Ideal and Resistive Magnetohydrodynamics

This book describes state-of-the-art theories and applications of ideal and resistive magnetohydrodynamics (MHD). It fills a gap in the market between existing textbooks and the latest cutting-edge research developments, reviewing the profound and pioneering contributions of the theory's founders and the latest developments and technologies.

This book guides readers from the foundations and first principles of MHD to then explore equilibrium theory from classic formulations to modern innovations in coil design, stellarator symmetry, and X-point physics.

Clear explanations connect advanced ideal and resistive stability theory to real fusion challenges, including interchange, ballooning, and Alfvén modes. The book also unpacks key nonlinear phenomena such as magnetic islands, tearing modes, sawtooth, and edge transport dynamics.

Concluding with a forward–looking assessment of tokamak, stellarator, and mirror concepts, this volume offers a deep yet accessible review of the field for anyone seeking to understand the MHD principles that underpin controlled nuclear fusion.

This book will appeal to graduate students and researchers in plasma physics. Although it focuses mainly on magnetically confined fusion plasma physics, it will also be useful to students and researchers in adjoining fields, including space plasma physics.

Key features:
- One of only a handful of books dedicated entirely to advanced ideal and resistive MHD theory for magnetically confined plasmas
- Fills a gap between existing textbooks and cutting-edge research in plasma physics
- Self-contained, covering advanced theory with detailed mathematical treatments

Dr. Linjin Zheng is a research scientist at the Institute for Fusion Studies, The University of Texas at Austin (USA). He is a theoretical physicist for controlled thermonuclear fusion plasmas. He earned his MS degree from the University of Science and Technology of China and PhD from the Institute of Physics – Beijing, Chinese Academy of Sciences. He has published more than 100 scientific papers in *Physics Review Letters, Physics Letters, Nuclear Fusion, Physics of Plasmas,* and at major conferences. He has edited one book and authored five more, covering ideal and resistive MHD, multifluids, and kinetic theories. Dr. Zheng's major contributions with his colleagues include the reformulation of gyrokinetic theory, renormalization of Lie transform theory, development of perpendicular magnetohydrodynamic theory, invention of the free boundary ballooning representation, solution of the X point equilibrium, discoveries of the second toroidal Alfvén eigenmodes and the current interchange tearing modes, etc. With his colleagues, he also developed the tokamak equilibrium code ATEQ and stability codes AEGIS and AEGIS-K.

Ideal and Resistive Magnetohydrodynamics

Linjin Zheng

CRC Press
Taylor & Francis Group
Boca Raton London New York

CRC Press is an imprint of the
Taylor & Francis Group, an **informa** business

Designed cover image: Zheng

First edition published 2027
by CRC Press
2385 NW Executive Center Drive, Suite 320, Boca Raton FL 33431

and by CRC Press
4 Park Square, Milton Park, Abingdon, Oxon, OX14 4RN

CRC Press is an imprint of Taylor & Francis Group, LLC

ISBN: 978-1-032-86206-4 (hbk)
ISBN: 978-1-032-86271-2 (pbk)
ISBN: 978-1-003-52213-3 (ebk)

DOI: 10.1201/9781003522133

Typeset in Nimbus Roman font
by KnowledgeWorks Global Ltd.

Publisher's note: This book has been prepared from camera-ready copy provided by the author.

This volume is dedicated with love
to my wife and children

Contents

Preface

This book describes the advanced ideal and resistive magnetohydrodynamics (MHD) theory. It aims at filling the gap between the textbooks and up-to-date research development. It focuses mainly on magnetically confined fusion plasma physics. Nevertheless, it is also useful to other fields in plasma physics, such as space plasma physics, etc.

The physics of magnetically confined plasmas is complex. It deals with a many-charged-particle system with a long correlation length. Charged particles are often confined in a complicated magnetic field configuration. The ideal MHD theory is unique in its simplicity and beauty. It taps the key feature of charged particle motions in a magnetic field: the particles tend to move together with the magnetic field lines in the perpendicular direction. However, even in the MHD description, the theory can still be complicated. After decades of pioneering efforts, many profound and elegant theories were developed in this field. The book is intended to give them an in-depth review.

The book starts with the fundamentals of magnetohydrodynamics theory by introducing the MHD theory from the kinetic description. The MHD force vector, the energy principle, and the Euler-Lagrange formulation of MHD are then described. Using the general MHD formalism, the basic MHD waves and instabilities are discussed. Afterward, the non-ideal MHD effects are considered with emphasis on the resistivity effects, although two-fluid and kinetic modifications are also outlined. The advanced theoretical analyses are accompanied by fundamental and easy-to-understand physics pictures.

In the MHD equilibrium, besides the conventional topics such as the flux coordinates and the Grad-Shafranov equation, it also addresses how to determine the coil structure or a feedback control system from a given fixed boundary equilibrium. The advanced topics, such as the quasi-symmetric stellarator concept, the Lagrangian and Hamiltonian descriptions of magnetic field line flow, the X-point equilibrium, etc., are also described.

In the ideal MHD stability theory, both cylindrical and toroidal theories are described. The MHD mode spectrum, the accumulation points, and continuum damping are discussed. The global mode description and the theoretical formalisms for interchange, ballooning, and toroidal Alfvén eigenmodes are also detailed. The theories are advanced. Nevertheless, the equilibrium with a large aspect ratio and circular cross-section is used to show the advanced concepts directly. The ideal MHD theories in the cylindrical and toroidal geometries are extended to the resistive MHD case. Especially, the resistive MHD theories by A. H. Glasser, J. M. Greene, and J. L. Johnson and the resistive ballooning mode theory are detailed.

A broad overview of the nonlinear theories is also given, covering the magnetic island and stochastic field, the nonlinear tearing modes, neoclassical tearing modes, sawtooth oscillations, current interchange tearing modes, etc. The couplings of the

MHD modes to the external structure and the inductive electric field are also discussed in connection to the edge-localized modes and the transport barrier induced by the off-axis heating. Besides, the blob transport, the edge harmonic oscillations, and the Sweet-Parker model in the forced reconnection are also addressed.

In conclusive remarks, the prospects of three major magnetic confinements, tokamak, stellarator, and mirror, are discussed, together with the importance of MHD formulation. It is hoped that the review of ideal and resistive MHD in this book will contribute to a better understanding of plasma physics to conquer the challenges mankind faces in controlling nuclear fusion.

Dr. Linjin Zheng
Institute for Fusion Studies
Department of Physics
The University of Texas at Austin
Austin, Texas, USA

Acknowledgments

This book was written when the author worked at the Institute for Fusion Studies, The University of Texas at Austin, with support from the U.S. Department of Energy, Office of Fusion Energy Science (Grant No. DE-FG02-04ER54742). Many research results in this book have benefited from the collaborations with colleagues in this institute. The author is also grateful for collaborations with many other colleagues in this field. He especially thanks Professors R. D. Hazeltine and M. Tessarotto for looking over the manuscript and providing valuable suggestions and comments.

Author

Dr Linjin Zheng is a theoretical physicist for controlled thermonuclear fusion plasmas. He earned his MS degree from the University of Science and Technology of China and PhD from the Institute of Physics – Beijing, Chinese Academy of Sciences. He is currently a research scientist at the Institute for Fusion Studies, The University of Texas at Austin. He published more than 100 scientific papers in *Physics Review Letters*, *Physics Letters*, *Nuclear Fusion*, *Physics of Plasmas*, and at major conferences. Dr. Zheng edited one book and authored five more covering ideal and resistive MHD, multifluids, and kinetic theories. His major contributions with his colleagues include the reformulation of gyrokinetic theory, renormalization of Lie transform theory, development of perpendicular magnetohydrodynamic theory, invention of the free boundary ballooning representation, solution of the X point equilibrium, discoveries of the second toroidal Alfvén eigenmodes and the current interchange tearing modes, etc. With his colleagues, he also developed the tokamak equilibrium code ATEQ and stability codes AEGIS and AEGIS-K.

1 Introduction

The term magnetohydrodynamics (MHD) includes three parts: magneto means magnetic field, hydro means water or fluid, and dynamics means the motion under the action of forces. Historically, the MHD description of electrically conducting fluids in a magnetic field was first developed by Hannes Alfvén to study the Alfvén waves, for which he received the Nobel Prize in Physics in 1970 [1]. It is Hannes Alfvén who later introduced the term magnetohydrodynamics. Here, the "hydro" refers particularly to the many charged particle system, also known as plasma. Unlike conventional fluids, like water or gas, plasma is usually hot. It therefore cannot be confined by a solid vessel. Because they are conductive, the magnetic field is usually used to confine them. MHD theory has applications in multiple fields, including magnetically confined fusion, space physics, geophysics, astrophysics, etc. The current book focuses on its application in controlled magnetic fusion [2, 3], although the theory can also apply to other fields.

Fusion reactions combine light atomic nuclei, for example, deuterium and tritium, into heavier particles, for example, an alpha particle (Helium-4 nucleus) and a neutron. The energy is released due to the mass loss in such a nuclear reaction. The energy yield is given by the kinetic energy of the reaction products. The fuel atoms are fully ionized in this case. In order for the light nuclei to collide to overcome the electrostatic repulsion between the nuclei, the fuel nuclei must have high kinetic energy. In terms of temperature, it must be about hundreds of millions of degrees. To have a net energy yield, the density has to be sufficiently high and the confinement time has to be sufficiently long to meet the so-called Lawson criterion (triple product) [4]. The magnetic field is used to confine such hot plasmas. The MHD theory then becomes the fundamental tool to study this kind of system.

The combination of the terms "magneto" and "hydro" reflects the key difference between magnetically confined plasma and liquid or gas. Liquid or gas is described as a fluid since strong collisions make the nearby particles move together as a fluid cell. A local thermal equilibrium is established in the fluid cell. The magnetically confined plasma becomes fluid-like because the magnetic field plays a role in particle localization. Charged particles tend to gyrate around a magnetic field line. This makes charged particles become frozen in the magnetic field line. It is the presence of a magnetic field that makes charged particles move together like a fluid cell. Therefore, it is a magnetofluid or "magnetohydro".

We need also to note that the charged-particle gyration occurs in the direction perpendicular to the magnetic field lines. Therefore, charged particles behave as being fluid-like only in the direction perpendicular to the magnetic field. In the direction parallel to the magnetic field, charged particles can move rather freely. The collisions are usually not strong enough to hold them to move collectively as a fluid cell, therefore, strictly speaking, the magnetohydrodynamics should be the perpendicular magnetohydrodynamics [5].

DOI: 10.1201/9781003522133-1

Nevertheless, even in the weak collision case, the resistivity effect due to the collision can play a significant role, as it can release charged particles from being frozen in the magnetic field. This book describes not only the ideal MHD without taking into account the collisional effects but also the resistive MHD with the collision effects included. Besides the collision effects, the two fluid and kinetic effects may also play a significant role in the parallel direction.

Since the magnetic field can only effectively confine the charged particles in the direction perpendicular to the magnetic field, various concepts are developed in controlled magnetic fusion to confine the plasmas to avoid the charged-particle losses along the direction parallel to the magnetic field. Bending them into a donut-like torus is a major effort, such as tokamak [6, 7], stellarator [8, 9], etc. There are also open field line systems, like the magnetic mirror [10], in which a stronger magnetic field is used to plug the end losses of charged particles. The review of various confinement concepts in the magnetically confined fusion can be found in reference [2].

The plasma physics research often ties the charged particle motions to the magnetic field. This leads the magnetically confined plasmas to have the long-range behavior. The local perturbation tends to excite the long-range waves. Furthermore, the magnetic confinement of plasmas is based on the balance of magnetic pressure and plasma thermal pressure. Although the balance is usually referred to as equilibrium in this field, it is totally different from thermal equilibrium. Thermal equilibrium implies the homogeneity of thermal parameters, such as density and temperature. This does not occur in the magnetically confined plasmas. Therefore, the force-balanced system can be in the high-energy state. The release of the energy can lead to the so-called instabilities. The excess energies can result from the configuration or velocity space. The MHD instability is usually related to the free energy from the configuration space. Equilibrium and stability are the two major topics in MHD theory. For controlled magnetic fusion, there are two other major fields: heating and transport [11, 12].

The current book focuses on the advanced theory of ideal and resistive MHD. It is, nevertheless, self-contained with the mathematical treatments detailed and the fundamental physics interpreted. It is intended to fill the gap between existing textbooks and cutting-edge research in plasma physics.

References

1. Alfvén H 1942 Existence of Electromagnetic-Hydrodynamic Waves. *Nature, 150*, 405–406. DOI:10.1038/150405d0. S2CID 4072220.
2. Teller E (ed) 1981 *Fusion* (New York: Academic).
3. Hazeltine R D, Meiss J D 1992 Plasma Confinement (New York: Addison-Wesley).
4. Lawson J D 1955 Some Criteria for a Power Producing Thermonuclear Reactor. *Proceedings of the Physical Society, Section B, 70 (1)*, 6–10. DOI:10.1088/0370-1301/70/1/303
5. Freidberg J P 1987 Ideal Magnetohydrodynamics (New York: Plenum Press).

6. Artsimovich L A 1958 Research on Controlled Thermonuclear Reactions in the USSR, *Proceedings of the Second United Nations International Conference on the Peaceful Uses of Atomic Energy*.
7. Wesson J 1997 Tokamaks (Clarendom Press, Oxford).
8. Spitzer L 1951 A proposed stellarator, *Technical Report*, Project Matterhorn.
9. Sagara A, Igitkhanov Y, Najmabadi F 2010. Review of stellarator/heliotron design issues towards MFE DEMO. *Fusion Engineering and Design, 85 (7–9),* 1336-1341. https://DOI.org/10.1016/j.fusengdes.2010.03.041
10. Post R. F. 1987 The magnetic mirror approach to fusion. *Nucl. Fusion 27,* 1579. DOI:10.1088/0029-5515/27/10/001
11. Hinton F L, Hazeltine R D 1976 Theory of plasma transport in toroidal confinement systems *Rev. Mod. Phys. 48*, 239. DOI: https://doi.org/10.1103/RevModPhys.48.239
12. Horton W 2012 Turbulent Transport in Magnetized Plasmas Illustrated Edition (Hackensack, NJ: World Scientific).

2 From kinetic to fluid descriptions

In this chapter, we discuss the foundation of MHD theory: from kinetic to fluid descriptions. This explains how the MHD equations are derived from the kinetic equations and what are the underlying physics considerations. Besides explaining the root of MHD theory, the theoretical framework to extend the MHD theory to the anisotropic case, two-fluid, and kinetic descriptions are also laid out. These theoretical frameworks form the basis for further research review in this book.

We first describe Braginskii's two-fluid theory [1] since the MHD theory originates from it. Given the limitations of the MHD description, we next proceed to further advanced theories. To consider the anisotropic case, we next introduce the Chew-Goldberger-Low (CGL) theory [2]. This is necessary for investigating the mirror type of confinements. Noting that the magnetic fluid concept can only be applicable in the direction perpendicular to the magnetic field lines, the perpendicular magnetofluid theory is subsequently introduced [3]. This represents the minimum modification of fluid description to include the parallel kinetic effects. We will also discuss the hybrid description with the kinetic closure. This makes the extended MHD contain the parallel electric field and wave-particle resonance effects.

After the general introduction of the foundation of fluid description, we return to construct the ideal and resistive MHD equations from Braginskii's two-fluid equations. This is preceded by deriving the generalized Ohm's law. The ideal and resistive MHD equations are then obtained. They will be used throughout this book.

2.1 BRAGINSKII'S TWO FLUID EQUATIONS

In this section, we outline the derivation of the Braginskii two-fluid equations from the Boltzmann equation [1]. One may refer to author's earlier book [4] for detailed review. Braginskii's two-fluid equations form the basis for the development of ideal and resistive MHD equations. The fluid equations are obtained from the moment equations of kinetic equation, the Boltzmann equation. Instead of directly taking the moment integrations over the Boltzmann equation, we first transform the Boltzmann equation to the center of mass and then derive the moment equations. This simplifies a lot. Three moment equations will be derived: the density, momentum, and energy equations. These moment equations will be used to construct the MHD equations later in Section 2.5.

DOI: 10.1201/9781003522133-2

2.1.1 BOLTZMANN EQUATION IN THE CENTER OF MASS

In this subsection, we briefly introduce the Boltzmann equation and transform it to the center of mass [5]. Since we focus on the Braginskii theory, the details about the Boltzmann equation and collisional operator are referred to other plasma physics books [6, 7].

We consider a classical many-particle system governed by Newton's second law of motion. It is noted that Newton's second law is a second-order differential equation in time t. The particle motion is determined by the initial position $\mathbf{x}$ and velocity $\mathbf{v}$. For N particles, the total number of initial conditions becomes $6N$. Therefore, the state of a many-particle system with N particles is determined by the probability density function $f_N(\mathbf{x}_1, \mathbf{v}_1, \cdots, \mathbf{x}_N, \mathbf{v}_N)$. One may use either the BBGKY or Klimontovich formalisms to describe the many-particle statistical problem in the $6N$-dimentional phase space [8, 9].

Instead, the Boltzmann equation formalism uses the particle distribution functions in the six-dimensional phase space: $f_a(t, \mathbf{x}, \mathbf{v})$, where the subscript a is introduced to label the particle species with $a = i, e$ denoting the ion and electron species, respectively. This reduces the description in the $6N$ dimensional phase space to the 6 dimensional one. The particle correlation effects are then treated as the collisional operator in the Boltzmann formalism. In this formalism, $f(t, \mathbf{x}, \mathbf{v})d\mathbf{x}d\mathbf{v}$ is the number of particles in the six-dimensional volume element $d\mathbf{x}d\mathbf{v}$.

The Boltzmann equation is as follows [5]

$$\frac{\partial f_a}{\partial t} + \mathbf{v}_a \cdot \frac{\partial f_a}{\partial \mathbf{x}} + \mathbf{a}_a \cdot \frac{\partial f_a}{\partial \mathbf{v}} = C_a, \tag{2.1}$$

where $\mathbf{a}_a$ is the acceleration, which can be related to the force by Newton's second law of motion. In the case with the electromagnetic field $\mathbf{E}$ and $\mathbf{B}$, one has

$$\mathbf{a}_a = \frac{e_a}{m_a} (\mathbf{E} + \mathbf{v}_a \times \mathbf{B}),$$

where e_a denotes the charge and m_a the mass. In equation (2.1), C_a on the right-hand side is the so-called collisional operator. It accounts for the remaining effects of reducing the $6N$ dimensional description f_N to the six-dimensional one f [8, 9]. Because particles of species a can collide with each other or with other species, the collisional operator can be expressed

$$C_a = \sum_b C_{ab}(f_a, f_b),$$

where $C_{ab}(f_a, f_b)$ represents the change of the distribution function per unit time from the collisions of species a with species b with $b = a$ included as well.

In the Braginskii two-fluid theory, the Landau collisional operator [10] is used:

$$C_{ab}(f_a, f_b) = -\frac{2\pi\lambda e_a^2 e_b^2}{m_a} \frac{\partial}{\partial v_{a\beta}} \int \left\{ \frac{f_a(\mathbf{v}_a)}{m_b} \frac{\partial f_b(\mathbf{v}_b)}{\partial v_{b\gamma}} - \frac{f_b(\mathbf{v}_b)}{m_a} \frac{\partial f_a(\mathbf{v}_a)}{\partial v_{a\gamma}} \right\} U_{\beta\gamma} d\mathbf{v}_b, \tag{2.2}$$

where

$$U_{\beta\gamma} = \frac{1}{u^3}\left(u^2\delta_{\beta\gamma} - u_\beta u_\gamma\right), \quad \text{with } u_\beta = v_{a\beta} - v_{b\beta}$$

and λ is the so-called Coulomb logarithm, which equals to the logarithm of the ratio of maximum to minimum impact parameters $\lambda = \log(P_{\max}/P_{\min})$. Here and later on, the Einstein summation convention with the same indices is implied for the same indices,

Braginskii's fluid theory considers a collision-dominant plasma. It is assumed that the mean free path is much smaller than the macroscopic scale length and the collision time is shorter than the time scale. In this case, the local thermal equilibrium is established with the Maxwellian distribution:

$$f_a^0 = \frac{n_a}{(2\pi T_a/m_a)^{3/2}} e^{-\frac{m_a}{2T_a}(\mathbf{v}_a - \mathbf{V}_a)^2}, \tag{2.3}$$

where the temperature is expressed in the energy unit so that the Boltzmann constant is left out and $\mathbf{V}$ is the average velocity. The random velocity is defined as $\mathbf{v}_a' = \mathbf{v}_a - \mathbf{V}_a$.

The kinetic equation (2.1) can be transformed to the center of mass, $\mathbf{V}_a$, for electron and ion species with $a = i, e$. We first note that the kinetic equation (2.1) can be rewritten with the random velocity $\mathbf{v}_a' = \mathbf{v}_a - \mathbf{V}_a$ as follows

$$\frac{\partial f_a}{\partial t} + \frac{\partial}{\partial x_\beta}(v_\beta f_a) + \frac{\partial}{\partial v_\beta}\left(\frac{F_{a\beta}}{m_a} f_a\right) = C_a. \tag{2.4}$$

Without causing ambiguity, the prime denoting the random velocity is dropped for brevity. We further note that

$$\begin{aligned}
\frac{\partial f_a}{\partial t} &= \frac{\partial f_a}{\partial t} - \frac{\partial V_{a\beta}}{\partial t}\frac{\partial f_a}{\partial v_{a\beta}}, \\
\frac{\partial}{\partial x_\beta}(v_\beta f_a) &= v_\beta \frac{\partial f_a}{\partial x_\beta}, \\
&= v_{a\beta}\frac{\partial f_a}{\partial x_\beta} + V_{a\beta}\frac{\partial f_a}{\partial x_\beta} - v_{a\alpha}\frac{\partial V_{a\beta}}{\partial x_\alpha}\frac{\partial f_a}{\partial v_{a\beta}} - V_{a\alpha}\frac{\partial V_{a\beta}}{\partial x_\alpha}\frac{\partial f_a}{\partial v_{a\beta}}, \\
\frac{\partial}{\partial v_\beta}\left(\frac{F_{a\beta}}{m_a} f_a\right) &= \frac{e_a}{m_a}\frac{\partial}{\partial v_{a\beta}}\left(E_\beta f_a + \frac{e_{\beta\gamma\alpha} v_\gamma B_\beta}{c} f_a\right) \\
&= \frac{e_a}{m_a}\left\{E_\beta \frac{\partial f_a}{\partial v_{a\beta}} + \frac{e_{\beta\gamma\alpha} V_{a\gamma} B_\alpha}{c}\frac{\partial f_a}{\partial v_{a\beta}} + \frac{e_{\beta\gamma\alpha} v_{a\gamma} B_\alpha}{c}\frac{\partial f_a}{\partial v_{a\beta}}\right\}.
\end{aligned}$$

Using these results equation (2.4) is then transformed to

$$\begin{aligned}
&\frac{d_a f_a}{dt} + v_{a\beta}\frac{\partial f_a}{\partial x_\beta} - \left(\frac{e_a}{m_a} E_{a\beta}^* - \frac{d_a V_{a\beta}}{dt}\right)\frac{\partial f_a}{\partial v_{a\beta}} \\
= &\frac{\partial V_{a\gamma}}{\partial x_\beta} v_{\alpha\beta}\frac{\partial f_\alpha}{\partial v_{a\gamma}} + \frac{e_a}{m_a} e_{\beta\gamma\alpha} v_{a\gamma} B_\alpha \frac{\partial f_a}{\partial v_{a\beta}} - \sum_b C_{ab}(f_a, f_b).
\end{aligned} \tag{2.5}$$

where

$$\frac{d_a}{dt} \equiv \frac{\partial}{\partial t} + \mathbf{V}_a \cdot \nabla, \quad \mathbf{E}_a^* \equiv \mathbf{E} + \mathbf{V}_a \times \mathbf{B},$$

and the Landau collisional operator is unchanged due to the boost invariance of the collisional operator. Equation (2.5) will be used to derive the moment equations in the following subsections.

2.1.2 GENERAL MOMENT EQUATIONS

With the Boltzmann equation in the center of mass derived in the previous subsection, we will derive the general moment equation in this subsection.

For any phase space quantity $\eta_a(\mathbf{x}_a, \mathbf{v}_a)$, the corresponding moment of the distribution function is defined as follows

$$\langle \eta_a \rangle = \frac{1}{n_a} \int \eta_a f_a d\mathbf{v}_a,$$

where $n_\alpha = \langle 1 \rangle$ is simply the fluid density. Multiplying equation (2.1) with η_a and integrating it over the velocity space, one can derive the general moment equation. Noting that

$$\begin{aligned}
\left\langle \eta_a \frac{d_a f_a}{dt} \right\rangle &= \frac{d_a \langle \eta_a f_a \rangle}{dt}, \\
\left\langle \eta_a v_{a\beta} \frac{\partial f_a}{\partial x_\beta} \right\rangle &= \frac{\partial \langle \eta_a v_{a\beta} f_a \rangle}{\partial x_\beta}, \\
\left\langle \eta_a \left(\frac{e_a}{m_a} E^*_{a\beta} - \frac{d_a V_{a\beta}}{dt} \right) \frac{\partial f_a}{\partial v_{a\beta}} \right\rangle &= -\left(\frac{e_a}{m_a} E^*_{a\beta} - \frac{d_a V_{a\beta}}{dt} \right) \left\langle \frac{\partial \eta_a}{\partial v_{a\beta}} f_a \right\rangle, \\
-\left\langle \eta_a \frac{\partial V_{a\gamma}}{\partial x_\beta} v_{\alpha\beta} \frac{\partial f_\alpha}{\partial v_{a\gamma}} \right\rangle &= \frac{\partial V_{a\gamma}}{\partial x_\beta} \left\langle \frac{\partial \eta_a v_{\alpha\beta}}{\partial v_{a\gamma}} f_\alpha \right\rangle, \\
-\left\langle \eta_a \frac{e_a}{m_a} e_{\beta\gamma\alpha} v_{a\gamma} B_\alpha \frac{\partial f_a}{\partial v_{a\beta}} \right\rangle &= e_{\beta\gamma\alpha} \frac{e_a}{m_a} B_\alpha \left\langle \frac{\partial \eta_a v_{a\gamma}}{\partial v_{a\beta}} f_a \right\rangle,
\end{aligned}$$

one obtains the general moment equation:

$$\begin{aligned}
&\underbrace{\frac{d_a \langle \eta_a f_a \rangle}{dt}}_{1} + \underbrace{\frac{\partial \langle \eta_a v_{a\beta} f_a \rangle}{\partial x_\beta}}_{2} \underbrace{- \left(\frac{e_a}{m_a} E^*_{a\beta} - \frac{d_a V_{a\beta}}{dt} \right) \left\langle \frac{\partial \eta_a}{\partial v_{a\beta}} f_a \right\rangle}_{3} \\
&+ \underbrace{\frac{\partial V_{a\gamma}}{\partial x_\beta} \left\langle \frac{\partial \eta_a v_{\alpha\beta}}{\partial v_{a\gamma}} f_\alpha \right\rangle}_{4} + \underbrace{e_{\beta\gamma\alpha} \frac{e_a}{m_a} B_\alpha \left\langle \frac{\partial \eta_a v_{a\gamma}}{\partial v_{a\beta}} f_a \right\rangle}_{5} \\
&= \underbrace{\sum_b \langle \eta_a C_{ab}(f_a, f_b) \rangle}_{6}. \qquad (2.6)
\end{aligned}$$

Using equation (2.6), one can derive the fluid equations, such as continuity, momentum, and energy transport equations, in the following subsections. To explain the further derivations, each term in equation (2.6) is numbered.

2.1.3 CONTINUITY EQUATION

Letting $\eta_a = 1$ with $a = e, i$ in equation (2.6), one obtains the continuity equations for electrons and ions as follows:

Note that terms 2, 3, 5, and 6 in equation (2.6) vanish. Terms 1 and 4 yield the continuity equations for the ion and electron species

$$\frac{\partial n_e}{\partial t} + \frac{\partial}{\partial x_\alpha}(n_e V_{e\alpha}) = 0, \tag{2.7}$$

$$\frac{\partial n_i}{\partial t} + \frac{\partial}{\partial x_\alpha}(n_i V_{i\alpha}) = 0, \tag{2.8}$$

where the fluid velocity is defined as follows

$$\mathbf{v}_a = \frac{1}{n_a}\int d\mathbf{x}d\mathbf{v}\; v_a f_a$$

with $a = i, e$.

2.1.4 MOMENTUM EQUATION

To derive the momentum equation, we let $\eta_a = m_a \mathbf{v}_a$ with $a = e, i$ in equation (2.6). Note that terms 1, 4, and 5 vanish. Term 2 can be simplified by introducing

$$\begin{aligned} T_a &= \frac{1}{3} m_a \left\langle v'^2 \right\rangle, \\ \pi_{a\alpha\beta} &= \left\langle v'_{a\alpha} v'_{a\beta} - \frac{1}{3} v'^2_a \delta_{\alpha\beta} \right\rangle, \end{aligned} \tag{2.9}$$

and term 6 gives

$$R_{a\alpha} = \sum_b R_{ab\alpha} = \sum_b \int m_a v'_{a\alpha} C_{ab} d\mathbf{v}_a \tag{2.10}$$

with

$$R_{ei\alpha} = -R_{ie\alpha}. \tag{2.11}$$

Terms 2, 3, and 6 yield the momentum equations fot the ion and electron species:

$$n_e m_e \frac{d_e V_{e\alpha}}{dt} = -\frac{\partial p_e}{\partial x_\alpha} - \frac{\partial \pi_{e\alpha\beta}}{\partial x_\beta} - n_e e(E_\alpha + e_{\alpha\beta\gamma} V_{e\beta} B_\gamma) + R_{e\alpha}, \tag{2.12}$$

$$n_i m_i \frac{d_i V_{i\alpha}}{dt} = -\frac{\partial p_i}{\partial x_\alpha} - \frac{\partial \pi_{i\alpha\beta}}{\partial x_\beta} + n_i e(E_\alpha + e_{\alpha\beta\gamma} V_{i\beta} B_\gamma) + R_{i\alpha}, \tag{2.13}$$

where the pressure $p_a = n_a T_a$.

2.1.5 ENERGY TRANSPORT EQUATION

To derive the energy transport equation, we let $\eta_a = \frac{1}{2} m_a v_a^2$ with $a = e, i$ in equation (2.6). Note that Terms 3 and 5 vanish.

$$Term\ 1 = \frac{3}{2}T, \tag{2.14}$$

$$term\ 2 = n_a m_a \left\langle \frac{1}{2} v'^2_a v_{a\alpha} \right\rangle = q_{a\alpha}, \tag{2.15}$$

$$Term\ 4 = \frac{\partial V_{a\gamma}}{\partial x_\beta} \left\langle \frac{\partial \eta_a v_{\alpha\beta}}{\partial v_{a\gamma}} f_\alpha \right\rangle = \frac{\partial V_{a\gamma}}{\partial x_\beta} \left(p_a \delta_{\gamma\beta} + \pi_{a\gamma\beta} \right) \tag{2.16}$$

$$Term\ 6 = Q_a = \sum_b Q_{ab} = \sum_b \int \frac{m_a v'^2_a}{2} C_{ab} d\mathbf{v}_a. \tag{2.17}$$

Here, q_a is the so-called heat conductivity, $\pi_{a\alpha\beta}$ denotes the pressure tensor, and Q_a represents the heat generation due to the collision. Using equations (2.14)–(2.17), one obtains the energy equations for electrons and ions as follows:

$$n_e \frac{\partial}{\partial t}\left(\frac{3}{2}T_e\right) + n_e V_{e\alpha} \frac{\partial}{\partial x_\beta}\left(\frac{3}{2}T_e\right)$$
$$+ \frac{\partial V_{e\gamma}}{\partial x_\beta}\left(p_e \delta_{\gamma\beta} + \pi_{e\gamma\beta}\right) + \frac{\partial}{\partial x_\beta} q_{e\beta} = Q_e, \tag{2.18}$$

$$n_i \frac{\partial}{\partial t}\left(\frac{3}{2}T_i\right) + n_i V_{i\beta} \frac{\partial}{\partial x_\beta}\left(\frac{3}{2}n_i T_i\right)$$
$$+ \frac{\partial V_{i\gamma}}{\partial x_\beta}\left(p_i \delta_{\gamma\beta} + \pi_{i\gamma\beta}\right) + \frac{\partial}{\partial x_\beta} q_{i\beta} = Q_i. \tag{2.19}$$

Sometimes, the flow velocity is added to the thermal kinetic energy to express the energy equation. From the moment equations (2.12) and (2.13), one obtains

$$n_e \frac{\partial}{\partial t}\left(\frac{m_e}{2}V_e^2\right) + n_e V_{e\beta} \frac{\partial}{\partial x_\beta}\left(\frac{m_e}{2}V_e^2\right)$$
$$+ V_{e\alpha} \frac{\partial}{\partial x_\beta}\left(p_e \delta_{\alpha\beta} + \pi_{e\alpha\beta}\right) = e_e n_e E_\beta V_{e\beta} + R_{e\beta} V_{e\beta}, \tag{2.20}$$

$$n_i \frac{\partial}{\partial t}\left(\frac{m_i i}{2}V_i^2\right) + n_i V_{i\beta} \frac{\partial}{\partial x_\beta}\left(\frac{m_i}{2}V_i^2\right)$$
$$+ V_{i\alpha} \frac{\partial}{\partial x_\beta}\left(p_i \delta_{\alpha\beta} + \pi_{i\alpha\beta}\right) = e_i n_i E_\beta V_{i\beta} + R_{i\beta} V_{i\beta}, \tag{2.21}$$

Combining equations (2.18)–(2.19) and equations (2.20)–(2.21) and using the continuity equations (2.7)–(2.8) yield

$$\frac{\partial}{\partial t}\left(\frac{m_e n_e}{2}V_e^2+\frac{3}{2}n_eT_e\right)+\frac{\partial}{\partial x_\beta}\left[\left(\frac{m_e n_e}{2}V_e^2+\frac{3}{2}n_eT_e\right)V_{e\beta}\right.$$

$$\left.+\left(p_e\delta_{\alpha\beta}+\pi_{e\alpha\beta}\right)V_{e\alpha}+q_{e\beta}\right]=e_en_eE_\beta V_{e\beta}+R_{e\beta}V_{e\beta}+Q_e, \tag{2.22}$$

$$\frac{\partial}{\partial t}\left(\frac{m_i n_i}{2}V_i^2+\frac{3}{2}n_iT_i\right)+\frac{\partial}{\partial x_\beta}\left[\left(\frac{m_i n_i}{2}V_i^2+\frac{3}{2}n_iT_i\right)V_{i\beta}\right.$$

$$\left.+\left(p_i\delta_{\alpha\beta}+\pi_{i\alpha\beta}\right)V_{i\alpha}+q_{i\beta}\right]=e_in_iE_\beta V_{i\beta}+R_{i\beta}V_{i\beta}+Q_i. \tag{2.23}$$

From the energy equations, one can also derive the entropy equations by defining

$$s_a\equiv\ln\left(T_a^{3/2}/n_a\right)=\ln\left(p_a^{3/2}/n_a^{5/2}\right), \tag{2.24}$$

which actually represents the entropy per particle within an unimportant constant. Again, with the continuity equations (2.7) and (2.8) being used, the entropy equations can be obtained from the energy transport equations (2.22) and (2.23)

$$\begin{aligned} n_eT_e\frac{d_es_e}{dt} &= T_e\left[\frac{\partial n_es_e}{\partial t}+\partial_\beta\left(n_es_eV_{e\beta}\right)\right] \\ &= -\frac{\partial q_{e\alpha}}{\partial x_\alpha}-\pi_{i\alpha\beta}\frac{\partial V_{e\alpha}}{\partial x_\beta}+Q_e, \end{aligned} \tag{2.25}$$

$$\begin{aligned} n_iT_i\frac{d_is_i}{dt} &= T_i\left[\frac{\partial n_is_i}{\partial t}+\partial_\beta\left(n_is_iV_{i\beta}\right)\right] \\ &= -\frac{\partial q_{i\alpha}}{\partial x_\alpha}-\pi_{i\alpha\beta}\frac{\partial V_{i\alpha}}{\partial x_\beta}+Q_i. \end{aligned} \tag{2.26}$$

In the MHD description, one often considers the adiabatic case, i.e., $\mathbf{q}=0$ or more generally the right-hand sides of equations (2.25) and (2.26) vanish. From equations (2.25), (2.26), and (2.24), the energy equations become simply

$$\frac{d}{dt}\left(\frac{p_e}{n_e^\Gamma}\right) = 0, \tag{2.27}$$

$$\frac{d}{dt}\left(\frac{p_e}{n_e^\Gamma}\right) = 0, \tag{2.28}$$

where $\Gamma=5/3$ is the ratio of specific heats. They are the so-called adiabatic energy equations in the MHD description.

2.1.6 CLOSURE AND TRANSPORT COEFFICIENTS

The moment equations have a hierarchical structure. The equation of lower moment depends on the higher-order moment. To apply the hierarchy of moment equations, one needs to limit the number of moments to be considered and find out the closure for the moment equations. In the framework of Braginskii's fluid equations, the three lowest moments—densities n_a, velocities $\mathbf{V}_a$, and temperatures T_a for electron and ion fluids—are considered. Instead of deriving the higher-order moment equations, the stress tensor $\pi_{\alpha\beta}$, the heat flux $\mathbf{q}$, the momentum transfer $\mathbf{R}$, and the heat generation Q are derived. In the derivations, it is assumed that the collision frequency and particle gyrofrequency are much higher than the frequency of interest, and the particle mean free path and the Larmor radius are much shorter than the scale length of the problems to be investigated. Their derivations are detailed in the Braginskii's work [1] and further explained in reference [4]. Here, we just give the results for application.

The electron and ion collision times are defined as follows:

$$\tau_e = \frac{3\sqrt{m_e}T_e^{3/2}}{4\sqrt{2\pi}\lambda e^4 Z^2 n_i}, \tag{2.29}$$

$$\tau_i = \frac{3\sqrt{m_i}T_i^{3/2}}{4\sqrt{\pi}\lambda e^4 Z^4 n_i}. \tag{2.30}$$

The momentum transfer $\mathbf{R}_{ei}$ is obtained as follows:

$$\mathbf{R}_{ei} = -\mathbf{R}_{ie} \equiv \mathbf{R} = \mathbf{R}_u + \mathbf{R}_T.$$

The friction force is given by

$$\mathbf{R}_u = -\frac{m_e n_e}{\tau_e}\left(0.5\mathbf{u}_\parallel + \mathbf{u}_\perp\right) = en\left(\frac{\mathbf{j}_\parallel}{\sigma_\parallel} + \frac{\mathbf{j}_\perp}{\sigma_\perp}\right), \tag{2.31}$$

where $\mathbf{u} = \mathbf{V}_e - \mathbf{V}_i$ and the electrical conductivities are

$$\sigma_\perp = \frac{e^2 n_e \tau_e}{m_e}, \quad \sigma_\parallel = 1.96\sigma_\perp. \tag{2.32}$$

The resistivity is just the inverse of conductivity. The thermal force is as follows:

$$\mathbf{R}_T = -0.71 n_e \nabla_\parallel T_e - \frac{3}{2}\frac{n_e}{\omega_{ce}\tau_e}\mathbf{b} \times \nabla T_e. \tag{2.33}$$

In the presence of strong magnetic field, the stress tensor is given in tensor notation as follows:

$$\pi_{\alpha\beta} = -\eta_0 W_{0\alpha\beta} - \eta_1 W_{1\alpha\beta} - \eta_2 W_{2\alpha\beta} + \eta_3 W_{3\alpha\beta} + \eta_4 W_{4\alpha\beta} \tag{2.34}$$

where

$$
\begin{aligned}
W_{0\alpha\beta} &= \frac{3}{2}\left(h_\alpha h_\beta - \frac{1}{2}\delta_{\alpha\beta}\right)\left(h_\mu h_\nu - \frac{1}{3}\delta_{\mu\nu}\right)W_{\mu\nu},\\
W_{1\alpha\beta} &= \left(\delta^\perp_{\alpha\mu}\delta^\perp_{\beta\nu} + \frac{1}{2}\delta^\perp_{\alpha\beta}h_\mu h_\nu\right)W_{\mu\nu},\\
W_{2\alpha\beta} &= \left(\delta^\perp_{\alpha\mu}h_\beta h_\nu + \delta^\perp_{\beta\nu}h_\alpha h_\mu\right)W_{\mu\nu},\\
W_{3\alpha\beta} &= \frac{1}{2}\left(\delta^\perp_{\alpha\mu}e_{\beta\gamma\nu} + \delta^\perp_{\beta\nu}e_{\alpha\gamma\mu}\right)h_\gamma W_{\mu\nu},\\
W_{4\alpha\beta} &= \left(h_\alpha h_m u e_{\beta\gamma\nu} + h_\beta h_\nu e_{\alpha\gamma\mu}\right)h_\gamma W_{\mu\nu},
\end{aligned}
$$

where $\delta^\perp_{\alpha\beta} = \delta_{\alpha\beta} - h_\alpha h_\beta$ and $e_{\alpha\beta\gamma}$ is the alternating tensor. The electron heat flux is given as follows:

$$
\mathbf{q}^e = \mathbf{q}^e_u + \mathbf{q}^e_T, \tag{2.35}
$$

where the collisional heat flux is

$$
\mathbf{q}^e_u = 0.71 n_e T_e \mathbf{u}_\parallel + \frac{3}{2}\frac{n_e T_e}{\omega_e \tau_e}\mathbf{b}\times\mathbf{u} \tag{2.36}
$$

and the thermal conductivity

$$
\mathbf{q}^e_T = -\kappa^e_\parallel \nabla_\parallel T_e - \kappa^e_\perp \nabla_\perp T_e - \frac{5}{2}\frac{n_e T_e}{eB}\mathbf{b}\times\nabla T_e \tag{2.37}
$$

with

$$
\begin{aligned}
\kappa^e_\parallel &= 3.16\frac{n_e T_e \tau_e}{m_e},\\
\kappa^e_\perp &= 4.66\frac{n_e T_e}{m_e \omega_{ce}^2 \tau_e}.
\end{aligned}
$$

Because of the large difference between ion and electron masses, the collision heat flux from electrons to ions is negligible. The ion heat flux is therefore contributed mainly by the thermal gradient heat flux

$$
\mathbf{q}^i = \mathbf{q}^i_T = -\kappa^i_\parallel \nabla_\parallel T_i - \kappa^i_\perp \nabla_\perp T_i + \frac{5}{2}\frac{n_i T_i}{ZeB}\mathbf{b}\times\nabla T_i, \tag{2.38}
$$

where

$$
\begin{aligned}
\kappa^e_\parallel &= 3.9\frac{n_i T_i \tau_i}{m_i},\\
\kappa^i_\perp &= 2\frac{n_i T_i}{m_i \omega_{ci}^2 \tau_i}.
\end{aligned}
$$

The heat acquired by the ions due to the collision with electrons is

$$
Q_i = Q_\Delta = \frac{3m_e}{m_i}\frac{n_e}{\tau_e}(T_e - T_i). \tag{2.39}
$$

The heat acquired by the electrons from collisions with ions is then

$$Q_e = -\mathbf{R}\cdot\mathbf{u} - Q_\Delta = \frac{j_\parallel^2}{\sigma_\parallel} + \frac{j_\perp^2}{\sigma_\perp} + \frac{1}{en_e}\mathbf{j}\cdot\mathbf{R}_T - \frac{3m_e}{m_i}\frac{n_e}{\tau_e}(T_e - T_i). \quad (2.40)$$

2.2 CHEW-GOLDBERGER-LOW THEORY

The Chew-Goldberger-Low theory (CGL, [2]) is a direct extension of the conventional ideal MHD formulation to the case with the nonisotropic pressure. More importantly, the CGL theory is constructed with the Vlasov equation [11] so that it is sometimes referred to as collisionless MHD. The CGL theory is well-known. However, its derivation is not trivial, and some details are omitted in the original paper [2]. This leads to some further efforts to demonstrate it, for example, in reference [12]. Here, we describe the theory in the two-fluid framework. The CGL theory is also useful for space plasma physics and magnetic fusion research.

The CGL theory started with the Vlasov equation [11]

$$\frac{\partial f}{\partial t} + \mathbf{v}\cdot\frac{\partial f}{\partial \mathbf{x}} + \frac{e}{m}(\mathbf{E}+\mathbf{v}\times\mathbf{B})\cdot\frac{\partial f}{\partial \mathbf{v}} = 0, \quad (2.41)$$

The expansion in powers of m/e is used to solve equation (2.41). The distribution function is expanded as follows:

$$f = f_0 + f_1 + f_2 + \cdots.$$

The Vlasov equation (2.41) is then expanded order by order as follows:

$$(\mathbf{E}+\mathbf{v}\times\mathbf{B})\cdot\frac{\partial f_0}{\partial \mathbf{v}} = 0, \quad (2.42)$$

$$\frac{\partial f_0}{\partial t} + \mathbf{v}\cdot\frac{\partial f_0}{\partial \mathbf{x}} + \frac{e}{m}(\mathbf{E}+\mathbf{v}\times\mathbf{B})\cdot\frac{\partial f_1}{\partial \mathbf{v}} = 0, \quad (2.43)$$

$$\frac{\partial f_1}{\partial t} + \mathbf{v}\cdot\frac{\partial f_1}{\partial \mathbf{x}} + \frac{e}{m}(\mathbf{E}+\mathbf{v}\times\mathbf{B})\cdot\frac{\partial f_2}{\partial \mathbf{v}} = 0, \quad (2.44)$$

$$\cdots$$

In these expansions, it is assumed that the parallel electric field $\mathbf{E}_\parallel = 0$ so that

$$\mathbf{E} = -\boldsymbol{\alpha}\times\mathbf{B} \quad \text{with} \quad \boldsymbol{\alpha} = \mathbf{E}\times\mathbf{B}/B^2.$$

It is also assumed that $\mathbf{E}_{\parallel 0} = 0$. This assumption leads equation (2.42) to become

$$\mathbf{v}'\times\mathbf{B}\cdot\frac{\partial f_0}{\partial \mathbf{v}} = 0, \quad \text{where} \quad \mathbf{v}' = \mathbf{v} - \boldsymbol{\alpha}. \quad (2.45)$$

From equation (2.45) one can see that the lowest order distribution function should be

$$f_0((\mathbf{v}-\boldsymbol{\alpha})^2, \mathbf{v}\cdot\mathbf{B}, \mathbf{r}, t).$$

It applies both for the electrons and ions with species index is suppressed.

Next, let us derive the moment equations of the first-order kinetic equation, equation (2.43). Noting that $\int d\mathbf{v}$, $\partial/\partial t$, and $\partial/\partial x_i$ are commutable and

$$\int d\mathbf{v} e_{\alpha\beta\gamma} v_\beta B_\gamma \frac{\partial}{\partial v_\alpha} f_1 = -\int d\mathbf{v} e_{\alpha\beta\gamma} \delta_{\alpha\beta} B_\gamma f_1 = 0,$$

one obtains

$$\frac{\partial n}{\partial t} + \frac{\partial}{\partial x_\alpha}(n u_\alpha) = 0, \tag{2.46}$$

where

$$\begin{aligned} n &= \int d\mathbf{v} f_0, \\ \mathbf{u} &= \int d\mathbf{v} \mathbf{v} f_0. \end{aligned}$$

Equation (2.46) applies both for the electrons and ions with species index is suppressed.

To derive the momentum equation, we multiply equation (2.43) with $\eta_a = m_a \mathbf{v}_a$ for both species and integrate it over velocity space. Noting that

$$\begin{aligned} \int d\mathbf{v} v_\alpha \partial_t f_0 &= \partial_t (n u_\alpha) \\ &= n\partial_t u_\alpha + u_\alpha \partial_t n = n \partial_t u_\alpha \underbrace{-u_\alpha \partial\left(n u_\beta\right)}_{x1}, \\ \int d\mathbf{v} v_\alpha v_\beta \partial_\beta f_0 &= \partial_\beta\left(n \langle v_\beta v_\alpha \rangle\right) \\ &= \partial_\beta\left[n\left\langle (v'_\beta + u_\beta)(v'_\alpha + u_\alpha)\right\rangle\right] \\ &= \partial_\beta\left(n\left\langle v'_\beta v'_\alpha + v'_\beta u_\alpha + v'_\alpha u_\beta + u_\beta u_\alpha\right\rangle\right) \\ &= \partial_\beta\left(n\left\langle v'_\beta v'_\alpha\right\rangle\right) + n u_\beta \partial_\beta u_\alpha + \underbrace{u_\alpha \partial_\beta\left(n u_\beta\right)}_{x1}, \\ nm\langle \mathbf{v}'\mathbf{v}' \rangle &= P_\parallel \mathbf{nn} + P_\perp\left(\overleftrightarrow{I} - \mathbf{nn}\right) \equiv \overleftrightarrow{P}, \\ e\int d\mathbf{v} v_\delta e_{\alpha\beta\gamma} v_\beta B_\gamma \frac{\partial}{\partial v_\alpha} f_1 &= -e\int d\mathbf{v} e_{\delta\beta\gamma} v_\beta B_\gamma f_1 = e_{\delta\beta\gamma} j_\beta B_\gamma, \end{aligned}$$

where for brevity we denote $\partial_\alpha = \partial/\partial x_\alpha$ and $\partial_t = \partial/\partial t$, one obtains the momentum transport equations for electrons and ions as follows:

$$n_e m_e \frac{d_e u_{e\alpha}}{dt} = -\frac{\partial P_{e\alpha\beta}}{\partial x_\beta} - n_e e E_\alpha + e_{\alpha\beta\gamma} j_{e\beta} B_\gamma, \tag{2.47}$$

$$n_i m_i \frac{d_i u_{i\alpha}}{dt} = -\frac{\partial P_{i\alpha\beta}}{\partial x_\beta} + n_i e E_\alpha + e_{\alpha\beta\gamma} j_{i\beta} B_\gamma, \tag{2.48}$$

Here, it has been noted that f_0 does not contribute to the ion and electron current densities $\mathbf{j}_{i,e}$.

To derive the energy transport equation, we multiply equation (2.43) with $m(\mathbf{v}-\mathbf{u})(\mathbf{v}-\mathbf{u})$ for both species and integrate it over velocity space. Note that

$$\begin{aligned}
&\int d\mathbf{v} m(\mathbf{v}-\mathbf{u})(\mathbf{v}-\mathbf{u})\frac{\partial f_0}{\partial t} \\
= \;& m\frac{\partial}{\partial t}\int d\mathbf{v}(\mathbf{v}-\mathbf{u})(\mathbf{v}-\mathbf{u})f_0 - 2m\frac{\partial u_\alpha}{\partial t}\int d\mathbf{v}(\mathbf{v}-\mathbf{u})f_0 \\
- \;& \frac{\partial \overleftrightarrow{P}}{\partial t},
\end{aligned} \tag{2.49}$$

$$\begin{aligned}
&\int d\mathbf{v} m(\mathbf{v}-\mathbf{u})(\mathbf{v}-\mathbf{u})\mathbf{v}\cdot\nabla f_0 \\
= \;& \int d\mathbf{v} m(\mathbf{v}-\mathbf{u})(\mathbf{v}-\mathbf{u})(\mathbf{v}-\mathbf{u}+\mathbf{u})\cdot\nabla f_0 \\
= \;& \int d\mathbf{v} m(\mathbf{v}-\mathbf{u})(\mathbf{v}-\mathbf{u})(\mathbf{v}-\mathbf{u})\cdot\nabla f_0 + \int d\mathbf{v} m(\mathbf{v}-\mathbf{u})(\mathbf{v}-\mathbf{u})\mathbf{u}\cdot\nabla f_0 \\
= \;& \nabla\cdot\int d\mathbf{v} m(\mathbf{v}-\mathbf{u})(\mathbf{v}-\mathbf{u})(\mathbf{v}-\mathbf{u})f_0 - \nabla\cdot\mathbf{u}\int d\mathbf{v} m(\mathbf{v}-\mathbf{u})(\mathbf{v}-\mathbf{u})f_0 \\
& -\left(\int d\mathbf{v} m(v_\alpha-u_\alpha)(v_\beta-u_\beta)f_0\right)\partial_\alpha\mathbf{u} - \left(\int d\mathbf{v} m(v_\alpha-u_\alpha)(v_\beta-u_\beta)f_0\right)\partial_\beta\mathbf{u} \\
& +\mathbf{u}\cdot\nabla\int d\mathbf{v} m(\mathbf{v}-\mathbf{u})(\mathbf{v}-\mathbf{u})f_0 \\
= \;& \nabla\cdot\mathbf{q} - \overleftrightarrow{P}\nabla\cdot\mathbf{u} - \left(\overleftrightarrow{P} + \overleftrightarrow{P}^T\right)\cdot\nabla\mathbf{u} + \mathbf{u}\cdot\nabla\overleftrightarrow{P},
\end{aligned}$$

and

$$\int d\mathbf{v}(\mathbf{v}-\mathbf{u})(\mathbf{v}-\mathbf{u})\mathbf{E}\cdot\nabla_{\mathbf{v}} f_1 = -2\mathbf{E}\cdot\int d\mathbf{v}(\mathbf{v}-\mathbf{u})f_0 = 0.$$

Here, it has been noted that $\int d\mathbf{v}(\mathbf{v}-\mathbf{u})f_1$ vanishes in the transverse direction and $\mathbf{E}_\parallel = 0$. It is further noted that

$$\begin{aligned}
&\int d\mathbf{v}(\mathbf{v}-\mathbf{u})(\mathbf{v}-\mathbf{u})\mathbf{v}\times\mathbf{B}\cdot\nabla_{\mathbf{v}} f_0 \\
= \;& \int d\mathbf{v}(v_\delta-u_\delta)(v_\lambda-u_\lambda)e_{\alpha\beta\gamma}v_\beta B_\gamma\partial_{v_\alpha} f_0 \\
= \;& -e_{\alpha\beta\gamma}B_\gamma\int d\mathbf{v} f_1\left[\delta_{\alpha\delta}(v_\lambda-u_\lambda)+\delta_{\alpha\lambda}(v_\delta-u_\delta)\right]v_\beta \\
= \;& -e_{\alpha\beta\gamma}B_\gamma\int d\mathbf{v} f_1\left[\delta_{\alpha\delta}(v_\lambda-u_\lambda)+\delta_{\alpha\lambda}(v_\delta-u_\delta)\right](v_\beta-u_\beta) \\
= \;& \mathbf{B}\times\left(\overleftrightarrow{P} + \overleftrightarrow{P}^T\right).
\end{aligned} \tag{2.50}$$

Using equations (2.49)–(2.50) one finally obtains

$$\frac{d\overleftrightarrow{P}}{dt} = -\overleftrightarrow{P}\nabla\cdot\mathbf{u} - \left(\overleftrightarrow{P} + \overleftrightarrow{P}^T\right)\cdot\nabla\mathbf{u} - \nabla\cdot\mathbf{q} + \frac{e}{m}\mathbf{B}\times\left(\overleftrightarrow{P} + \overleftrightarrow{P}^T\right).$$

By projecting onto the two directions $\mathbf{bb} = \delta_{\alpha 3}\delta_{\beta 3}$ and $\overleftrightarrow{I} - \mathbf{bb} = \delta_{\alpha\beta} - \delta_{\alpha 3}\delta_{\beta 3}$, one obtains

$$\frac{dP_n}{dt} = -P_n\nabla\cdot\mathbf{u} - 2P_n(\mathbf{b}\cdot\nabla\mathbf{u})\cdot\mathbf{b} - \nabla\cdot[(q_n+q_s)\mathbf{b}] - 2\mathbf{b}\cdot\nabla q_s, \quad (2.51)$$

$$\frac{dP_s}{dt} = -sP_n\nabla\cdot\mathbf{u} + P_s(\mathbf{b}\cdot\nabla\mathbf{u})\cdot\mathbf{b} - \nabla\cdot(q_s\mathbf{b}) - q_s\nabla\cdot\mathbf{b}, \quad (2.52)$$

where the species index is given by $a = i, e$ and

$$q_{\alpha\beta\gamma} = q_n b_\alpha b_\beta b_\gamma + q_s\left(\delta_{\alpha\beta}b_\gamma + \delta_{\gamma\alpha}b_\beta + \delta_{\beta\gamma}b_\alpha\right). \quad (2.53)$$

In the adiabatic case $\mathbf{q} = 0$, one has

$$\frac{d}{dt}\left(\frac{P_n B^2}{\rho^3}\right) = 0, \quad (2.54)$$

$$\frac{d}{dt}\left(\frac{P_s}{\rho B}\right) = 0. \quad (2.55)$$

In summary, the continuity equations in equation (2.46), the momentum equations in equations (2.47) and (2.48), and the energy equations in equations (2.51) and (2.52) (or equations (2.54) and (2.55)) consist of the complete set of CGL equations in the two-fluid description. Following the CGL formalism, the parallel electric field effect has been neglected. In principle, one can add it back using the formalism described in subsection 3.6.2.

It is straightforward to derive the single fluid version of the CGL equations from the two fluid ones using the quasineutrality condition $Zn_i = n_e$, where Z is the ion charge number. The set of equations is summarized as follows

$$\frac{\partial n}{\partial t} + \frac{\partial}{\partial x_\alpha}(nu_{i\alpha}) = 0, \quad (2.56)$$

$$n_i m_i \frac{d_i u_{i\alpha}}{dt} = -\frac{\partial P_{\alpha\beta}}{\partial x_\beta} + e_{\alpha\beta\gamma} j_\beta B_\gamma, \quad (2.57)$$

$$\frac{d}{dt}\left(\frac{P_n B^2}{\rho^3}\right) = 0, \quad (2.58)$$

$$\frac{d}{dt}\left(\frac{P_s}{\rho B}\right) = 0. \quad (2.59)$$

Summation of equations (2.12) and (2.13) leads to a single fluid equation with anisotropic pressure

$$n_i m_i \frac{d_i u_{i\alpha}}{dt} = -\frac{\partial P_{\alpha\beta}}{\partial x_\beta} + e_{\alpha\beta\gamma} j_\beta B_\gamma, \quad (2.60)$$

Here, in the derivation of the single-fluid momentum equation (2.57), the electron inertia contribution has been neglected noting that its mass is small. To constitute a complete set of equations, the Maxwell equations need to be included with $\mu_0 \mathbf{j} = \nabla \times \mathbf{B}$ and the quasineutrality condition applied.

2.3 PERPENDICULAR MAGNETOFLUID EQUATIONS

In this section, we outline the perpendicular magnetofluid theory [3]. The detailed review can be found in reference [4]. Magnetically confined plasmas tend to be weakly collisional or collisionless. Otherwise, the magnetic field cannot even confine them due to the collisional transport in the perpendicular direction. Since the collision cannot hold charged particles to move together along the field line as a fluid cell, one cannot employ the fluid parameters, like the parallel fluid velocity, to describe the parallel motion. Nevertheless, because of gyration, the fluid description is still rather good in the direction perpendicular to the magnetic field. These features enable the perpendicular magnetofluid description of magnetically confined plasmas. It is the most straightforward extension of conventional fluid theories, such as the Braginskii two-fluid theory. Like Braginskii's two-fluid theory, the perpendicular magnetofluid can account for the FLR effects of second order.

We first outline the derivations of the kinetic equation in the field-line-aligned coordinates and next derive the general perpendicular moment equations. Afterward, the continuity, momentum, and energy equations will be derived. The results will be used to discuss the extension of MHD theory described in this book.

2.3.1 VLASOV EQUATION IN THE FIELD-LINE-ALIGNED COORDINATES

In this subsection, the Vlasov equation is expressed in the field-line-aligned velocity coordinates [3]. This paves the way for deriving the perpendicular magnetofluid equations. In magnetically confined plasmas, charged particles gyrate around magnetic field lines in the direction perpendicular to the magnetic field while moving rather freely along the field lines. To embrace this anisotropic feature, it is preferable to have a velocity space coordinate system that is aligned to the magnetic field line.

Since we focus on explaining the kinetic effects on the ideal MHD theory, the collisionless case is considered. The Vlasov equation [11] is used as the starting equation:

$$\frac{\partial f}{\partial t} + \mathbf{v} \cdot \frac{\partial f}{\partial \mathbf{x}} + \frac{e}{m} \mathbf{a} \cdot \frac{\partial f}{\partial \mathbf{v}} = 0, \tag{2.61}$$

where f is the distribution function, which depends on the particle coordinates $\mathbf{x}$, velocity $\mathbf{v}$, and time t, and $\mathbf{a} = \mathbf{E} + \mathbf{v} \times \mathbf{B}$. The subscript a for particle species is suppressed for brevity.

We intend to transform the Vlasov equation (2.61) from $(\mathbf{x}, \mathbf{v})$ to $(\mathbf{x}; \mathbf{v}_\perp, \mathbf{v}_\parallel)$ coordinate systems. Note that the magnetic field $\mathbf{B}$ depends on $\mathbf{x}$ and t. The perpendicular and parallel components of velocity $(\mathbf{v}_\perp, \mathbf{v}_\parallel)$ become dependent on $\mathbf{x}$ and t although the complete velocity $\mathbf{v}$ itself is not. We introduce the field-line-aligned coordinate system with the right-handed basis $(\hat{\mathbf{e}}_1, \hat{\mathbf{e}}_2, \hat{\mathbf{e}}_3)$, where $\hat{\mathbf{e}}_{1,2,3}$ are unit vectors, $\hat{\mathbf{e}}_3 = \mathbf{B}/B$, $\hat{\mathbf{e}}_1$ is perpendicular to the magnetic field, and $\hat{\mathbf{e}}_2 = \hat{\mathbf{e}}_3 \times \hat{\mathbf{e}}_1$. We also introduce the laboratory Cartesian coordinate system with the basis $(\mathbf{e}_1, \mathbf{e}_2, \mathbf{e}_3)$, where $\mathbf{e}_{1,2,3}$ are unit vectors. To describe the transformation between two coordinate systems, the transform matrix is defined as follows:

$$g_{\alpha\beta}(\mathbf{x}, t) = \hat{\mathbf{e}}_\alpha \cdot \mathbf{e}_\beta. \tag{2.62}$$

The velocity can be decomposed in the two coordinate systems as follows:

$$\begin{aligned}\mathbf{v} &= v_1\mathbf{e}_1 + v_2\mathbf{e}_2 + v_3\mathbf{e}_3 \\ &= \hat{v}_1\hat{\mathbf{e}}_1 + \hat{v}_2\hat{\mathbf{e}}_2 + \hat{v}_3\hat{\mathbf{e}}_3.\end{aligned} \tag{2.63}$$

Here, $\hat{v}_3$ is the parallel component while $\hat{v}_{1,2}$ are the perpendicular components. Using equation (2.62), one obtains

$$\hat{v}_\alpha = g_{\alpha\beta} v_\beta, \quad v_\alpha = g_{\beta\alpha}\hat{v}_\beta. \tag{2.64}$$

Before transforming to the field-line-aligned coordinates, we first derive the kinetic equation in the center of perpendicular fluid motion by introducing $\mathbf{v}' = \mathbf{v} - \mathbf{V}_\perp$. Note that

$$\begin{aligned}\frac{\partial f}{\partial t} &= \frac{\partial f}{\partial t} - \frac{\partial V_{a\beta}}{\partial t}\frac{\partial f}{\partial v_{a\beta}}, \\ v_\beta \frac{\partial f}{\partial x_\beta} &= v_{a\beta}\frac{\partial f_a}{\partial x_\beta} + V_{a\beta}\frac{\partial f_a}{\partial x_\beta} - v_{a\alpha}\frac{\partial V_{a\beta}}{\partial x_\alpha}\frac{\partial f_a}{\partial v_{a\beta}} - V_{a\alpha}\frac{\partial V_{a\beta}}{\partial x_\alpha}\frac{\partial f_a}{\partial v_{a\beta}}, \\ \frac{F_{a\beta}}{m_a}\frac{\partial f}{\partial v_\beta} &= \frac{e_a}{m_a}\left\{E_\beta\frac{\partial f_a}{\partial v_{a\beta}} + \frac{e_{\beta\gamma\alpha}V_{a\gamma}B_\alpha}{c}\frac{\partial f_a}{\partial v_{a\beta}} + \frac{e_{\beta\gamma\alpha}v_{a\gamma}B_\alpha}{c}\frac{\partial f_a}{\partial v_{a\beta}}\right\}.\end{aligned}$$

Using these results, one obtains

$$\begin{aligned}&\frac{df}{dt} + v'_\beta\frac{\partial f}{\partial x_\beta} + \left(\frac{e}{m}E^*_\beta - \frac{dV_{\perp\beta}}{dt}\right)\frac{\partial f}{\partial v'_\beta} \\ &= \frac{\partial V_{\perp\gamma}}{\partial x_\beta}v'_\beta\frac{\partial f_\alpha}{\partial v'_\gamma} - \frac{e}{m}e_{\beta\gamma\alpha}v'_\gamma B_\alpha\frac{\partial f}{\partial v'_\beta},\end{aligned} \tag{2.65}$$

where

$$\begin{aligned}\frac{d}{dt} &= \left.\frac{\partial}{\partial t}\right|_{\mathbf{x},\mathbf{v}} + \hat{V}_{\perp\alpha}\left.\frac{\partial}{\partial x_\alpha}\right|_{\mathbf{v},t}, \\ E^*_\alpha &\equiv E_\alpha + e_{\alpha\beta\gamma}V_{\perp\beta}\hat{B}_\gamma.\end{aligned}$$

Next, we transform equation (2.65) to the field-line-aligned coordinates. Note that

$$
\begin{aligned}
\frac{df}{dt} &= \left.\frac{\hat{d}f_{\hat{a}}}{dt}\right|_{\mathbf{x},\hat{\mathbf{v}}} + g_{\delta\nu}\hat{v}_\delta \frac{\hat{d}g_{\mu\nu}}{dt}\left.\frac{\partial f}{\partial \hat{v}_\mu}\right|_{\mathbf{x},t}, \\
\frac{\partial V_{\perp\gamma}}{\partial x_\beta} v_\beta \frac{\partial f}{\partial v_\gamma} &= \frac{\partial g_{l\gamma}\hat{V}_{\perp l}}{\partial x_\beta} g_{p\beta}\hat{v}_p g_{r\gamma}\left.\frac{\partial f}{\partial \hat{v}_r}\right|_{\mathbf{x},t}, \\
-\frac{e}{m}e_{\beta\gamma\alpha}v_\gamma B_\alpha \frac{\partial f}{\partial v_\beta} &= -\frac{e}{m}e_{\beta\gamma\alpha}g_{l\gamma}\hat{v}_l B_\alpha g_{r\beta}\left.\frac{\partial f}{\partial \hat{v}_r}\right|_{\mathbf{x},t}.
\end{aligned}
$$

where

$$
\frac{\hat{d}}{dt} \equiv \left.\frac{\partial}{\partial t}\right|_{\mathbf{x},\hat{\mathbf{v}}} + g_{\beta\alpha}\hat{V}_{\perp\beta}\left.\frac{\partial}{\partial x_\alpha}\right|_{\hat{\mathbf{v}},t}.
$$

Using these results, equation (2.65) is transformed to

$$
\begin{aligned}
&\frac{\hat{d}f}{dt} + g_{\beta\alpha}\hat{v}_\beta \left.\frac{\partial f}{\partial x_\alpha}\right|_{\hat{\mathbf{v}},t} + \left(\frac{e}{m}E^*_\beta - \frac{\hat{d}g_{l\beta}\hat{V}_{\perp l}}{dt}\right) g_{r\beta}\left.\frac{\partial f}{\partial \hat{v}_r}\right|_{\mathbf{x},t} \\
&= g_{p\beta}g_{r\gamma}\hat{v}_p \frac{\partial g_{l\gamma}\hat{V}_{\perp l}}{\partial x_\beta}\left.\frac{\partial f}{\partial \hat{v}_r}\right|_{\mathbf{x},t} - \frac{e}{m}e_{\beta\gamma\alpha}g_{l\gamma}\hat{v}_l B_\alpha g_{r\beta}\left.\frac{\partial f}{\partial \hat{v}_r}\right|_{\mathbf{x},t} \\
&\quad - \left(g_{\gamma\delta}\hat{v}_\gamma \frac{\hat{d}g_{\beta\delta}}{dt} + g_{\delta\alpha}g_{\gamma\mu}\hat{v}_\delta\hat{v}_\gamma \frac{\partial g_{\beta\mu}}{\partial x_\alpha}\right)\left.\frac{\partial f}{\partial \hat{v}_\beta}\right|_{\mathbf{x},t}, \qquad (2.66)
\end{aligned}
$$

where

$$
E^*_\alpha = E_\alpha + e_{\alpha\gamma\delta}g_{q\gamma}\hat{V}_{\perp q}B_\delta.
$$

Since equation (2.66) is given in the $(\mathbf{x},\hat{\mathbf{v}},t)$ coordinates, the variables $\mathbf{x}$, $\hat{\mathbf{v}}$, and t in this equation are independent, and their derivatives and integrations are commutable. We can then calculate its perpendicular moments straightforwardly.

Different from Braginskii's formulation in Section 2.1, it is assumed that the distribution function of lowest order is given by the perpendicular Maxwellian

$$
f^0 = n\frac{m}{2\pi T}\exp\left\{-\frac{m\hat{v}'^2_\perp/2}{T}\right\},
$$

instead of the full Maxwellian distribution function. As discussed in reference [4], this is consistent with the lowest-order equilibrium in the gyrokinetic theory, in which the lowest-order equilibrium distribution function is assumed to be a function of the magnetic moment μ.

2.3.2 GENERAL PERPENDICULAR MOMENT EQUATIONS

In this subsection, the general perpendicular moment equation is derived. The moments are constructed only in the direction perpendicular to the magnetic field, while the direction parallel to the magnetic field remains unrestricted. This treatment is different from the Braginskii formalism described in Section 2.1.

We first introduce the definitions of the perpendicular density

$$n(\mathbf{x},\hat{v}_3,t) = \int\int d\hat{v}_1 d\hat{v}_2 f \tag{2.67}$$

and the general perpendicular moment for a phase space variable $\eta(\mathbf{x},\hat{\mathbf{v}})$

$$\langle\eta\rangle = \frac{1}{n}\int\int d\hat{v}_1 d\hat{v}_2 \eta f.$$

The general moment equation is derived by multiplying the kinetic equation (2.66) with a phase-space variable η and integrating it over the velocity space perpendicular to the magnetic field. We note that

$$\begin{aligned}
\int \eta \left.\frac{\partial f}{\partial(x,t)}\right|_{\hat{\mathbf{v}}} d\hat{\mathbf{v}}_\perp &= \frac{\partial(n\langle\eta\rangle)}{\partial(x,t)} - n\left\langle\frac{\partial\eta}{\partial(x,t)}\right\rangle, \\
\int \eta \frac{\partial f}{\partial\hat{v}_\beta} d\hat{\mathbf{v}}_\perp &= -n\left\langle\frac{\partial\eta}{\partial\hat{v}_\beta}\right\rangle + \delta_{\beta 3}\frac{\partial n\langle\eta\rangle}{\partial\hat{v}_3}, \\
\int \eta\hat{v}_p \frac{\partial f}{\partial\hat{v}_\beta} d\hat{\mathbf{v}}_\perp &= -n\left(\left\langle\hat{v}_p\frac{\partial\eta}{\partial\hat{v}_\beta}\right\rangle + \langle\eta\rangle\delta_{p\beta}\right) + \delta_{\beta 3}\frac{\partial n\langle\eta\hat{v}_p\rangle}{\partial\hat{v}_3}, \\
\int \eta\hat{v}_p\hat{v}_q \frac{\partial f}{\partial\hat{v}_\beta} d\hat{\mathbf{v}}_\perp &= -n\left(\left\langle\hat{v}_p\hat{v}_q\frac{\partial\eta}{\partial\hat{v}_\beta}\right\rangle + \langle\eta\hat{v}_q\rangle\delta_{p\beta} + \langle\eta\hat{v}_p\rangle\delta_{q\beta}\right) \\
&\quad + \delta_{\beta 3}\frac{\partial n\langle\eta\hat{v}_p\hat{v}_q\rangle}{\partial\hat{v}_3}.
\end{aligned}$$

and further that

$$\frac{e}{m}e_{\beta\gamma\alpha}g_{l\gamma}B_\alpha g_{r\beta}\delta_{rl} = \frac{e}{m}e_{\beta\gamma\alpha}g_{r\gamma}B_\alpha g_{r\beta} = \frac{e}{m}e_{\beta\gamma\alpha}B_\alpha\delta_{\gamma\beta} = 0,$$

Using these results, one obtains the general moment equation from equation (2.66)

$$\begin{aligned}
&\frac{\partial}{\partial t}(n\langle\eta\rangle) - n\left\langle\frac{\partial\eta}{\partial t}\right\rangle + \frac{\partial}{\partial x_\alpha}\left(n\langle g_{\beta\alpha}\hat{V}_{\perp\beta}\eta\rangle\right) - n\left\langle\frac{\partial}{\partial x_\alpha}\left(g_{\beta\alpha}\hat{V}_{\perp\beta}\eta\right)\right\rangle \\
&+\frac{\partial}{\partial x_\alpha}\left(g_{\beta\alpha}n\langle\eta\hat{v}_\beta\rangle\right) - n\left\langle\hat{v}_\beta\frac{\partial}{\partial x_\alpha}(\eta g_{\beta\alpha})\right\rangle
\end{aligned}$$

$$
\begin{aligned}
& -\left(\frac{e}{m}E_\beta^* - \frac{\hat{d}g_{l\beta}\hat{V}_{\perp l}}{dt}\right) g_{r\beta} n \left\langle \frac{\partial \eta}{\partial \hat{v}_r} \right\rangle \\
= \quad & -g_{p\beta} g_{r\gamma} \frac{\partial g_{l\gamma}\hat{V}_{\perp l}}{\partial x_\beta} n \left(\left\langle \hat{v}_p \frac{\partial \eta}{\partial \hat{v}_r} \right\rangle + \langle \eta \rangle \delta_{pr} \right) + \frac{e}{m} e_{\beta\gamma\alpha} g_{l\gamma} B_\alpha g_{r\beta} n \left\langle \frac{\partial \eta}{\partial \hat{v}_r} \hat{v}_l \right\rangle \\
& + g_{p\delta} \frac{\hat{d}g_{\beta\delta}}{dt} n \left(\left\langle \hat{v}_p \frac{\partial \eta}{\partial \hat{v}_\beta} \right\rangle + \langle \eta \rangle \delta_{p\beta} \right) \\
& + g_{p\alpha} g_{q\mu} \frac{\partial g_{\beta\mu}}{\partial x_\alpha} n \left(\left\langle \hat{v}_p \hat{v}_q \frac{\partial \eta}{\partial \hat{v}_\beta} \right\rangle + \langle \eta \hat{v}_q \rangle \delta_{p\beta} \right) \\
& -\left(\frac{e}{m}E_\beta^* - \frac{\hat{d}g_{l\beta}\hat{V}_{\perp l}}{dt}\right) g_{3\beta} \frac{\partial n \langle \eta \rangle}{\partial \hat{v}_3} + g_{p\beta} g_{3\gamma} \frac{\partial g_{l\gamma}\hat{V}_{\perp l}}{\partial x_\beta} \frac{\partial n \langle \eta \hat{v}_p \rangle}{\partial \hat{v}_3} \\
& -g_{p\delta} \frac{\hat{d}g_{3\delta}}{dt} \frac{\partial n \langle \eta \hat{v}_p \rangle}{\partial \hat{v}_3} - g_{p\alpha} g_{q\mu} \frac{\partial g_{3\mu}}{\partial x_\alpha} \frac{\partial n \langle \eta \hat{v}_p \hat{v}_q \rangle}{\partial \hat{v}_3}. \qquad (2.68)
\end{aligned}
$$

Note here that the terms with a derivative with respect to $\hat{v}_3$ appear in the general moment equation (2.68). This is different from the Braginskii formulation. As shown in reference [4], one can integrate over $\hat{v}_3$ to recover the general moment equation (2.6) in Section 2.1.2 in the Braginskii formulation.

Let us now define the thermal parameters to be used in the individual moment equations. The density moment has been defined in equation (2.67). The velocity, temperature, and pressure are defined as follows:

$$0 = \langle \hat{v}_\alpha \rangle, \qquad (2.69)$$

$$T = \frac{m}{2} \langle \hat{v}_\alpha^2 \rangle, \qquad (2.70)$$

$$p = nT. \qquad (2.71)$$

The fourth-order thermal parameter, referred to as T-square scalar, needs also to be introduced in the perpendicular moment equations

$$D = \frac{m_a^2}{4} \langle \hat{v}^4 \rangle. \qquad (2.72)$$

We can use the perpendicular Maxwellian distribution function in equation (2.67) to examine these thermal parameters. Direct calculations yield.

$$n = \langle 1 \rangle_0 = \int\int d\hat{v}_1 d\hat{v}_2 f^0, \qquad (2.73)$$

$$0 = \langle v_{\perp\alpha} \rangle_0 = \frac{1}{n} \int\int d\hat{v}_1 d\hat{v}_2 v_{\perp\alpha} f^0, \qquad (2.74)$$

$$T = \left\langle \frac{1}{2} m \hat{v}_\perp^2 \right\rangle_0 = \frac{1}{n} \int\int d\hat{v}_1 d\hat{v}_2 \frac{1}{2} m \hat{v}_\perp^2 f^0 = T, \tag{2.75}$$

$$D = \left\langle \frac{1}{4} m^2 \hat{v}_\perp^4 \right\rangle_0 = \frac{1}{n} \int\int d\hat{v}_1 d\hat{v}_2 \frac{1}{4} m^2 \hat{v}_\perp^4 f^0 = 2T^2. \tag{2.76}$$

The next-order thermal parameters, such as viscosity $\pi_{\alpha\beta}$, heat flux q_α, etc., are also introduced as follows:

$$\pi_{\alpha\beta} = \rho \left\langle \hat{v}_{\perp\alpha} \hat{v}_{\perp\beta} - \frac{1}{2} \hat{v}_\perp^2 \delta_{\alpha\beta}^\perp \right\rangle. \tag{2.77}$$

$$q_\alpha = \frac{1}{2} \rho \left\langle \hat{v}_{\perp v}^2 \hat{v}_{\perp\alpha} \right\rangle, \tag{2.78}$$

$$\Theta_{\alpha\beta\gamma} = \rho \left\langle \left(\hat{v}_{\perp\alpha} \hat{v}_{\perp\beta} - \frac{1}{2} v_\perp'^2 \delta_{\alpha\beta}^\perp \right) \hat{v}_{\perp\gamma} \right\rangle, \tag{2.79}$$

$$\Xi_{\alpha\beta} = \frac{m^2}{2} \left\langle \hat{v}_\perp'^2 \left(\hat{v}_{\perp\alpha} \hat{v}_{\perp\beta} - \frac{1}{2} \hat{v}_\perp^2 \delta_{\alpha\beta}^\perp \right) \right\rangle. \tag{2.80}$$

Here, the new thermal parameters in the perpendicular magnetofluid description $\Theta_{\alpha\beta\gamma}$ and $\Xi_{\alpha\beta}$ are referred to as the heat flux tensor and T-square tensor, respectively. It can be proved that the perpendicular Maxwellian distribution function in equation (2.67) does not contribute to these quantities, i.e., $\pi_{\alpha\beta 0} = q_{\alpha 0} = \Theta_{\alpha\beta\gamma 0} = \Theta_{\alpha\beta\gamma 0} = 0$. The next-order distribution function can contribute to them, which are related to the FLR effects.

Using these definitions, one proves that

$$\rho \left\langle \hat{v}_{\perp\alpha} \hat{v}_{\perp\beta} \right\rangle = p\delta_{\alpha\beta}^\perp + \pi_{\alpha\beta}, \tag{2.81}$$

$$\rho \left\langle \hat{v}_{\perp\alpha} \hat{v}_{\perp\beta} \hat{v}_{\perp\gamma} \right\rangle = q_\gamma \delta_{\alpha\beta}^\perp + \Theta_{\alpha\beta\gamma}, \tag{2.82}$$

$$\frac{1}{2} m^2 \left\langle \hat{v}_\perp'^2 v_{\perp\alpha} \hat{v}_{\perp\beta} \right\rangle = D\delta_{\alpha\beta}^\perp + \Xi_{\alpha\beta}. \tag{2.83}$$

Here, $\delta_{\alpha\beta}^\perp$ $(\alpha, \beta < 3)$, which excludes the third component 3. Noting the definition of $\delta_{\alpha\beta}^\perp$, one can see that the pressure p in equation (2.81), q_α in equation (2.82), and the parameter D in equation (2.83) appear only in the perpendicular direction. Different from the isotropic case in the Braginskii two-fluid theory, equation (2.9), where the factor $1/3$ is introduced, the factor $1/2$ is introduced in the definitions in equations (2.77), (2.79), and (2.80) to make their traces to be zero.

Similar to the Barginskii theory, the following constraints are introduced

$$\langle 1 \rangle_1 = \langle \hat{v}_{\perp\alpha} \rangle_1 = \left\langle \hat{v}_\perp^2 \right\rangle_1 = \left\langle \hat{v}_\perp^4 \right\rangle_1 = 0, \tag{2.84}$$

where $\langle \cdots \rangle_1$ denotes the perpendicular velocity space average with the first-order distribution function f^1.

2.3.3 CONTINUITY EQUATION

In this subsection, we derive the continuity equation in the perpendicular magnetofluid formalism. The general moment equation (2.68) is used to derive the equation by letting $\eta = 1$.

We can reduce the general moment equation, equation (2.68), term by term as follows. The terms having no contributions to the continuity equation are not explained.

$$\begin{aligned}
&\frac{\partial}{\partial t}(n\langle\eta\rangle) - n\left\langle\frac{\partial\eta}{\partial t}\right\rangle + \frac{\partial}{\partial x_\alpha}\left(n\langle g_{\beta\alpha}\hat{V}_{\perp\beta}\eta\rangle\right) - n\left\langle\frac{\partial}{\partial x_\alpha}\left(g_{\beta\alpha}\hat{V}_{\perp\beta}\eta\right)\right\rangle \\
&= \frac{\partial n}{\partial t} + g_{\beta\alpha}\hat{V}_{\perp\beta}\frac{\partial n}{\partial x_\alpha},
\end{aligned}$$

$$\begin{aligned}
&\frac{\partial}{\partial x_\alpha}(g_{\beta\alpha}n\langle\eta\hat{v}_\beta\rangle) - n\left\langle\hat{v}_\beta\frac{\partial}{\partial x_\alpha}(\eta g_{\beta\alpha})\right\rangle \\
&= \frac{\partial}{\partial x_\alpha}(g_{3\alpha}n\hat{v}_3) - n\hat{v}_3\frac{\partial g_{3\alpha}}{\partial x_\alpha} = g_{3\alpha}\hat{v}_3\frac{\partial n}{\partial x_\alpha},
\end{aligned}$$

$$-g_{p\beta}g_{r\gamma}\frac{\partial g_{l\gamma}\hat{V}_{\perp l}}{\partial x_\beta}n\left(\left\langle\tilde{v}_p\frac{\partial\eta}{\partial\tilde{v}_r}\right\rangle + \langle\eta\rangle\delta_{pr}\right) = -g_{r\beta}g_{r\gamma}\frac{\partial g_{l\gamma}\hat{V}_{\perp l}}{\partial x_\beta}n = -\frac{\partial g_{l\beta}\hat{V}_{\perp l}}{\partial x_\beta}n,$$

$$g_{p\delta}\frac{\hat{d}g_{\beta\delta}}{dt}n\left(\left\langle\hat{v}_p\frac{\partial\eta}{\partial\hat{v}_\beta}\right\rangle + \langle\eta\rangle\delta_{p\beta}\right) = \frac{n}{2}\frac{\hat{d}g^2_{\beta\delta}}{dt} = 0,$$

$$g_{p\alpha}g_{q\mu}\frac{\partial g_{\beta\mu}}{\partial x_\alpha}n\left(\left\langle\hat{v}_p\hat{v}_q\frac{\partial\eta}{\partial\hat{v}_\beta}\right\rangle + \langle\eta\hat{v}_q\rangle\delta_{p\beta}\right) = -n\hat{v}_3\frac{\partial g_{3\alpha}}{\partial x_\alpha},$$

$$-\left(\frac{e}{m}E^*_\beta - \frac{\hat{d}g_{l\beta}\hat{V}_{\perp l}}{dt}\right)g_{3\beta}\frac{\partial n\langle\eta\rangle}{\partial\hat{v}_3} = -\left(\frac{e}{m}E^*_\beta - \frac{\hat{d}g_{l\beta}\hat{V}_{\perp l}}{dt}\right)g_{3\beta}\frac{\partial n}{\partial\hat{v}_3},$$

$$g_{p\beta}g_{3\gamma}\frac{\partial g_{l\gamma}\hat{V}_{\perp l}}{\partial x_\beta}\frac{\partial n\langle\hat{v}_p\eta\rangle}{\partial\hat{v}_3} = g_{3\beta}g_{3\gamma}\frac{\partial g_{l\gamma}\hat{V}_{\perp l}}{\partial x_\beta}\hat{v}_3\frac{\partial n}{\partial\hat{v}_3} + ng_{3\beta}g_{3\gamma}\frac{\partial g_{l\gamma}\hat{V}_{\perp l}}{\partial x_\beta},$$

$$-g_{p\delta}\frac{\hat{d}g_{3\delta}}{dt}\frac{\partial n\langle\eta\hat{v}_p\rangle}{\partial\hat{v}_3} = -g_{3\delta}\frac{\hat{d}g_{3\delta}}{dt}\frac{\partial n\langle\hat{v}_3\rangle}{\partial\hat{v}_3} = 0,$$

and

$$\begin{aligned}&-g_{p\alpha}g_{q\mu}\frac{\partial g_{3\mu}}{\partial x_\alpha}\frac{\partial n\langle\eta\hat{v}_p\hat{v}_q\rangle}{\partial\hat{v}_3}\\ =\;&-\frac{1}{m}\frac{\partial g_{3\alpha}}{\partial x_\alpha}\frac{\partial nT}{\partial\hat{v}_3}-\frac{1}{m}g_{p\alpha}g_{q\mu}\frac{\partial g_{3\mu}}{\partial x_\alpha}\frac{\partial\pi_{pq}}{\partial\hat{v}_3}.\end{aligned}$$

Using these results, we obtain the continuity equations:

$$\begin{aligned}&\frac{\partial n}{\partial t}+g_{\beta\alpha}\hat{V}_{\perp\beta}\frac{\partial n}{\partial x_\alpha}+g_{3\alpha}\hat{v}_3\frac{\partial n}{\partial x_\alpha}+n\frac{\partial g_{l\gamma}\hat{V}_{\perp l}}{\partial x_\beta}\left(\delta_{\beta\gamma}-g_{3\beta}g_{3\gamma}\right)\\ &+n\hat{v}_3\frac{\partial g_{3\alpha}}{\partial x_\alpha}+\frac{1}{m}\frac{\partial g_{3\alpha}}{\partial x_\alpha}\frac{\partial nT}{\partial\hat{v}_3}-g_{3\beta}g_{3\gamma}\frac{\partial g_{l\gamma}\hat{V}_{\perp l}}{\partial x_\beta}\hat{v}_3\frac{\partial n}{\partial\hat{v}_3}\\ =\;&-g_{3\beta}\frac{\partial n}{\partial\hat{v}_3}\left(\frac{e}{m}E^*_\beta-\frac{\hat{d}g_{l\beta}\hat{V}_{\perp l}}{dt}\right)-\frac{1}{m}g_{p\alpha}g_{q\mu}\frac{\partial g_{3\mu}}{\partial x_\alpha}\frac{\partial\pi_{pq}}{\partial\hat{v}_3}.\end{aligned}\tag{2.85}$$

In equation (2.85), the species index has been dropped. The equation applies for both the ion and electron species.

Let us briefly discuss this result. From equation (2.85), one can see that only the perpendicular fluid compression and convection appear in the continuity equation through $\hat{\mathbf{V}}_\perp$. The parallel dynamics remain to be fully kinetic through the parallel particle velocity $\hat{v}_3$. The FLR correction π_{pq} also appears in this continuity equation. As shown in reference [4], one can derive the dispersion relation for electrostatic modes using equation (2.85).

2.3.4 MOMENTUM TRANSPORT EQUATION

In this subsection, we derive the momentum transport equation in the perpendicular magnetofluid formalism. The general moment equation (2.68) is used to derive the equation by letting $\eta = mg_{\nu\lambda}\hat{v}_{\perp\nu}$. The term-by-term simplifications are given as follows.

First, it is noted that using equation (2.69), one can prove that

$$\frac{\partial}{\partial t}(n\langle\eta\rangle)-n\left\langle\frac{\partial\eta}{\partial t}\right\rangle+\frac{\partial}{\partial x_\alpha}\left(n\langle g_{\beta\alpha}\hat{V}_{\perp\beta}\eta\rangle\right)-n\left\langle\frac{\partial}{\partial x_\alpha}\left(g_{\beta\alpha}\hat{V}_{\perp\beta}\eta\right)\right\rangle = 0.$$

Next, we reduce the remaining terms sequentially. Note that

$$\begin{aligned}&\frac{\partial}{\partial x_\alpha}(g_{\beta\alpha}n\langle\eta\hat{v}_\beta\rangle)-n\left\langle\tilde{v}_\beta\frac{\partial}{\partial x_\alpha}(\eta g_{\beta\alpha})\right\rangle\\ =\;&(\delta_{\lambda\alpha}-g_{3\lambda}g_{3\alpha})\frac{\partial p}{\partial x_\alpha}+g_{\nu\lambda}g_{\beta\alpha}\frac{\partial\pi_{\beta\nu}}{\partial x_\alpha},\end{aligned}$$

$$
\begin{aligned}
& -\left(\frac{e}{m}E^*_\beta - \frac{\hat{d}g_{l\beta}\hat{V}_{\perp l}}{dt}\right) g_{r\beta} n \left\langle \frac{\partial \eta}{\partial \hat{v}_r} \right\rangle \\
= \quad & -\rho \left(\frac{e}{m}E^*_\beta - \frac{\hat{d}g_{l\beta}\hat{V}_{\perp l}}{dt}\right) \left(\delta_{\beta\lambda} - g_{3\beta}g_{3\lambda}\right),
\end{aligned}
$$

$$
\begin{aligned}
& g_{p\beta}g_{r\gamma}\frac{\partial g_{l\gamma}\hat{V}_{\perp l}}{\partial x_\beta} n \left(\left\langle \hat{v}_p \frac{\partial \eta}{\partial \hat{v}_r}\right\rangle + \langle \eta \rangle \delta_{pr}\right) \\
= \quad & \left(\delta_{\gamma\lambda} - g_{3\gamma}g_{3\lambda}\right) \rho \hat{v}_3 g_{3\beta} \frac{\partial g_{l\gamma}\hat{V}_{\perp l}}{\partial x_\beta},
\end{aligned}
$$

$$
-\frac{e}{m} e_{\beta\gamma\alpha} g_{l\gamma} B_\alpha g_{r\beta} n \left\langle \frac{\partial \eta}{\partial \hat{v}_r} \hat{v}_l \right\rangle = 0,
$$

$$
-g_{p\delta}\frac{\hat{d}g_{\beta\delta}}{dt} n \left(\left\langle \hat{v}_p \frac{\partial \eta}{\partial \hat{v}_\beta}\right\rangle + \langle \eta \rangle \delta_{p\beta}\right) = \left(\delta_{\delta\lambda} - g_{3\delta}g_{3\lambda}\right) \rho \hat{v}_3 \frac{\hat{d}g_{3\delta}}{dt},
$$

$$
\begin{aligned}
& -g_{p\alpha}g_{q\mu}\frac{\partial g_{\beta\mu}}{\partial x_\alpha} n \left(\left\langle \hat{v}_p \hat{v}_q \frac{\partial \eta}{\partial \hat{v}_\beta}\right\rangle + \langle \eta \hat{v}_q \rangle \delta_{p\beta}\right) \\
= \quad & \underbrace{-g_{3\alpha}\frac{\partial g_{3\lambda}}{\partial x_\alpha} p}_{x1} + g_{3\alpha}\hat{v}_3^2 \frac{\partial g_{3\lambda}}{\partial x_\alpha}\rho + \left(\frac{\partial g_{p\alpha}g_{q\lambda}}{\partial x_\alpha} + g_{p\alpha}g_{q\mu}g_{3\lambda}\frac{\partial g_{3\mu}}{\partial x_\alpha}\right)\pi_{pq},
\end{aligned}
$$

$$
\left(\frac{e}{m}E^*_\beta - \frac{\hat{d}g_{l\beta}\hat{V}_{\perp l}}{dt}\right) g_{3\beta} \frac{\partial n \langle \eta \rangle}{\partial \hat{v}_3} = 0,
$$

$$
\begin{aligned}
& \left(g_{p\delta}\frac{\hat{d}g_{3\delta}}{dt} - g_{p\beta}g_{3\gamma}\frac{\partial g_{l\gamma}\hat{V}_{\perp l}}{\partial x_\beta}\right) \frac{\partial n \langle \hat{v}_p \eta \rangle}{\partial \hat{v}_3} \\
= \quad & \frac{\hat{d}g_{3\lambda}}{dt}\frac{\partial p}{\partial \hat{v}_3} - \left(\delta_{\beta\lambda} - g_{3\beta}g_{3\lambda}\right) g_{3\gamma}\frac{\partial g_{l\gamma}\hat{V}_{\perp l}}{\partial x_\beta}\frac{\partial p}{\partial \hat{v}_3} \\
& + \left(g_{p\delta}\frac{\hat{d}g_{3\delta}}{dt} - g_{p\beta}g_{3\gamma}\frac{\partial g_{l\gamma}\hat{V}_{\perp l}}{\partial x_\beta}\right) g_{\nu\lambda}\frac{\partial \pi_{p\nu}}{\partial \hat{v}_3},
\end{aligned}
$$

and

$$
\begin{aligned}
& g_{p\alpha} g_{q\mu} \frac{\partial g_{3\mu}}{\partial x_\alpha} \frac{\partial n\langle \eta \hat{v}_p \hat{v}_q \rangle}{\partial \hat{v}_3} = m g_{p\alpha} g_{q\mu} \frac{\partial g_{3\mu}}{\partial x_\alpha} g_{\nu\lambda} \frac{\partial n \langle \hat{v}_p \hat{v}_q \hat{v}_{\perp\nu} \rangle}{\partial \hat{v}_3} \\
= \;& g_{\nu\lambda} \frac{\partial g_{3\alpha}}{\partial x_\alpha} \frac{\partial q_\nu}{\partial \hat{v}_3} + g_{p\alpha} g_{q\mu} g_{\nu\lambda} \frac{\partial g_{3\mu}}{\partial x_\alpha} \frac{\partial \Theta_{pq\nu}}{\partial \hat{v}_3} + \underbrace{g_{3\alpha} \frac{\partial g_{3\lambda}}{\partial x_\alpha} p}_{x1} + g_{3\alpha} g_{q\mu} g_{\nu\lambda} \frac{\partial g_{3\mu}}{\partial x_\alpha} \pi_{q\nu} \\
& + \hat{v}_3 g_{3\alpha} \frac{\partial g_{3\lambda}}{\partial x_\alpha} \frac{\partial p}{\partial \hat{v}_3} + \hat{v}_3 g_{3\alpha} g_{q\mu} g_{\nu\lambda} \frac{\partial g_{3\mu}}{\partial x_\alpha} \frac{\partial \pi_{q\nu}}{\partial \hat{v}_3}.
\end{aligned}
$$

Using these results and noting the cancellations between the terms with the same indices "xi", one finally obtains the momentum equation:

$$
\begin{aligned}
& (\delta_{\lambda\alpha} - g_{3\alpha} g_{3\lambda}) \left(neE^*_\alpha - \rho \frac{\hat{d} g_{l\alpha} \hat{V}_{\perp l}}{dt} - \frac{\partial p}{\partial x_\alpha} - \hat{v}_3 g_{3\beta} \frac{\partial g_{3\alpha}}{\partial x_\beta} \frac{\partial p}{\partial \hat{v}_3} \right. \\
& \left. - g_{3\beta} \hat{v}_3^2 \frac{\partial g_{3\alpha}}{\partial x_\beta} \rho + g_{3\gamma} \frac{\partial g_{l\gamma} \hat{V}_{\perp l}}{\partial x_\alpha} \frac{\partial p}{\partial \hat{v}_3} - \rho \hat{v}_3 g_{3\beta} \frac{\partial g_{l\alpha} \hat{V}_{\perp l}}{\partial x_\beta} - \rho \hat{v}_3 \frac{\hat{d} g_{3\alpha}}{dt} - \frac{\hat{d} g_{3\alpha}}{dt} \frac{\partial p}{\partial \hat{v}_3} \right) \\
= \;& g_{\nu\lambda} g_{\beta\alpha} \frac{\partial \pi_{\beta\nu}}{\partial x_\alpha} + g_{\nu\lambda} \left(g_{p\delta} \frac{\hat{d} g_{3\delta}}{dt} - g_{p\beta} g_{3\gamma} \frac{\partial g_{l\gamma} \hat{V}_{\perp l}}{\partial x_\beta} + \hat{v}_3 g_{3\alpha} g_{p\mu} \frac{\partial g_{3\mu}}{\partial x_\alpha} \right) \frac{\partial \pi_{p\nu}}{\partial \hat{v}_3} \\
& + \left(\frac{\partial g_{p\alpha} g_{q\lambda}}{\partial x_\alpha} + g_{p\alpha} g_{q\mu} g_{3\lambda} \frac{\partial g_{3\mu}}{\partial x_\alpha} + g_{3\alpha} g_{q\mu} g_{p\lambda} \frac{\partial g_{3\mu}}{\partial x_\alpha} \right) \pi_{pq} \\
& + g_{\nu\lambda} \frac{\partial g_{3\alpha}}{\partial x_\alpha} \frac{\partial q_\nu}{\partial \hat{v}_3} + g_{p\alpha} g_{q\mu} g_{\nu\lambda} \frac{\partial g_{3\mu}}{\partial x_\alpha} \frac{\partial \Theta_{pq\nu}}{\partial \hat{v}_3},
\end{aligned}
\tag{2.86}
$$

where $E^*_\alpha = E_\alpha + e_{\alpha\gamma\delta} g_{q\gamma} \hat{V}_q B_\delta$. In equation (2.86), the species index has been dropped. The equation applies both for the ion and electron species.

Let us briefly discuss the momentum transport equation (2.86). We first point out that (2.86) actually contains only two perpendicular components. This can be proved by applying the parallel-projection operator $g_{3\lambda}$, which shows that the parallel projection of equation (2.86) is a trivial equation.

2.3.5 ENERGY TRANSPORT EQUATION

In this subsection, we derive the energy transport equation in the perpendicular magnetofluid formalism. The general moment equation (2.68) is used to derive the equation by letting $\eta = m\hat{v}_{\perp\nu}\hat{v}_{\perp\nu}/2$.

The term-by-term simplifications are given as follows.

$$
\begin{aligned}
& \frac{\partial}{\partial t}(n\langle\eta\rangle) - n\left\langle \frac{\partial \eta}{\partial t} \right\rangle + \frac{\partial}{\partial x_\alpha}\left(n\langle g_{\beta\alpha} \hat{V}_{\perp\beta} \eta\rangle\right) - n\left\langle \frac{\partial}{\partial x_\alpha}\left(g_{\beta\alpha} \hat{V}_{\perp\beta} \eta\right)\right\rangle \\
= \;& \frac{\partial}{\partial t}(nT) + g_{\beta\alpha} \hat{V}_{\perp\beta} \frac{\partial}{\partial x_\alpha}(nT),
\end{aligned}
$$

$$\begin{aligned}
&\frac{\partial}{\partial x_\alpha}(g_{\beta\alpha} n \langle \eta \hat{v}_\beta \rangle) - n \left\langle \hat{v}_\beta \frac{\partial}{\partial x_\alpha}(\eta g_{\beta\alpha}) \right\rangle \\
= \; & \frac{\partial}{\partial x_\alpha}(g_{\beta\alpha} q_\beta) + \underbrace{\hat{v}_3 \frac{\partial}{\partial x_\alpha}(g_{3\alpha} p)}_{o1} - \underbrace{\rho \left\langle \frac{1}{2}\hat{v}_{\perp\nu}^2 \hat{v}_\beta \right\rangle \frac{\partial g_{\beta\alpha}}{\partial x_\alpha}}_{x2},
\end{aligned}$$

$$\begin{aligned}
& g_{q\alpha} g_{p\mu} \frac{\partial g_{\beta\mu}}{\partial x_\alpha} n \left(\left\langle \hat{v}_p \hat{v}_q \frac{\partial \eta}{\partial \hat{v}_\beta} \right\rangle + \langle \eta \hat{v}_q \rangle \delta_{p\beta} \right) \\
= \; & \underbrace{-\hat{v}_3 \frac{\partial g_{3\alpha}}{\partial x_\alpha} p}_{o1} - \hat{v}_3 g_{p\alpha} g_{q\mu} \frac{\partial g_{3\mu}}{\partial x_\alpha} \pi_{pq} - \underbrace{\frac{\partial g_{p\alpha}}{\partial x_\alpha} \rho \left(\left\langle \frac{1}{2}\hat{v}_{\perp\nu}^2 \hat{v}_p \right\rangle \right)}_{x2},
\end{aligned}$$

$$-\left(\frac{e}{m} E_\beta^* - \frac{\hat{d} g_{l\beta} \hat{V}_{\perp l}}{dt} \right) g_{r\beta} n \left\langle \frac{\partial \eta}{\partial \hat{v}_r} \right\rangle = 0,$$

$$\begin{aligned}
& -g_{p\beta} g_{r\gamma} \frac{\partial g_{l\gamma} \hat{V}_{\perp l}}{\partial x_\beta} n \left(\left\langle \hat{v}_p \frac{\partial \eta}{\partial \hat{v}_r} \right\rangle + \langle \eta \rangle \delta_{pr} \right) \\
= \; & -2 \frac{\partial g_{l\beta} \hat{V}_{\perp l}}{\partial x_\beta} p + \underbrace{g_{3\beta} g_{3\gamma} \frac{\partial g_{l\gamma} \hat{V}_{\perp l}}{\partial x_\beta} p}_{o3} - g_{p\beta} g_{r\gamma} \frac{\partial g_{l\gamma} \hat{V}_{\perp l}}{\partial x_\beta} \pi_{pr},
\end{aligned}$$

$$\frac{e}{m} e_{\beta\gamma\alpha} g_{l\gamma} B_\alpha g_{r\beta} n \left\langle \frac{\partial \eta}{\partial \hat{v}_r} \hat{v}_l \right\rangle = e e_{\beta\gamma\alpha} g_{l\gamma} B_\alpha g_{r\beta} n \langle \hat{v}_{\perp r} \hat{v}_{\perp l} \rangle = 0,$$

$$g_{p\delta} \frac{\hat{d} g_{\beta\delta}}{dt} n \left(\left\langle \hat{v}_p \frac{\partial \eta}{\partial \hat{v}_\beta} \right\rangle + \langle \eta \rangle \delta_{p\beta} \right) = 0,$$

$$-\left(\frac{e}{m} E_\beta^* - \frac{\hat{d} g_{l\beta} \hat{V}_{\perp l}}{dt} \right) g_{3\beta} \frac{\partial n \langle \eta \rangle}{\partial \hat{v}_3} = -\left(\frac{e}{m} E_\beta^* - \frac{\hat{d} g_{l\beta} \hat{V}_{\perp l}}{dt} \right) g_{3\beta} \frac{\partial p}{\partial \hat{v}_3},$$

$$\begin{aligned}
& g_{p\beta} g_{3\gamma} \frac{\partial g_{l\gamma} \hat{V}_{\perp l}}{\partial x_\beta} \frac{\partial n \langle \hat{v}_p \eta \rangle}{\partial \hat{v}_3} \\
= \; & g_{p\beta} g_{3\gamma} \frac{\partial g_{l\gamma} \hat{V}_{\perp l}}{\partial x_\beta} \frac{\partial q_p}{\partial \hat{v}_3} + \hat{v}_3 g_{3\beta} g_{3\gamma} \frac{\partial g_{l\gamma} \hat{V}_{\perp l}}{\partial x_\beta} \frac{\partial p}{\partial \hat{v}_3} + \underbrace{g_{3\beta} g_{3\gamma} \frac{\partial g_{l\gamma} \hat{V}_{\perp l}}{\partial x_\beta} p}_{o3},
\end{aligned}$$

$$-g_{p\delta}\frac{\hat{d}g_{3\delta}}{dt}\frac{\partial n\langle\hat{v}_p\eta\rangle}{\partial\hat{v}_3} = -g_{p\delta}\frac{\hat{d}g_{3\delta}}{dt}\frac{\partial q_p}{\partial\hat{v}_3},$$

and

$$\begin{aligned}
&-g_{p\alpha}g_{q\mu}\frac{\partial g_{3\mu}}{\partial x_\alpha}\frac{\partial n\langle\eta\hat{v}_p\hat{v}_q\rangle}{\partial\hat{v}_3}\\
&= -\frac{\partial g_{3\alpha}}{\partial x_\alpha}\frac{1}{m}\frac{\partial nD}{\partial\hat{v}_3} - g_{p\alpha}g_{q\mu}\frac{\partial g_{3\mu}}{\partial x_\alpha}\frac{1}{m}\frac{\partial n\Xi_{pq}}{\partial\hat{v}_3}\\
&\quad -\hat{v}_3 g_{3\alpha}g_{q\mu}\frac{\partial g_{3\mu}}{\partial x_\alpha}\frac{\partial q_q}{\partial\hat{v}_3} - g_{3\alpha}g_{q\mu}\frac{\partial g_{3\mu}}{\partial x_\alpha}q_q.
\end{aligned}$$

Using these results and noting the cancellations between the terms with the same indices "*xi*" and the combinations with the same indices "*oi*", one obtains the energy transport equation

$$\begin{aligned}
&\frac{\partial p}{\partial t} + g_{\beta\alpha}\hat{V}_{\perp\beta}\frac{\partial p}{\partial x_\alpha} + g_{3\alpha}\hat{v}_3\frac{\partial p}{\partial x_\alpha} + 2p\frac{\partial g_{l\gamma}\hat{V}_{\perp l}}{\partial x_\beta}(\delta_{\beta\gamma} - g_{3\beta}g_{3\gamma})\\
&+2\hat{v}_3\frac{\partial g_{3\alpha}}{\partial x_\alpha}p + \frac{\partial g_{3\alpha}}{\partial x_\alpha}\frac{1}{m}\frac{\partial nD}{\partial\hat{v}_3} - \hat{v}_3 g_{3\beta}g_{3\gamma}\frac{\partial g_{l\gamma}\hat{V}_{\perp l}}{\partial x_\beta}\frac{\partial p}{\partial\hat{v}_3}\\
&= -g_{3\alpha}\left(\frac{e}{m}E^*_\alpha - \frac{\hat{d}g_{l\alpha}\hat{V}_{\perp l}}{dt}\right)\frac{\partial p}{\partial\hat{v}_3} - g_{p\alpha}g_{q\mu}\left(\hat{v}_3\frac{\partial g_{3\mu}}{\partial x_\alpha} + \frac{\partial g_{l\mu}\hat{V}_{\perp l}}{\partial x_\alpha}\right)\pi_{pq}\\
&+\left(g_{p\beta}g_{3\gamma}\frac{\partial g_{l\gamma}\hat{V}_{\perp l}}{\partial x_\beta} - g_{p\delta}\frac{\hat{d}g_{3\delta}}{dt}\right)\frac{\partial q_p}{\partial\hat{v}_3} - \frac{\partial}{\partial x_\alpha}(g_{\beta\alpha}q_\beta)\\
&-\hat{v}_3 g_{3\alpha}g_{q\mu}\frac{\partial g_{3\mu}}{\partial x_\alpha}\frac{\partial q_q}{\partial\hat{v}_3} - g_{3\alpha}g_{q\mu}\frac{\partial g_{3\mu}}{\partial x_\alpha}q_q - g_{p\alpha}g_{q\mu}\frac{\partial g_{3\mu}}{\partial x_\alpha}\frac{1}{m}\frac{\partial n\Xi_{pq}}{\partial\hat{v}_3}. \qquad (2.87)
\end{aligned}$$

Using the continuity equation in equation (2.85), one can reduce equation (2.87) to

$$\begin{aligned}
&n\frac{\partial T}{\partial t} + ng_{\beta\alpha}\hat{V}_{\perp\beta}\frac{\partial T}{\partial x_\alpha} + ng_{3\alpha}\hat{v}_3\frac{\partial T}{\partial x_\alpha} + p\frac{\partial g_{l\gamma}\hat{V}_{\perp l}}{\partial x_\beta}(\delta_{\gamma\beta} - g_{3\beta}g_{3\gamma})\\
&+\hat{v}_3\frac{\partial g_{3\alpha}}{\partial x_\alpha}p + \frac{\partial g_{3\alpha}}{\partial x_\alpha}\frac{T^2}{m}\frac{\partial n}{\partial\hat{v}_3} + \frac{\partial g_{3\alpha}}{\partial x_\alpha}\frac{3p}{m}\frac{\partial T}{\partial\hat{v}_3} - \hat{v}_3 g_{3\beta}g_{3\gamma}\frac{\partial g_{l\gamma}\hat{V}_{\perp l}}{\partial x_\beta}n\frac{\partial T}{\partial\hat{v}_3}\\
&= -g_{3\alpha}\left(\frac{e}{m}E^*_\alpha - \frac{\hat{d}g_{l\alpha}\hat{V}_{\perp l}}{dt}\right)n\frac{\partial T}{\partial\hat{v}_3} + \frac{T}{m}g_{p\alpha}g_{q\mu}\frac{\partial g_{3\mu}}{\partial x_\alpha}\frac{\partial\pi_{pq}}{\partial\hat{v}_3}\\
&-g_{p\alpha}g_{q\mu}\left(\hat{v}_3\frac{\partial g_{3\mu}}{\partial x_\alpha} + \frac{\partial g_{l\mu}\hat{V}_{\perp l}}{\partial x_\alpha}\right)\pi_{pq} + \left(g_{p\beta}g_{3\gamma}\frac{\partial g_{l\gamma}\hat{V}_{\perp l}}{\partial x_\beta} - g_{p\delta}\frac{\hat{d}g_{3\delta}}{dt}\right)\frac{\partial q_p}{\partial\hat{v}_3}\\
&-\frac{\partial}{\partial x_\alpha}(g_{\beta\alpha}q_\beta) - \hat{v}_3 g_{3\alpha}g_{q\mu}\frac{\partial g_{3\mu}}{\partial x_\alpha}\frac{\partial q_q}{\partial\hat{v}_3} - g_{3\alpha}g_{q\mu}\frac{\partial g_{3\mu}}{\partial x_\alpha}q_q\\
&-g_{p\alpha}g_{q\mu}\frac{\partial g_{3\mu}}{\partial x_\alpha}\frac{1}{m}\frac{\partial n\Xi_{pq}}{\partial\hat{v}_3}. \qquad (2.88)
\end{aligned}$$

In equation (2.88), the species index has been dropped. The equation applies both for the ion and electron species.

2.3.6 CLOSURE AND TRANSPORT COEFFICIENTS

Like Braginskii's fluid theory, the moment equations in the perpendicular magnetofluid description also have a hierarchical structure. The equation of the lower moment depends on the higher-order moment. To apply the hierarchy of moment equations, one needs to limit the number of moments to be considered and find the closure for the moment equations. In the framework of Braginskii's fluid equations, it is assumed that the collision frequency and particle gyrofrequency are much higher than the frequency of interest, and the particle mean free path and the Larmor radius are much shorter than the scale length of the problems to be investigated. In the perpendicular magnetofluid description, the collisionless case is considered. But, one can still keep the ordering about the gyration of charged particles, i.e., the particle gyrofrequency is much higher than the frequency of interest, and the Larmor radius is much shorter than the scale length of the problems to be investigated. Detailed derivations are given in references [3] and [4]. Here, we just summarize the results. Since only the ion Larmor radius effects needed to be kept, only the transport coefficients for ion species are described.

The gyroviscous pressure tensor is as follows:

$$\pi_{11} = -\pi_{22} = -\frac{nT}{2\omega_c} W_{12},$$

$$\pi_{12} = \pi_{21} = \frac{nT}{2\omega_c}\frac{1}{2}\left(W_{11} - W_{22}\right),$$

and the rest of the components tied to the parallel direction are zero. Here, the rate-of-strain tensor is given as follows:

$$\begin{aligned} W_{\mu\nu} &= W_{\mu\nu}^{\perp} + W_{\mu\nu}^{\parallel}, \\ W_{\mu\nu}^{\perp} &= g_{\mu\alpha} g_{\nu\beta} \frac{\partial g_{l\beta} \hat{V}_{\perp l}}{\partial x_\alpha} + g_{\nu\alpha} g_{\mu\beta} \frac{\partial g_{l\beta} \hat{V}_{\perp l}}{\partial x_\alpha} - \delta_{\mu\nu}^{\perp} \frac{\partial g_{l\alpha} \hat{V}_{\perp l}}{\partial x_\alpha}, \\ W_{\mu\nu}^{\parallel} &= \left(g_{\mu\alpha} g_{\nu\beta} \frac{\partial g_{3\beta}}{\partial x_\alpha} + g_{\nu\alpha} g_{\mu\beta} \frac{\partial g_{3\beta}}{\partial x_\alpha} - \delta_{\mu\nu}^{\perp} \frac{\partial g_{3\alpha}}{\partial x_\alpha} \right) \left(\hat{v}_3 + \frac{T}{m} \frac{\partial \ln n}{\partial \hat{v}_3} \right). \end{aligned}$$

The gyroviscous tensor $\boldsymbol{\pi}$ can also be expressed in the index notation:

$$\pi_{\alpha\beta} = \eta_3 W_{3\alpha\beta},$$

where

$$\begin{aligned} \eta_3 &= nT/2\omega_c, \\ W_{3\alpha\beta} &= \frac{1}{2}\left(\delta_{\alpha\mu}^{\perp} e_{\beta\gamma\nu} + \delta_{\beta\nu}^{\perp} e_{\alpha\gamma\mu} \right) b_\gamma W_{\mu\nu}. \end{aligned}$$

The heat flux vector is as follows:

$$\mathbf{q} = \frac{2nT}{m\omega_c}\mathbf{b} \times \nabla T.$$

The heat flux tensor is given as follows:

$$\begin{aligned}
\Theta_{111} &= \frac{nT^2}{m\omega_c^2}\hat{\mathbf{e}}_1 \times \boldsymbol{\omega_c} \cdot \nabla \ln T, \\
\Theta_{112} &= -\frac{nT^2}{m\omega_c^2}\hat{\mathbf{e}}_2 \times \boldsymbol{\omega_c} \cdot \nabla \ln T, \\
\Theta_{121} &= -\Theta_{112}, \\
\Theta_{122} &= \Theta_{111}, \\
\Theta_{211} &= -\Theta_{112}, \\
\Theta_{212} &= \Theta_{111}, \\
\Theta_{221} &= -\Theta_{111}, \\
\Theta_{222} &= -\Theta_{112},
\end{aligned}$$

and the rest of the components tied to the parallel direction are zero. The T-square tensor is as follows:

$$\begin{aligned}
\Xi_{11} &= -\Xi_{22} = \frac{3T^2}{2\omega_c}W_{12}, \\
\Xi_{12} &= \Xi_{21} = \frac{3T^2}{2\omega_c}\frac{1}{2}\left(W_{11} - W_{22}\right),
\end{aligned}$$

and the rest of the components tied to the parallel direction are zero.

To constitute the complete set of equations, the Maxwell equations also need to be included.

2.4 HYBRID DESCRIPTION, KINETIC CLOSURE

We have described the fluid theories. Their general feature is that the closure of the moment equation hierarchy is needed. The kinetic equation is solved with suitable ordering considerations to construct the so-called transport coefficients. There are other possibilities for closure simply by letting the kinetic equation be part of the basic set of equations. The higher moments are determined by the solution of the kinetic equation in this case. It is often referred to as the hybrid description. There are two approaches: the pressure-moment approach and the current-moment approach. The pressure moment approach is the conventional approach, and the current moment approach is a more sophisticated one in that it avoids the usage of the parallel fluid velocity. A fluid description of particle parallel velocities oversimplifies the wave-particle resonance to acoustic resonance.

2.4.1 PRESSURE-MOMENT CLOSURE APPROACH

In this section, we outline the conventional pressure-moment-closure approach in the fluid-kinetic hybrid theory [13]. This is a traditional hybrid formalism. The fluid moment equations need closure, since the low-order moment equations involve the higher-order moments. In the Braginskii fluid theory, a high collision frequency is assumed to get the local Maxwellian solution of the kinetic equation. The local Maxwellian solution is then used to compute various transport coefficients to close the hierarchy of fluid moment equations. Since magnetically confined plasmas are usually weakly collisional, the hybrid models are proposed without relying on the solution of the kinetic equation. In the pressure-moment closure approach, only density and velocity moment equations are introduced, while directly using the kinetic equation to describe the pressure moment.

The continuity and momentum equations in the Braginskii two-fluid theory are described in Section 2.1 in equations (2.7), (2.8) and equations (2.12), (2.13) as follows:

$$\frac{\partial n_a}{\partial t} + \frac{\partial}{\partial x_\alpha}(n_a V_{a\alpha}) = 0, \tag{2.89}$$

$$n_a m_a \frac{d_a V_{a\alpha}}{dt} = -\frac{\partial P_{a\alpha\beta}}{\partial x_\beta} - n_a e_a (E_\alpha + e_{\alpha\beta\gamma} V_{a\beta} B_\gamma) + R_{a\alpha}, \tag{2.90}$$

where

$$P_{a\alpha\beta} = m_a \int v'_{a\alpha} v'_{a\beta} f_a d\mathbf{v}_a, \tag{2.91}$$

$$R_{a\alpha} = \sum_b R_{ab\alpha} = \sum_b \int m_a v'_{a\alpha} C_{ab} d\mathbf{v}_a, \tag{2.92}$$

$$\frac{d_a}{dt} \equiv \frac{\partial}{\partial t} + \mathbf{V}_a \cdot \nabla. \tag{2.93}$$

Here, the particle species index $a = e, i$ for electrons and ions. The momentum equations are coupled to the electric and magnetic fields, which are governed by the Maxwell equations

$$\nabla \cdot \mathbf{E} = \frac{1}{\varepsilon_0} \sum_{a=i,e} n_a e_a, \tag{2.94}$$

$$\nabla \times \mathbf{E} = -\frac{\partial \mathbf{B}}{\partial t}, \tag{2.95}$$

$$\nabla \cdot \mathbf{B} = 0, \tag{2.96}$$

$$\nabla \times \mathbf{B} = \mu_0 \sum_{a=i,e} n_a e_a \mathbf{V}_a. \tag{2.97}$$

The set of equations (2.89)–(2.97) involves only the density and velocity moments. As soon as the distribution function is given, the pressure and friction in equations (2.91) and (2.92) are determined so that the system becomes complete. For this, one just needs to introduce the kinetic equations in equation (2.5)

$$\frac{d_a f_a}{dt} + v_{a\beta}\frac{\partial f_a}{\partial x_\beta} - \left(\frac{e_a}{m_a}E^*_{a\beta} - \frac{d_a V_{a\beta}}{dt}\right)\frac{\partial f_a}{\partial v_{a\beta}}$$
$$= \frac{\partial V_{a\gamma}}{\partial x_\beta} v_{\alpha\beta}\frac{\partial f_\alpha}{\partial v_{a\gamma}} + \frac{e_a}{m_a} e_{\beta\gamma\alpha} v_{a\gamma} B_\alpha \frac{\partial f_a}{\partial v_{a\beta}} - \sum_b C_{ab}(f_a, f_b). \tag{2.98}$$

where $\mathbf{v}$ is the relative velocity with respect to the velocity of mass center $\mathbf{V}_a$, and

$$\mathbf{E}^*_a \equiv \mathbf{E} + \mathbf{V}_a \times \mathbf{B}.$$

Here, the closure is addressed in the two-fluid formalism. One can introduce the mass center velocity to convert the description to the single fluid description, such as the construction of MHD equations in Section 2.5.

Here, however, we should point out that the pressure moment formalism does not completely resolve the wave-particle resonance issue. This can be seen from solving the density using the continuity equation

$$\delta n_a = -i\frac{n\nabla \cdot \delta\mathbf{V}}{\omega - k_\parallel \mathbf{V}_\parallel}, \tag{2.99}$$

where $k_\parallel$ is the parallel wavenumber. In equation (2.99), one can see that $\omega - k_\parallel \mathbf{V}_\parallel = 0$ is the resonance condition. This is a fluid cell resonance instead of individual wave-particle resonance. The so-called Landau fluid model is introduced in this field to address this issue [14]. The perpendicular magnetofluid theory in reference [3] is another effort to solve this issue.

2.4.2 CURRENT MOMENT APPROACH

Let us now describe the current-moment approach for hybrid description. The main consideration is as follows. One can decompose the kinetic equation using the Fourier representation in gyrophase α as follows: $\delta f = \sum \delta f_l \exp\{il\alpha\}$. In the conventional pressure-moment approach, the density and moment equations are constructed. In the current-moment approach, instead, the gyrophase-dependent distribution function δf_l $(l \neq 0)$ is used to compute the perpendicular current. Since the gyrofrequency is much larger than other frequencies, one can solve the kinetic equation order by order for δf_l $(l \neq 0)$. Ampere's law,

$$\mu_0 \delta\mathbf{j} = \nabla \times \delta\mathbf{B}, \tag{2.100}$$

is used to construct the complete set of equations.

The approach has been described in reference [15]. In ideal MHD, one has in the lowest order $\delta\mathbf{E} = \delta\mathbf{v} \times \mathbf{B}$ assuming that the equilibrium velocity vanishes. Again, here $\delta\mathbf{v}_\perp$ can be determined by the gyrophase-dependent distribution function δf_l $(l \neq 0)$.

Since the current moment approach has been reviewed in the author's earlier book in describing the gyrokinetic theory [16], here, let us outline the main procedures to explain how the ideal MHD is recovered in this approach. For simplicity, the equilibrium electric field is ignored. To simplify the explanation, sometimes the linearization process is used, which will appear in the next chapter.

The Vlasov equation is as follows [11]

$$\frac{\partial f}{\partial t} + \mathbf{v} \cdot \nabla f + \frac{e}{m} (\mathbf{E} + \mathbf{v} \times \mathbf{B}) \cdot \nabla_{\mathbf{v}} f = 0. \tag{2.101}$$

Using velocity space cylinder coordinate to prove

$$\frac{e}{m} \mathbf{v} \times \mathbf{B} \cdot \nabla_{\mathbf{v}} f = -\Omega \frac{\partial f}{\partial \alpha}$$

Note $\mathbf{v} = v_\perp (\mathbf{e}_1 \cos\alpha + \mathbf{e}_2 \sin\alpha) + v_\parallel \mathbf{b}$ Equation (2.102) needs to be treated by the gyrophase-averaged and gyrophase-dependent parts. Note that the gyrofrequency is dominant. We therefore obtain that the distribution function in leading order is gyrophase independent.

$$\frac{\partial \bar{f}}{\partial t} + \mathbf{v} \cdot \nabla \bar{f} + \frac{e}{m} \mathbf{E} \cdot \nabla_{\mathbf{v}} \bar{f} + \frac{e}{m} \mathbf{v} \times \mathbf{B} \cdot \nabla_{\mathbf{v}} \tilde{f} = 0 \tag{2.102}$$

where $\bar{f} = \langle f \rangle_\alpha$ and $\tilde{f} = f - \bar{f}$. Solution of the average part of equation (2.102) with the FLR effects included has been described in the gyrokinetic theory [15]. Here, we first focus on the gyrophase-dependent part in order to recover the MHD equation in the hybrid model.

We first solve the contribution from the second term in equation (2.102). The gyrophase-dependent part can be solved as follows:

$$\Omega \frac{\partial \tilde{f}}{\partial \alpha} = \mathbf{v}_\perp \cdot \nabla f, \; \rightarrow \; \tilde{f} = \frac{1}{\Omega} \mathbf{b} \times \mathbf{v} \cdot \nabla \bar{f}. \tag{2.103}$$

It's contribution to the perpendicular current density is evaluated as follows:

$$\begin{aligned} \mathbf{j}_\perp^{\nabla p} &= \sum_{e,i} e \int d\mathbf{v} \mathbf{v}_\perp \frac{1}{\Omega} \mathbf{b} \times \mathbf{v} \cdot \nabla \bar{f} \\ &= e \frac{1}{\Omega} e_{\alpha\beta\gamma} b_\gamma \partial_\alpha \int d\mathbf{v} v_{\perp\delta} v_{\perp\beta} \bar{f} \\ &= e \frac{1}{\Omega} e_{\alpha\beta\gamma} b_\gamma \partial_\alpha \left(P_{\delta\beta} + n V_{\mathbf{E}\delta} V_{\mathbf{E}\beta} \right). \end{aligned}$$

Here, we have extracted the $\mathbf{E} \times \mathbf{B}$ drift velocity from the velocity definition by defining

$$\mathbf{v} = \mathbf{V}_\mathbf{E} + \mathbf{v}' \quad \text{with} \quad \mathbf{V}_E = \frac{\mathbf{E} \times \mathbf{B}}{B^2}.$$

Consequently, the distribution function is transformed to $\hat{f}$ and equation (2.101) is transformed to

$$\frac{\partial \hat{f}}{\partial t} + \mathbf{v_E} \cdot \nabla \hat{f} + \mathbf{v}' \cdot \nabla \hat{f} + \frac{e}{m} \mathbf{v}' \times \mathbf{B} \cdot \nabla_{\mathbf{v}'} \hat{f} = 0. \tag{2.104}$$

Therefore, the pressure tensor is defined as

$$P_{\delta\beta} = \int d\mathbf{v}' v'_{\perp\delta} v'_{\perp\beta} \bar{\hat{f}}.$$

The gyrophase-averaged distribution function is required to calculate the pressure tensor. We can average equation (2.104) and obtain

$$\frac{\partial \bar{\hat{f}}}{\partial t} + \mathbf{v_E} \cdot \nabla \bar{\hat{f}} + v'_{\parallel} \mathbf{B} \cdot \nabla \bar{\hat{f}} = -\frac{e}{m} \left\langle \mathbf{v}' \times \mathbf{B} \cdot \nabla_{\mathbf{v}'} \left(\frac{1}{\Omega} \mathbf{b} \times \mathbf{v} \cdot \nabla \bar{f}. \right) \right\rangle, \tag{2.105}$$

where the solution in equation (2.103) has been used. To see the exact picture, we introduce the linearization *a posteriori*, $\partial/\partial t = -i\omega$, $\delta \mathbf{v_E} = i\omega \boldsymbol{\xi}$, and $\delta \mathbf{B} = \nabla \times (\boldsymbol{\xi} \times \mathbf{B})$. The left-hand side of equation (2.105) becomes

$$\begin{aligned} & \frac{\partial \bar{\hat{f}}}{\partial t} + \mathbf{v_E} \cdot \nabla \bar{\hat{f}} + v'_{\parallel} \mathbf{B} \cdot \nabla \bar{\hat{f}} \\ = \; & -i\omega \delta \bar{\hat{f}} + i\omega \boldsymbol{\xi} \cdot \nabla F_0 + v'_{\parallel} \nabla \times (\boldsymbol{\xi} \times \mathbf{B}) \cdot \nabla F_0 + v'_{\parallel} \mathbf{B} \cdot \nabla \delta \bar{\hat{f}} \\ = \; & -i\omega \delta \bar{\hat{f}} + i\omega \boldsymbol{\xi} \cdot \nabla F_0 + v'_{\parallel} \mathbf{B} \cdot \nabla \boldsymbol{\xi} \cdot \nabla F_0 + v'_{\parallel} \mathbf{B} \cdot \nabla \delta \bar{\hat{f}} \\ = \; & -i\omega \Delta \bar{\hat{f}} + v'_{\parallel} \mathbf{B} \cdot \nabla \Delta \bar{\hat{f}}, \end{aligned} \tag{2.106}$$

where

$$\delta \bar{\hat{f}} = -\boldsymbol{\xi} \cdot \nabla F_0 + \Delta \bar{\hat{f}}. \tag{2.107}$$

Comparing with the perturbed pressure expression in the linear ideal MHD, $\delta p = -\boldsymbol{\xi} \cdot \nabla p - \Gamma p \nabla \cdot \boldsymbol{\xi}$, one can see that the distribution function in equation (2.107) has a similar feature. Equation (2.107) shows that the wave-particle resonance is recovered in this hybrid formulation. The right-hand side of equation (2.105) corresponds to the magnetic field perturbation.

Let us now go back to the third term in equation (2.102). It is treated as follows:

$$\Omega \frac{\partial \tilde{f}}{\partial \alpha} = \frac{e}{m} \mathbf{E} \cdot \nabla_{\mathbf{v}} \bar{f} = 2 \frac{e}{m} \mathbf{E} \cdot \mathbf{v}_{\perp} \frac{\partial \bar{f}}{\partial v_{\perp}^2}$$

The $\mathbf{E}_{\parallel}$ goes to the average equation. One obtains

$$\tilde{f} = \frac{2}{B} \mathbf{b} \times \mathbf{v}_{\perp} \cdot \mathbf{E} \frac{\partial \bar{f}}{\partial v_{\perp}^2} \tag{2.108}$$

Using the quasineutrality condition, one can prove that $\tilde{f}$ in equation (2.108) does not contribute to the current density $\mathbf{j}$. Inserting $\tilde{f}$ in equation (2.108) into the first term in equation (2.102) yields

$$\Omega\frac{\partial \tilde{f}}{\partial \alpha} = \frac{\partial}{\partial t}\left(\frac{2}{B}\mathbf{b}\times\mathbf{v}\cdot\mathbf{E}\frac{\partial \bar{f}}{\partial v_\perp^2}\right).$$

One obtains

$$\tilde{f} = \frac{1}{\Omega}\frac{\partial}{\partial t}\left(\mathbf{v}_\perp\cdot\frac{\mathbf{E}}{B}\frac{\partial \bar{f}}{\partial v_\perp^2}\right).$$

The perpendicular current

$$\begin{aligned}\mathbf{j}_\perp^{inertia} &= \sum_{i,e}\frac{2m}{B^2}\frac{\partial}{\partial t}\int d\mathbf{v}\mathbf{v}_\perp\mathbf{v}_\perp\cdot\frac{\mathbf{E}_\perp}{B}\frac{\partial \bar{f}}{\partial v_\perp^2}\\ &= \frac{\partial \rho\mathbf{V}_\mathbf{E}}{\partial t}\times\frac{\mathbf{B}}{B^2}.\end{aligned} \tag{2.109}$$

The total perpendicular current density is then given by

$$\mathbf{j}_\perp = \mathbf{j}_\perp^{\nabla p} + \mathbf{j}_\perp^{inertia}. \tag{2.110}$$

To construct the eigenmode equations, one can use Ampere's law in equation (2.100) to represent $\mathbf{j}_\perp$ on the left-hand side of equation (2.110). Because $\nabla\cdot\mathbf{B}=0$, there are only two independent components of $\mathbf{B}$. Since equation (2.110) has two components, they are fully determined. One can also use the so-called quasineutrality condition $\nabla\cdot\mathbf{j}=0$ to derive the so-called vorticity equation to replace one component of equation (2.110). The result is as follows:

$$\begin{aligned}&\mathbf{B}\cdot\nabla\frac{\mathbf{j}\cdot\mathbf{B}}{B^2}+\nabla\cdot\frac{\mathbf{B}\times\nabla\cdot\overleftrightarrow{P}}{B^2}\\ +\;&\nabla\cdot\frac{\mathbf{B}}{B^2}\times\left(\frac{\partial\rho\mathbf{V}_\mathbf{E}}{\partial t}+\mathbf{V}_\mathbf{E}\nabla\cdot(\rho\mathbf{V}_\mathbf{E})+\rho\mathbf{V}_\mathbf{E}\cdot\nabla\mathbf{V}_\mathbf{E}\right)=0.\end{aligned} \tag{2.111}$$

Here, $\mathbf{j}_\parallel$ can be determined by Ampere's law $\mu_0\mathbf{j}\cdot\mathbf{B}=\mathbf{B}\cdot\nabla\times\mathbf{B}$. These explain how the gyrokinetic theory in the lowest order is recovered without employing the guiding center transform to identify the ideal MHD roots. The fluid resonance is replaced by the wave-particle resonance in equation (2.106).

2.5 GENERALIZED OHM'S LAW AND MHD EQUATIONS

After giving a broad review of how the fluid framework evolves from the kinetic one, in this section we narrow the scope to the MHD theory. This includes the derivation of the generalized Ohm's law and the presentation of MHD equations. They will be the focus of this book.

We start with Braginskii's two fluid equations described in Section 2.1: the continuity equations

$$\begin{aligned}\frac{\partial n_e}{\partial t}+\frac{\partial}{\partial x_\alpha}(n_e V_{e\alpha}) &= 0, &(2.112)\\ \frac{\partial n_i}{\partial t}+\frac{\partial}{\partial x_\alpha}(n_i V_{i\alpha}) &= 0, &(2.113)\end{aligned}$$

the momentum equations

$$\begin{aligned}n_e m_e \frac{d_e V_{e\alpha}}{dt} &= -\frac{\partial p_e}{\partial x_\alpha} - n_e e(E_\alpha + e_{\alpha\beta\gamma}V_{e\beta}B_\gamma) + R_{e\alpha}, &(2.114)\\ n_i m_i \frac{d_i V_{i\alpha}}{dt} &= -\frac{\partial p_i}{\partial x_\alpha} + n_i e(E_\alpha + e_{\alpha\beta\gamma}V_{i\beta}B_\gamma) + R_{i\alpha}, &(2.115)\end{aligned}$$

and the adiabatic energy equations

$$\begin{aligned}\frac{d}{dt}\left(\frac{p_e}{n_e^\Gamma}\right) &= 0, &(2.116)\\ \frac{d}{dt}\left(\frac{p_e}{n_e^\Gamma}\right) &= 0 &(2.117)\end{aligned}$$

for electron and ion species, where the pressure $p_a = n_a T_a$ and the ratio of specific heats $\Gamma = 5/3$. Here, the pressure tensor, the viscosity, etc. have been neglected.

We first derive the generalized Ohm's law. Multiplying equation (2.114) with e/m_e and equation (2.115) with e/m_i, and subtracting the resulting equations yield

$$\begin{aligned}&ne\frac{\partial}{\partial t}(\mathbf{V}_i - \mathbf{V}_e) - ne\left(\mathbf{V}_i \cdot \nabla \mathbf{V}_i - \mathbf{V}_e \cdot \nabla \mathbf{V}_e\right)\\ =\; & -\frac{e}{m_i}\nabla p_i + \frac{e}{m_i}\nabla p_e + ne^2\left(\frac{1}{m_e}+\frac{1}{m_i}\right)\mathbf{E}\\ & + ne^2\left(\frac{m_e\mathbf{V}_i + m_i\mathbf{V}_e}{m_e m_i}\right)\times\mathbf{B} - e\left(\frac{1}{m_e}+\frac{1}{m_i}\right)\mathbf{R}_{ei}.\end{aligned}$$

Noting that $\mathbf{j} = ne(\mathbf{V}_i - \mathbf{V}_e)$, $p = p_i + p_e$, $nm_{i,e}V_{i,e}^2 \ll p_{i,e}$, and $m_i \gg m_e$, one obtains the generalized Ohm's law

$$\frac{m_e}{ne^2}\frac{\partial \mathbf{j}}{\partial t} = \mathbf{E} + \mathbf{V}\times\mathbf{B} + \frac{1}{2ne}\nabla p - \frac{1}{ne}\mathbf{j}\times\mathbf{B} - \eta\mathbf{j},$$

where $\eta = m_e \nu_{ei}/ne^2$ is the resistivity and the velocity of the center of mass

$$\mathbf{V} = \frac{m_i V_i + m_e V_e}{m_i m_e}.$$

The left term is referred to as the electron inertia effect; the third to fifth terms on the right are respectively the pressure drive term, the Hall effect, and the resistive term.

Their relative orderings can be estimated as follows:

$$\left(\frac{\omega}{\omega_p}\right)^2\left(\frac{c}{U}\right)^2 : 1 : 1 : \left(\frac{\omega}{\omega_{ci}}\right)\left(\frac{c_s}{U}\right)^2$$
$$: \left(\frac{\omega}{\omega_p}\right)\left(\frac{\omega_{ce}}{\omega_p}\right)\left(\frac{c}{U}\right)^2 : \left(\frac{\omega}{\omega_p}\right)^2\left(\frac{\nu_{ei}}{\omega_p}\right)\left(\frac{c}{U}\right)^2, \tag{2.118}$$

where $U \sim E/B$, $\omega_p = (ne^2/\varepsilon_0 m_e)^{1/2}$ is the plasma frequency, $c = (\mu_0\varepsilon_o)^{-1/2}$ is the speed of light, and $\omega_c = eB/m$ is the gyrofrequency. In most ordering schemes, one has the so-called Ohm's law

$$\mathbf{E} + \mathbf{V} \times \mathbf{B} = \eta \mathbf{j}. \tag{2.119}$$

Using the two fluid equations (2.112)–(2.117) and introducing the velocity of the center of mass $\mathbf{V}$, one can construct the single fluid equations, *i.e.*, the MHD equations

$$\frac{\partial \rho}{\partial t} + \nabla \cdot (\rho \mathbf{V}) = 0, \tag{2.120}$$
$$\rho \frac{d\mathbf{V}}{dt} = -\nabla p + \mathbf{j} \times \mathbf{B}, \tag{2.121}$$
$$\frac{d}{dt}\left(p\rho^{-\Gamma}\right) = 0. \tag{2.122}$$

To complete the set of equations, the Maxwell equations are also required. Together with the Ohm's law in equation (2.119), one has

$$\nabla \times \mathbf{E} = -\frac{\partial \mathbf{B}}{\partial t}, \tag{2.123}$$
$$\nabla \times \mathbf{B} = \mu_0 \mathbf{j}, \tag{2.124}$$
$$\nabla \cdot \mathbf{B} = 0, \tag{2.125}$$
$$\mathbf{E} + \mathbf{V} \times \mathbf{B} = \eta \mathbf{j}. \tag{2.126}$$

Here, the displacement current $\varepsilon_0 \partial \mathbf{E}/\partial t$ has been neglected in equation (2.124), and $\nabla \cdot \mathbf{B} = 0$ in equation (2.125) may be thought to be included in equation (2.123).

Equations (2.120)–(2.126) are the complete set of MHD equations. If $\eta = 0$, the framework is referred to as the ideal MHD; otherwise, the resistive MHD. The set of equations will be used in this book to describe ideal and resistive MHD theory.

2.6 CONCLUSIONS AND DISCUSSION

In this chapter, we discussed the foundation of MHD theory: from kinetic to fluid descriptions. Describing this is not only for understanding how the ideal and resistive MHD theories are derived from the kinetic equations but also for further studies to modify the MHD analyses with the anisotropic pressure, the two-fluid, and kinetic modifications.

In deriving the fluid equations, as well as the MHD equation, we follow strict and regular procedures. First, the Braginskii two-fluid equations are derived, followed by the derivation of moment equations. The derivation of moment equations is based on the description of the center of mass. It simplifies a lot, and also the transport coefficients in the Braginskii theory are actually derived in this frame.

Next, we introduced the Chew-Goldberger-Low (CGL) theory in the two-fluid framework. This is necessary for investigating the mirror type of confinements. In space plasmas, for example, the Earth's magnetic field is a mirror-field-like. In magnetic fusion, the mirror device is one of the most important confinement concepts with symmetry.

Further on, the perpendicular magnetofluid theory was briefly outlined. It represents the minimum and necessary modifications of the fluid description of plasmas. Charged particle motion along the magnetic field is not localized as a fluid cell. The trapped-particle effects and wave-particle resonance make the parallel fluid description inapplicable. The fluid description, nonetheless, is good in the perpendicular direction. Introducing it helps to understand the nature of fluid description in plasma physics.

In return, we discussed the fluid-kinetic-hybrid theory. It is aimed at explaining how to modify the fluid theory with the kinetic effects. We are not interested in systematically describing the kinetic theory. Instead, it is focused on revealing the roots of kinetic effects in the fluid or MHD descriptions. This helps to build the physics and mathematical pictures in considering the further extension of the MHD theory described in this book.

In the end, the generalized Ohm's law and MHD equations are derived. They will be used in this book for further analyses.

References

1. Braginskii S I 1966 Transport processes *Reviews of Plasma Physics*, vol. 1 ed. by Leontovich M (New York: Consultants Bureau) pp 205–311.
2. Chew G F, Goldberger M L, Low F E 1956 *Proc. R. Soc. London, Ser. A 236,* 112–118.
3. Zheng L 2020 Perpendicular magnetofluid theory for magnetically confined plasmas in the collisionless limit *Phys. Plasmas 27*, 052503. https://doi.org/10.1063/1.5144676
4. Zheng L 2020 Magnetically Confined Fusion Plasma Physics, Volume 2, Multifluid Theory (London: IOP Publishing). DOI: 10.1088/978-0-7503- 3575-1.
5. Lerner R G, Trigg G L 1991 Encyclopaedia of Physics (2nd Edition), VHC publishers, ISBN (Verlagsgesellschaft) 3-527-26954-1, ISBN (VHC Inc.) 0-89573-752-3.
6. Chapman S, Cowling T G 1953 Mathematical theory of nonuniform gases. (Cambridge, Cambridge University Press).
7. Hazeltine R D, Meiss J D 1992 Plasma Confinement (New York: Addison-Wesley).
8. Bogoliubov N N 1946 Kinetic Equations *Journal of Physics USSR, 10 (3)*, 265–274.
9. Klimontovich Yu L 1966 The statistical theory of non-equilibrium process in a plasma (Pergamon).
10. Landau L D 1937 Kinetic equation for the Coulomb effect. *Sov. Phys. JETP 7*, 203–209.

11. Vlasov A A 1938 On vibration properties of electron gas. *J. Exp. Theor. Phys. (in Russian). 8 (3)*, 291. doi:10.1070/PU1968v010n06ABEH003709. S2CID 122952713.
12. Hunana P, Tenerani A, Zank G P, *et. al.* 2019 An introductory guide to fluid models with anisotropic temperatures. Part 1. CGL description and collisionless fluid hierarchy. *J. Plasma Phys., 85,* 205850602. DOI: https://doi.org/10.1017/S0022377819000801
13. Kulsrud R M 1983 MHD description of plasma Basic Plasma Physics I (Handbook of Plasma Physics) ed A A Galeev and R N Sudan (Amsterdam: North-Holland) 1-31 pp
14. Hammet G W, Perkins F W 1990 Fluid moment models for Landau damping with application to the ion-temperature-gradient instability *Phys. Rev. Lett. 64*, 3019. https://doi.org/10.1103/PhysRevLett.64.3019
15. Zheng L J, Kotschenreuther M T, Van Dam J W 2007 Revisiting linear gyrokinetics to recover ideal magnetohydrodynamics and missing finite Larmor radius effects. *Phys. Plasmas 14*, 072505. https://doi.org/10.1063/1.2746811.
16. Zheng L 2022 Magnetically confined fusion plasma physics, Vol. 3, Kinetic theory (London: IOP Publishing). https://doi.org/10.1088/978-0-7503-3779-3.

3 Fundamentals of MHD theory

With the MHD equations constructed in the previous chapter, Chapter 2, we focus on describing the fundamentals of MHD theory in this chapter. Three subjects are addressed: the general linear MHD formalism, the basic waves and instabilities, and the non-ideal MHD extension. They represent many pioneering contributions in this field.

First, the MHD equations are linearized for equilibrium and stability analyses. The field line displacement will be introduced so that the so-called MHD force operator is derived. The energy principle will be constructed and the self-adjoint property of the force operator will be proved. Besides the direct approach, the variational theory based on the Eulerian and Lagrangian formulations of MHD are described. They are fundamental, but not trivial. Their applications extend beyond the MHD theory, especially the variational formulation of MHD. The kinetic energy principle in the guiding-center description is based on this formalism.

Next, the fundamental features of linear MHD are analyzed based on the proper decomposition of MHD force operator and energy sources. The equations for three basic MHD waves: the compressional Alfvén modes, the acoustic or sound waves, and the shear Alfvén waves, are derived and discussed. The two key MHD instability drives, the current-driven kink and pressure-gradient-driven modes, are identified using the energy principle. The analyses are general and intuitive, but no geometry simplification is resorted.

Furthermore, we also discuss the nonideal MHD extensions. The resistivity effects are first described. The resistivity causes the ideal MHD frozen-in law be released. Consequently, two types of resistive MHD modes appear: the resistive interchange and tearing modes. Although ideal and resistive MHD theories are good and elegant, they have limitations. In the direction parallel to the magnetic field line, charged particles can move rather freely. This makes fluid description in the parallel direction is not generally applicable. In the perpendicular direction, the so-called finite-Larmor-radius effects can also lead charged particles to drift away from the field lines. The two fluid and kinetic modifications of MHD are then outlined. It clarifies the MHD roots of two-fluid and kinetic descriptions.

3.1 LINEARIZATION OF MHD EQUATIONS AND MHD FORCE OPERATOR

We have described the MHD equations in Chapter 2. To proceed, we discuss first the linearization of MHD equations in this section [1, 2] The construction of the so-called MHD force operator will be described afterward.

DOI: 10.1201/9781003522133-3

The set of MHD equations has been constructed in equations (2.120)–(2.126) as follows

$$\begin{aligned}
\frac{\partial \rho}{\partial t} + \nabla \cdot (\rho \mathbf{v}) &= 0 && (3.1)\\
\rho \frac{d\mathbf{v}}{dt} &= -\nabla p + \mathbf{j} \times \mathbf{B} && (3.2)\\
\frac{d}{dt}\left(p\rho^{-\Gamma}\right) &= 0. && (3.3)\\
\nabla \times \mathbf{E} &= -\frac{\partial \mathbf{B}}{\partial t}, && (3.4)\\
\nabla \times \mathbf{B} &= \mu_0 \mathbf{j} && (3.5)\\
\nabla \cdot \mathbf{B} &= 0, && (3.6)\\
\mathbf{E} + \mathbf{v} \times \mathbf{B} &= 0, && (3.7)
\end{aligned}$$

where $d/dt \equiv \partial/\partial t + \mathbf{v} \cdot \nabla$. Combining equations (3.1) and (3.3) yields

$$\frac{dp}{dt} = -\Gamma p \nabla \cdot \mathbf{v}. \tag{3.8}$$

Equations (3.1)–(3.8) can be linearized by expanding the unknowns into the equilibrium and perturbed parts, for example, $\mathbf{v} = \mathbf{v}_0 + \mathbf{v}_1$, where the subscript 0 and 1 are introduced to represent the equilibrium and perturbed quantities. The equilibrium equations become

$$\begin{aligned}
\mathbf{j}_0 \times \mathbf{B}_0 &= \nabla P_0 + \rho_0 \mathbf{v}_0 \cdot \nabla \mathbf{v}_0, && (3.9)\\
\nabla \times \mathbf{B}_0 &= \mu_0 \mathbf{j}_0, && (3.10)\\
\nabla \cdot \mathbf{B}_0 &= 0, && (3.11)\\
\mathbf{E}_0 &= \mathbf{v}_0 \times \mathbf{B}_0. && (3.12)
\end{aligned}$$

Here, we have kept the equilibrium velocity. It is only important when the velocity is of the same order as the ion thermal velocity. Also, the resistivity effect is left out in the equilibrium since it is usually only important in the singular layer analyses in the first-order analyses. The linearized equations are as follows:

$$\rho \frac{\partial \mathbf{v}_1}{\partial t} + \rho \mathbf{v} \cdot \nabla \mathbf{v}_1 + \rho \mathbf{v}_1 \cdot \nabla \mathbf{v} + \rho_1 \mathbf{v} \cdot \nabla \mathbf{v} = -\nabla p_1 + \mathbf{j} \times \mathbf{B}_1 + \mathbf{j}_1 \times \mathbf{B}, \tag{3.13}$$

$$\frac{\partial \rho_1}{\partial t} = -\nabla \cdot (\rho \mathbf{v}_1) - \nabla \cdot (\rho_1 \mathbf{v}), \tag{3.14}$$

$$\frac{\partial p_1}{\partial t} + \mathbf{v} \cdot \nabla p = -\mathbf{v}_1 \cdot \nabla p - \Gamma p \nabla \cdot \mathbf{v}_1 - \Gamma p_1 \nabla \cdot \mathbf{v}, \tag{3.15}$$

$$\mathbf{E}_1 + \mathbf{v}_1 \times \mathbf{B} = \eta \mathbf{j}_1, \tag{3.16}$$

$$\frac{\partial \mathbf{B}_1}{\partial t} = -\nabla \times \mathbf{E}_1 \tag{3.17}$$

$$\nabla \cdot \mathbf{B}_1 = 0, \tag{3.18}$$

$$\frac{\partial \mathbf{j}_1}{\partial t} = \frac{1}{\mu_0} \nabla \times \frac{\partial \mathbf{B}_1}{\partial t}. \tag{3.19}$$

Here, we have dropped the subscript 0 for equilibrium quantities.

It is assumed that in the equilibrium, $\mathbf{E}_0 = 0$, $\mathbf{v}_0 = 0$. In the ideal MHD case, applying $\partial/\partial t$ on equation (3.13) yields

$$\rho\frac{\partial^2 \mathbf{v}_1}{\partial t^2} = \mathbf{j}\times\nabla\times(\mathbf{v}_1\times\mathbf{B})+\frac{1}{\mu_0}[\nabla\times(\nabla\times(\mathbf{v}_1\times\mathbf{B})]\times\mathbf{B}$$
$$+\nabla(\mathbf{v}_1\cdot\nabla p)+\Gamma\nabla(p\nabla\cdot\mathbf{v}_1) \quad (3.20)$$

Introducing the field line displacement $\boldsymbol{\xi}$ as follows:

$$\mathbf{v}_1(\mathbf{r},t) = \frac{\partial}{\partial t}\boldsymbol{\xi}(\mathbf{r},t),$$

where $\mathbf{r} = \mathbf{r}_0 + \boldsymbol{\xi}$, equastion (3.15)–(3.17) are reduced to

$$\rho_1 = -\nabla\cdot(\rho\boldsymbol{\xi}),$$
$$p_1 = -\boldsymbol{\xi}\cdot\nabla p - \Gamma p\nabla\cdot\boldsymbol{\xi},$$
$$\mathbf{B}_1 = \nabla\times(\boldsymbol{\xi}\times\mathbf{B}) \equiv \mathbf{Q}.$$

Using these results, the linearized ideal MHD equations become

$$\rho_m\gamma^2\boldsymbol{\xi} = \delta\mathbf{J}\times\mathbf{B}+\mathbf{J}\times\delta\mathbf{B}-\nabla\delta P, \quad (3.21)$$
$$\delta\mathbf{B} = \nabla\times\boldsymbol{\xi}\times\mathbf{B}, \quad (3.22)$$
$$\mu_0\delta\mathbf{J} = \nabla\times\delta\mathbf{B}, \quad (3.23)$$
$$\delta P = -\boldsymbol{\xi}\cdot\nabla P - \Gamma P\nabla\cdot\boldsymbol{\xi}. \quad (3.24)$$

The momentum equation (3.13) then becomes

$$\rho\frac{\partial^2\boldsymbol{\xi}}{\partial t^2} = \mathbf{F}(\boldsymbol{\xi}), \quad (3.25)$$

where

$$\mathbf{F}(\boldsymbol{\xi}) = \frac{1}{\mu_0}(\nabla\times\mathbf{Q})\times\mathbf{B}+\frac{1}{\mu_0}(\nabla\times\mathbf{B})\times\mathbf{Q}+\nabla(\boldsymbol{\xi}\cdot\nabla p)+\Gamma\nabla(p\nabla\cdot\boldsymbol{\xi})$$
$$= -\nabla\left(\frac{\mathbf{B}\cdot\mathbf{Q}}{\mu_0}+p_1\right)+\frac{1}{\mu_0}(\mathbf{B}\cdot\nabla\mathbf{Q}+\mathbf{Q}\cdot\nabla\mathbf{B}). \quad (3.26)$$

The operator **F** is referred to as the ideal MHD force operator.

3.2 BOUNDARY CONDITIONS

The linearized MHD equation (3.25) needs to be solved with the boundary conditions. Two types of equations are considered: plasma contacts a conducting wall or faces a vacuum region.

We first consider the case with plasma contacting a conducting wall directly. Noting that the electric field in the conducting wall vanishes and the tangential component of the electric field is continuous on the plasma-wall interface, one has the boundary condition that the normal electric field has to vanish on the interface in this case, i.e.,

$$\mathbf{E} \times \mathbf{n} = (\dot{\boldsymbol{\xi}} \times \mathbf{B}) \times \mathbf{n} = \mathbf{B}(\dot{\boldsymbol{\xi}} \cdot \mathbf{n}) - \dot{\boldsymbol{\xi}}(\mathbf{B} \cdot \mathbf{n}), \tag{3.27}$$

where $\mathbf{n}$ is the normal vector of the plasma-wall interface. In most cases, the boundary condition in equation (3.27) just implies that $\boldsymbol{\xi} = 0$. In the case that the magnetic field is tangential to the wall, we then have

$$\boldsymbol{\xi} \cdot \mathbf{n} = 0.$$

Next, we consider the case with the interface with a vacuum. Two boundary conditions are required. The first is the radial force balance. From equations (3.2) and (3.5), one has the force balance equation in the lowest order

$$p(\mathbf{r}_0) + \frac{B_i^2(r_0)}{2\mu_0} = \frac{B_o^2(r_0)}{2\mu_0}, \tag{3.28}$$

where $\mathbf{r}_0$ represents the plasma-vacuum interface and the subscripts i and o denote, respectively, the plasma and vacuum regions. In the first order, one needs to consider that the plasma interface is perturbed as well. Supposing the perturbed interface is denoted by $\mathbf{r} = \mathbf{r}_0 + \mathbf{r}_1$, one has in the first order

$$p(\mathbf{r}) + p_1(\mathbf{r}) + \frac{(\mathbf{B}_i(\mathbf{r}) + \mathbf{B}_{1i}(\mathbf{r}))^2}{2\mu_0} = \frac{(\mathbf{B}_e(\mathbf{r}) + \mathbf{B}_{1e}(\mathbf{r}))^2}{2\mu_0}. \tag{3.29}$$

Expanding the equilibrium quantities as follows

$$\begin{aligned} p(\mathbf{r}) &= p(\mathbf{r}_0) + \boldsymbol{\xi} \cdot \nabla p(r_0), \\ B^2(\mathbf{r}) &= B^2(\mathbf{r}_0) + \boldsymbol{\xi} \cdot \nabla B^2(r_0), \end{aligned} \tag{3.30}$$

and using equation (3.21), the boundary condition in equation (3.29) becomes

$$\begin{aligned} &-\Gamma p(\mathbf{r}_0)\nabla \cdot \boldsymbol{\xi} + \frac{\mathbf{B}_{0i}(\mathbf{r}_0) \cdot \mathbf{B}_{1i}(\mathbf{r}_0)}{2\mu_0} + \frac{\xi_n}{2\mu_0}\frac{\partial}{\partial n}B_{0i}^2(r_0) \\ = &\ \frac{\mathbf{B}_e(\mathbf{r}_0) \cdot \mathbf{B}_{1e}(\mathbf{r}_0)}{2\mu_0} + \frac{\xi_n}{2\mu_0}\frac{\partial}{\partial n}B_{0e}^2(r_0). \end{aligned} \tag{3.31}$$

The other boundary conditions can be found from the continuity of the electromagnetic field. First, we note that at the boundary on the plasma side, one has

$$\mathbf{E}_1 + \mathbf{v}_1 \times \mathbf{B} = 0 \text{ and } \mathbf{B}_1 = \nabla \times (\boldsymbol{\xi} \times \mathbf{B})$$

Note that $\nabla \cdot \mathbf{B}_1 = 0$ requires the normal component of the magnetic field to be continuous. One then obtains the boundary condition

$$\mathbf{n} \cdot \mathbf{B}_{1e} = \mathbf{n} \cdot \nabla \times (\boldsymbol{\xi} \times \mathbf{B}_e). \tag{3.32}$$

In the vacuum region, the electromagnetic field can be represented by the vector potential **A**,

$$\begin{aligned} \mathbf{E}_{1e} &= -\frac{\partial \mathbf{A}}{\partial t}, \\ \mathbf{B}_{1e} &= \nabla \times \mathbf{A}, \\ \nabla \cdot \mathbf{A} &= 0. \end{aligned}$$

Because the vacuum region is current-free, one has

$$\nabla \times \nabla \times \mathbf{A} = -\nabla^2 \mathbf{A} = 0.$$

Here, we point out that one cannot use alone the scalar potential φ to describe the magnetic field, although one has $\nabla \times \mathbf{B} = 0$ in the vacuum. This is because $B_\phi = -\nabla\varphi$ does not have a periodic solution for φ if the toroidal field $\mathbf{B}_\phi$ is finite.

The boundary condition in equation (3.32) becomes

$$\mathbf{n} \times \mathbf{A} = -\xi_n \mathbf{B}_{0e} \tag{3.33}$$

In magnetic fusion devices, the vacuum region often faces plasma on one side and a conducting or resistive wall on the other side. If the vacuum region meets a conducting wall, one has on the other side

$$\mathbf{n} \times \mathbf{A} = 0.$$

The resistive-wall boundary condition will be discussed when studying the resistive wall modes later.

3.3 ENERGY PRINCIPLE

In this section, the ideal MHD energy principle will be derived. The self-adjointness of the force operator will be proved afterward. In the end, the local energy conservation law with the resistive MHD will be derived. The energy principle is derived in this section directly. They will be revisited with the Eulerian and Lagrangian formulations in the next section, Section 3.4.

3.3.1 MHD ENERGY PRINCIPLE

In this subsection, we will derive the ideal MHD energy principle directly from the linearized MHD equation, equation (3.25), with the MHD force operator given in equation (3.26). The boundary conditions derived in Section 3.2 will be used to treat the surface contributions.

To derive the energy principle, we multiply the MHD equation (3.25) with η and integrate it over the plasma region. The term-by-term treatments are given as follows.

The first term in equation (3.26) is reduced as follows:

$$
\begin{aligned}
& \int d\mathbf{r}\boldsymbol{\eta} \cdot \frac{1}{\mu_0}(\nabla \times \mathbf{Q}) \times \mathbf{B} \\
= & -\frac{1}{\mu_0}\int d\mathbf{r}(\nabla \times \boldsymbol{\eta} \times \mathbf{B}) \cdot \mathbf{Q} + \frac{1}{\mu_0}\int d\mathbf{r}\nabla \cdot [\mathbf{Q} \times (\boldsymbol{\eta} \times \mathbf{B})] \\
= & -\frac{1}{\mu_0}\int d\mathbf{r}(\nabla \times \eta \times \mathbf{B}) \cdot \mathbf{Q} + \frac{1}{\mu_0}\oint d\mathbf{s} \cdot [\mathbf{Q} \times (\eta \times \mathbf{B})] \\
= & -\frac{1}{\mu_0}\int d\mathbf{r}(\nabla \times \eta \times \mathbf{B}) \cdot \mathbf{Q} + \frac{1}{\mu_0}\oint d\mathbf{s} \cdot \boldsymbol{\eta}\,(\mathbf{Q} \cdot \mathbf{B}).
\end{aligned} \tag{3.34}
$$

Here, in the last step, we have noted that $d\mathbf{s} \cdot \mathbf{B} = 0$. The second term in equation (3.26) is reduced as follows:

$$
\int d\mathbf{r}\boldsymbol{\eta} \cdot \frac{1}{\mu_0}(\nabla \times \mathbf{B}) \times \mathbf{Q} = -\frac{1}{\mu_0}\int d\mathbf{r}\mathbf{j} \cdot (\boldsymbol{\eta} \times \mathbf{Q}).
$$

The third and fourth terms in equation (3.26) are reduced as follows:

$$
\begin{aligned}
& \int d\mathbf{r}\boldsymbol{\eta} \cdot [\nabla(\boldsymbol{\xi} \cdot \nabla p) + \Gamma\nabla(p\nabla \cdot \boldsymbol{\xi})] \\
= & -\int d\mathbf{r}(\nabla \cdot \boldsymbol{\eta})(\boldsymbol{\xi} \cdot \nabla p + \Gamma p\nabla \cdot \boldsymbol{\xi}) + \int d\mathbf{r}\nabla \cdot [\boldsymbol{\eta} \cdot (\boldsymbol{\xi} \cdot \nabla p + \Gamma p\nabla \cdot \boldsymbol{\xi})] \\
= & -\int d\mathbf{r}(\nabla \cdot \boldsymbol{\eta})(\boldsymbol{\xi} \cdot \nabla p + \Gamma p\nabla \cdot \boldsymbol{\xi}) + \oint d\mathbf{s} \cdot \boldsymbol{\eta}\,(\boldsymbol{\xi} \cdot \nabla p + \Gamma p\nabla \cdot \boldsymbol{\xi}).
\end{aligned} \tag{3.35}
$$

Now, let us use the boundary conditions in subsection 3.2 to further reduce the surface integrations in equations (3.34) and (3.35). Assuming $\boldsymbol{\eta}$ satisfies the same boundary conditions as $\boldsymbol{\xi}$ and using equation (3.33), one can prove that

$$
\begin{aligned}
\oint d\mathbf{s} \cdot \boldsymbol{\eta}\mathbf{B} \cdot \mathbf{Q} &= -\oint (n \times \mathbf{A}) \cdot (\nabla \times \mathbf{A}) ds \\
&= -\oint (\mathbf{A} \times \nabla \times \mathbf{A}) \cdot d\mathbf{s} \\
&= \int_{V_e} \nabla \cdot (\mathbf{A} \times \nabla \times \mathbf{A}) d\mathbf{r} \\
&= \int_{V_e} [(\nabla \times \mathbf{A}) \cdot (\nabla \times \mathbf{A}) - \mathbf{A} \cdot (\nabla \times \mathbf{Q})]\, d\mathbf{r} \\
&= \int_{V_e} (\nabla \times \mathbf{A}) \cdot (\nabla \times \mathbf{A}) d\mathbf{r} = \int_{V_e} B_1^2 d\mathbf{r}.
\end{aligned}
$$

Here, V_e denotes the vacuum region, and in the last step, $\nabla \times \mathbf{Q} = \mathbf{j}_1 = 0$ in the vacuum has been considered. Using the boundary condition in equation (3.31), one obtains

$$
\begin{aligned}
& \oint d\mathbf{s} \cdot \boldsymbol{\eta}\,[(-\boldsymbol{\xi} \cdot \nabla p - \Gamma p\nabla \cdot \boldsymbol{\xi})] \\
= & \oint d\mathbf{s} \cdot \boldsymbol{\eta}\left[-\xi_n \frac{\partial}{\partial n}\left(p_0 + \frac{B_{0i}^2}{2\mu_0} - \frac{B_{0i}^2}{2\mu_0}\right)\right].
\end{aligned} \tag{3.36}
$$

Using equations (3.34)–(3.36) and regarding η as $\boldsymbol{\xi}$, one obtains

$$\delta W_k = \delta W(\boldsymbol{\xi},\boldsymbol{\xi}), \tag{3.37}$$

where

$$\delta W_k = \frac{1}{2}\int \rho_m \left(\frac{\partial \boldsymbol{\xi}}{\partial t}\right)^2 d\mathbf{r}, = -\frac{\omega^2}{2}\int \rho_m |\boldsymbol{\xi}|^2 d\mathbf{r}, \tag{3.38}$$

$$\delta W = \delta W_p + \delta W_s + \delta W_v. \tag{3.39}$$

Here, the plasma energy is given by

$$\delta W_p = \frac{1}{2}\int d\mathbf{r}\left\{\frac{\mathbf{Q}^2}{\mu_0} + \mathbf{j}\cdot(\boldsymbol{\xi}\times\mathbf{Q}) + (\boldsymbol{\xi}\cdot\nabla p)\nabla\cdot\boldsymbol{\xi} + \Gamma p(\nabla\cdot\boldsymbol{\xi})^2\right\}. \tag{3.40}$$

and the surface and vacuum energies are respectively given as follows:

$$\delta W_s = -\frac{1}{2}\oint ds \xi_n^2 \frac{\partial}{\partial n}\left(p_0 + \frac{B_{0i}^2}{2\mu_0} - \frac{B_{0i}^2}{2\mu_0}\right), \tag{3.41}$$

$$\delta W_v = \frac{1}{2\mu_0}\int_V (\nabla\times\mathbf{A})\cdot(\nabla\times\mathbf{A})d\mathbf{r} = \frac{1}{2\mu_0}\int_V B_1^2 d\mathbf{r}. \tag{3.42}$$

In equation (3.38), the assumption that the time dependence of perturbations is of the type $\exp\{-i\omega t\}$ has been used.

3.3.2 SELF-ADJOINTNESS OF THE MHD FORCE VECTOR

With the derivation of the energy principle in subsection 3.3.1, we will prove the self-adjointness of the MHD force operator in equation (3.26). The proof is based on the procedure in reference [1]. It is simpler. The alternative direct proof of the self-adjointness can be found in other MHD books, for example, references [2], [3], and [4].

The self-adjointness of the force vector is defined as follows:

$$\int d\mathbf{r}\boldsymbol{\eta}\cdot\mathbf{F}(\boldsymbol{\xi}) = \int d\mathbf{r}\boldsymbol{\xi}\cdot\mathbf{F}(\boldsymbol{\eta}), \tag{3.43}$$

Applying d/dt to the energy principle in equation (3.37), one obtains

$$-\frac{d\delta W_k}{dt} + \frac{d\delta W}{dt} = 0. \tag{3.44}$$

Note that

$$\frac{d\delta W_k}{dt} = \frac{1}{2}\int d\mathbf{r}2\rho_m \frac{\partial\boldsymbol{\xi}}{\partial t}\cdot\frac{\partial^2\boldsymbol{\xi}}{\partial t^2} = \int d\mathbf{r}\rho_m \frac{\partial\boldsymbol{\xi}}{\partial t}\cdot\frac{\partial^2\boldsymbol{\xi}}{\partial t^2} = \int d\mathbf{r}\dot{\boldsymbol{\xi}}\cdot\mathbf{F}(\boldsymbol{\xi}),$$

$$\frac{d\delta W}{dt} = \delta W(\dot{\boldsymbol{\xi}},\boldsymbol{\xi}) + \delta W(\boldsymbol{\xi},\dot{\boldsymbol{\xi}}),$$

$$\delta W(\dot{\boldsymbol{\xi}},\boldsymbol{\xi}) = \frac{1}{2}\int d\mathbf{r}\dot{\boldsymbol{\xi}}\cdot\mathbf{F}(\boldsymbol{\xi}),$$

where an overdot is used to denote the time derivative of the function. Inserting them into equation (3.44) yields

$$W(\dot{\boldsymbol{\xi}},\boldsymbol{\xi}) = W(\boldsymbol{\xi},\dot{\boldsymbol{\xi}}).$$

Regarding $\dot{\boldsymbol{\xi}}$ as $\boldsymbol{\eta}$, the self-adjointness of $\mathbf{F}$ in equation (3.43) is proved.

Noting equation (3.38) when the time dependence of perturbations is assumed to be $\exp\{-i\omega t\}$, the energy principle becomes

$$\omega^2 = \frac{\int d\mathbf{r}\boldsymbol{\xi}^* \cdot \mathbf{F}(\boldsymbol{\xi})}{\int d\mathbf{r}\rho|\boldsymbol{\xi}|^2}. \tag{3.45}$$

Using the self-adjointness property in equation (3.43), one can prove that $(\omega^2)^* = \omega^2$, i.e., the MHD eigenvalue ω^2 is real. Equation (3.45) can be used to determine the stability by the minimization of the MHD energy.

3.3.3 MHD ENERGY CONSERVATION LAW

The linear ideal MHD energy principle has been obtained in equation (3.37). In this subsection, we derive the general MHD energy conservation law with the resistivity and plasma flow taken into account. The general energy conservation law is obtained by individually deriving the electromagnetic and kinetic/thermal energy balance equations and then combining them together.

From Faraday's and Ampere's laws in equations (3.4) and (3.5), one obtains the electromagnetic energy balance equation

$$\frac{1}{\mu_0}\nabla \cdot (\mathbf{E} \times \mathbf{B}) + \frac{\partial}{\partial t}\left(\frac{B^2}{2\mu_0}\right) + \mathbf{E} \cdot \mathbf{j} = 0. \tag{3.46}$$

Using the Ohm's law in equation (3.7), one obtains

$$\mathbf{E} \cdot \mathbf{j} = \eta \mathbf{j}^2 - (\mathbf{j} \times \mathbf{B}) \cdot \mathbf{v}. \tag{3.47}$$

Inserting equation (3.47) to equation (3.46) yields

$$\frac{1}{\mu_0}\nabla \cdot (\mathbf{E} \times \mathbf{B}) + \frac{\partial}{\partial t}\left(\frac{B^2}{2\mu_0}\right) + \eta \mathbf{j}^2 - (\mathbf{j} \times \mathbf{B}) \cdot \mathbf{v} = 0. \tag{3.48}$$

From the momentum equation (3.2), one obtains the kinetic and thermal energy balance equation

$$(\mathbf{j} \times \mathbf{B}) \cdot \mathbf{v} = \frac{\partial}{\partial t}\left(\frac{\rho_m v^2}{2}\right) + \nabla \cdot \left(\frac{\rho_m v^2}{2}\mathbf{v}\right). + \mathbf{v} \cdot \nabla p. \tag{3.49}$$

Here, it has been noted that $\mathbf{v} \cdot \nabla v^2/2 = \mathbf{v} \cdot (\mathbf{v} \cdot \nabla \mathbf{v})$. To further reduce the last term, the adiabatic pressure equation (3.8) is used to prove that

$$\mathbf{v} \cdot \nabla p = \frac{\partial}{\partial t}\left(\frac{p}{\Gamma - 1}\right) + \nabla \cdot \left(\frac{\Gamma p}{\Gamma - 1}\right)\mathbf{v}. \tag{3.50}$$

Combining equations (3.49) and (3.50), one obtains the total energy conservation eqaution

$$\frac{\partial}{\partial t}\left(\frac{\rho_m \mathbf{v}^2}{2}+\frac{p}{\Gamma-1}+\frac{B^2}{2\mu_0}\right)$$
$$+\eta \mathbf{j}^2+\frac{1}{\mu_0}\nabla\cdot(\mathbf{E}\times\mathbf{B})+\nabla\cdot\left[\left(\frac{\rho_m \mathbf{v}^2}{2}+\frac{\Gamma p}{\Gamma-1}\right)\mathbf{v}\right]=0. \tag{3.51}$$

Equation (3.51) is interpreted as follows. The total energy, the first term, consists of, respectively, the kinetic energy of fluid motion, plasma thermal energy, and magnetic field energy. The variation of the total energy is compensated by the following sources. The second term in equation (3.51) represents the Ohmic heating and the third term results from the Poynting vector. Noting that $\Gamma p/(\Gamma-1)=p/(\Gamma-1)+p$, one can see that the last term in equation (3.51) contains two parts: the thermal energy flux and the work done by the thermal pressure. In the kinetic description, the pressure is related to the moment $(3/2)nT=\int(v^2/2)f d\mathbf{v}$.

In the ideal MHD case with $\mathbf{v}_0=0$ and $\eta=0$, equation (3.51) can be integrated over the configuration space and gives rise to the global energy balance equation

$$\frac{\partial}{\partial t}\int d\mathbf{r}\left(\frac{\rho \mathbf{v}^2}{2}+\frac{p}{\Gamma-1}+\frac{B^2}{2\mu_0}\right)$$
$$+\frac{1}{\mu_0}\oint d\mathbf{s}\cdot[(\mathbf{B}\times\mathbf{v})\times\mathbf{B}]+\oint d\mathbf{s}\cdot\frac{\Gamma p}{\Gamma-1}\mathbf{v}=0, \tag{3.52}$$

where the ideal MHD frozen-in law has been used to simplify the Poynting vector and only the terms up to the second order are retained. Equation (3.52) can alternatively be used to derive the linear ideal MHD energy principle. In the next section, Section 3.4, we will discuss it in the Euler and Lagrange variational formalisms.

3.4 VARIATIONAL FORMULATION OF LINEAR MHD

In this section, we describe the Eulerian and Lagrangian descriptions of MHD. The Eulerian and Lagrangian descriptions are widely used in fluid theory. Their applications in the MHD theory are still limited. Because the MHD frozen-in law causes the magnetic field to be compressed like a fluid, the variational description of ideal MHD becomes complicated. Many efforts have been made in this field, for example, references [5, 6, 7, 8, 9, 10]. Here, we focus on reviewing the systematic description of Eulerian and Lagrangian theory in references [5], [6], and [10]. The formalism was also reviewed in reference [11] for the guiding center theory in references [5] with the hybrid Euler-Lagrange formalism and only the fixed boundary case being considered. The Lagrangian energy expression in the fixed boundary case is given in reference [6]. The complete extension for including the boundary contributions in the Lagrangian description is given in reference [10].

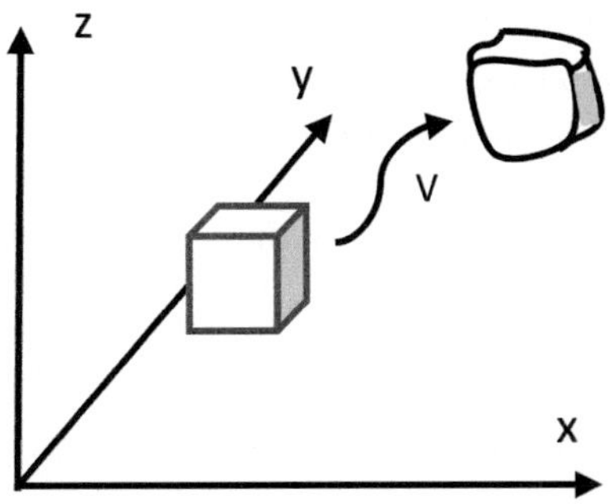

Figure 3.1 The Eulerian and Lagrangian descriptions of fluid mechanics. Reproduced from reference [10] with the permission of AIP Publishing.

3.4.1 EULERIAN AND LAGRANGIAN SPECIFICATIONS OF THE MHD FLOW FIELD

In this subsection, the Eulerian and Lagrangian specifications of the MHD flow field are described [5, 6, 10]. The reciprocal relation between the volume element and density variations is explained. While the topic may be apparent in conventional fluid theory, it is not obvious in the ideal MHD case.

As shown in figure 3.1, the Eulerian specification is based on the observer who remains still and observes changes of the fluid parameters at the fixed point, while in the Lagrangian description, the observer moves with the fluid cell to observe its movement. The Eulerian specification views the system as a field, while the Lagrangian specification as a collection of moving objects. Mathematically, supposing $\mathbf{v}$ is the fluid-cell velocity, in the Eulerian specification, one has

$$\frac{d\mathbf{v}}{dt} = -\mathbf{v} \cdot \nabla \mathbf{v}; \tag{3.53}$$

while, in the Lagrangian specification,

$$\frac{d\mathbf{v}}{dt} = 0. \tag{3.54}$$

Because it is MHD, besides the flow field, one also needs to specify the magnetic field $\mathbf{B}$. Since the Maxwell equations describe the electromagnetic "field" variations, the Maxwell equations naturally belong to the Eulerian description. To be specific, let us now explain how to describe the variation of magnetic field in two descriptions. In the Eulerian description, one has

$$\frac{d\mathbf{B}}{dt} = \frac{\partial \mathbf{B}}{\partial t}, \tag{3.55}$$

while in the Lagrangian description, one has

$$\frac{d\mathbf{B}}{dt} = \frac{\partial \mathbf{B}}{\partial t} + \mathbf{v}\cdot\nabla\mathbf{B}. \tag{3.56}$$

Because observing along with the moving fluid cell, the Lagrangian description includes the convective contribution, $\mathbf{v}\cdot\nabla\mathbf{B}$. The specifications in equations (3.55) and (3.56) also apply for the scalar pressure.

In an infinitesimal time interval δt, the fluid velocity can be represented by the displacement

$$\boldsymbol{\xi} = \mathbf{v}\delta t. \tag{3.57}$$

This applies for both the Eulerian and Lagrangian descriptions. Introducing the displacement in equation (3.57), equation (3.52) is reduced to

$$\delta W_k = \delta W_v + \delta W_{s0}, \tag{3.58}$$

where

$$\delta W_k = -\omega^2 \int d\mathbf{r} \frac{\rho \boldsymbol{\xi}^2}{2}, \tag{3.59}$$

$$\delta W_v = \int d\mathbf{r} \left(\frac{p}{\Gamma - 1} + \frac{B^2}{2\mu_0} \right), \tag{3.60}$$

$$\begin{aligned} \delta W_{s0} &= \oint d\mathbf{s} \cdot \left[\frac{1}{\mu_0} (\mathbf{B}\times\boldsymbol{\xi})\times\mathbf{B} + \frac{\Gamma p}{\Gamma - 1} \right] \mathbf{v} \\ &= \oint d\mathbf{s} \cdot \left[\left(\frac{p}{\Gamma - 1} + \frac{B^2}{2\mu_0} \right) \boldsymbol{\xi} + \left(p + \frac{B^2}{2\mu_0} \right) \boldsymbol{\xi} \right], \end{aligned} \tag{3.61}$$

where the time dependence of perturbations is assumed to be of the type $\exp\{-i\omega t\}$. Here, the Poynting vector in the second expression of δW_{s0} in equation (3.61) has been reduced. Consequently, the energy flux is split into two parts: the thermal and magnetic energies and the work done by the thermal and magnetic pressures. We introduce some further notations. As will be seen that δW_v can also give rise to the boundary contribution. This is denoted as δW_{sv} so that $\delta W_v = \delta W_p + \delta W_{sv}$ and $\delta W_s = \delta W_{s0} + \delta W_{sv}$. We will minimize $\delta W_v + \delta W_{s0}$ in equation (3.61) order by order to develop the equilibrium and stability theories in both the Eulerian and Lagrangian descriptions.

In general, one needs to deal with the energy integral in the Eulerian and Lagrangian variational formulations

$$\delta W = \int d\mathbf{r}\varepsilon. \tag{3.62}$$

For a displacement $\boldsymbol{\xi}$, the energy per volume can be expanded as follows:

$$\begin{aligned} \varepsilon(\mathbf{x}+\boldsymbol{\xi}) &= \varepsilon^{(0)} + \varepsilon^{(1)} + \varepsilon^{(2)} + \cdots \\ &= \varepsilon(\mathbf{x}) + \left(\frac{d\varepsilon}{dt}\right)_{\mathbf{x}} \delta t + \frac{1}{2}\left(\frac{d^2\varepsilon}{dt^2}\right)_{\mathbf{x}} (\delta t)^2 + \cdots. \end{aligned} \tag{3.63}$$

Note here that due to the Taylor expansion, there is a factor $1/2$ for the second order $\varepsilon^{(2)}$. The superscript (0) for the lowest order is often dropped for brevity. Thus, in the Eulerian description, the energy integral in equation (3.62) is expanded as follows:

$$\delta W^E = \int d\mathbf{r} \left(\varepsilon^{(0)} + \varepsilon^{(1)} + \varepsilon^{(2)} + \cdots \right). \tag{3.64}$$

Here, the superscript E is introduced to indicate the Eulerian description. Because it is in the Eulerian specification, the volume element in equation (3.64) remains unchanged.

In the Lagrangian description, however, the volume element varies due to the fluid motion. This can be calculated by the transformation of the Jacobian under the coordinate transform $\mathbf{x}' = \mathbf{x} + \boldsymbol{\xi}$,

$$\begin{aligned} \mathscr{J}(\mathbf{x}'|\mathbf{x}) &= \frac{\partial(x',y',z')}{\partial(x,y,z)} = \begin{vmatrix} 1+\frac{\partial \xi_x}{\partial x} & \frac{\partial \xi_x}{\partial y} & \frac{\partial \xi_x}{\partial z} \\ \frac{\partial \xi_y}{\partial x} & 1+\frac{\partial \xi_y}{\partial y} & \frac{\partial \xi_y}{\partial z} \\ \frac{\partial \xi_z}{\partial x} & \frac{\partial \xi_z}{\partial y} & 1+\frac{\partial \xi_x}{\partial z} \end{vmatrix} \\ &= 1 + \boldsymbol{\nabla}\cdot\boldsymbol{\xi} + \frac{1}{2}\left[(\boldsymbol{\nabla}\cdot\boldsymbol{\xi})^2 - \boldsymbol{\nabla}\boldsymbol{\xi} : \boldsymbol{\nabla}\boldsymbol{\xi}\right] + \cdots \\ &\equiv 1 + \delta^{(1)} + \delta^{(2)} + \cdots . \end{aligned} \tag{3.65}$$

where

$$\delta^{(1)} = \boldsymbol{\nabla}\cdot\boldsymbol{\xi}, \tag{3.66}$$

$$\delta^{(2)} = \frac{1}{2}\left[(\boldsymbol{\nabla}\cdot\boldsymbol{\xi})^2 - \boldsymbol{\nabla}\boldsymbol{\xi} : \boldsymbol{\nabla}\boldsymbol{\xi}\right]. \tag{3.67}$$

One can expect that the variation of the volume element in the Lagrangian description is reciprocal to the density variation. This can be proved as follows. Using the continuity equation (3.1), one obtains

$$\frac{d\rho}{dt} = \frac{\partial \rho}{\partial t} + \mathbf{v}\cdot\boldsymbol{\nabla}\rho = -\mathbf{v}\cdot\boldsymbol{\nabla}\rho - \rho\boldsymbol{\nabla}\cdot\mathbf{v} + \mathbf{v}\cdot\boldsymbol{\nabla}p = -\rho\boldsymbol{\nabla}\cdot\mathbf{v}.$$

Introducing $\boldsymbol{\xi}$ in equation (3.57), this gives the density variation in the first order

$$\rho^{(1)} = -\rho(\boldsymbol{\nabla}\cdot\boldsymbol{\xi}). \tag{3.68}$$

Carrying out the second derivative, one obtains

$$\begin{aligned} \frac{d^2\rho}{dt^2} &= -\frac{d\rho}{dt}\boldsymbol{\nabla}\cdot\mathbf{v} - \rho\boldsymbol{\nabla}\cdot\frac{\partial \mathbf{v}}{\partial t} - \rho\mathbf{v}\cdot\boldsymbol{\nabla}\boldsymbol{\nabla}\cdot\mathbf{v} \\ &= -\gamma\frac{d\rho}{dt}\boldsymbol{\nabla}\cdot\mathbf{v} - \rho\boldsymbol{\nabla}\cdot\frac{\partial \mathbf{v}}{\partial t} - \rho\boldsymbol{\nabla}\cdot(\mathbf{v}\cdot\boldsymbol{\nabla}\mathbf{v}) + \rho\boldsymbol{\nabla}\mathbf{v} : \boldsymbol{\nabla}\mathbf{v}) \\ &= \rho(\boldsymbol{\nabla}\cdot\mathbf{v})^2 + \rho\boldsymbol{\nabla}\mathbf{v} : \boldsymbol{\nabla}\mathbf{v}, \end{aligned}$$

where equation (3.54) has been used. Introducing $\boldsymbol{\xi}$ in equation (3.57), one obtains the density variation in the second order

$$\rho^{(2)} = \frac{\rho}{2}\left[(\nabla\cdot\boldsymbol{\xi})^2+\nabla\boldsymbol{\xi}:\nabla\boldsymbol{\xi}\right]. \tag{3.69}$$

The density variations in equations (3.68) and (3.69) can be used to verify the reciprocal relation between the volume element and density variations in the Lagrangian description as follows:

$$\begin{aligned} d\mathbf{r} &= d\mathbf{r}'\frac{\rho}{\rho\left\{1-\nabla\cdot\boldsymbol{\xi}+\frac{1}{2}[(\nabla\cdot\boldsymbol{\xi})^2+\nabla\boldsymbol{\xi}:\nabla\boldsymbol{\xi}]\right\}} \\ &= d\mathbf{r}'\left\{1+\nabla\cdot\boldsymbol{\xi}+(\nabla\cdot\boldsymbol{\xi})^2-\frac{1}{2}\left[(\nabla\cdot\boldsymbol{\xi})^2+\nabla\boldsymbol{\xi}:\nabla\boldsymbol{\xi}\right]\right\} \\ &= d\mathbf{r}'\left\{1+\nabla\cdot\boldsymbol{\xi}+\frac{1}{2}\left[(\nabla\cdot\boldsymbol{\xi})^2-\nabla\boldsymbol{\xi}:\nabla\boldsymbol{\xi}\right]\right\}, \end{aligned} \tag{3.70}$$

which recovers the Lagrangian volume element variation equation (3.65). The reciprocal relation justifies the physical picture of the Lagrangian description in ideal MHD: tracking the motion of individual fluid parcels.

Including the volume element variation, the energy variation in equation (3.62) in the Lagrangian description becomes

$$\delta W^L = \int d\mathbf{r}\left(1+\delta^{(1)}+\delta^{(2)}+\cdots\right)\left(\varepsilon^{(0)}+\varepsilon^{(1)}+\varepsilon^{(2)}+\cdots\right).$$

Here, the superscript L is introduced to indicate the Lagrangian description.

Besides the volume energy, one also needs to determine the surface fluxes as required for the Lagrangian formulation of the energy conservation law in equation (3.61). For a general vector $\mathbf{A}$, one has

$$\begin{aligned} \delta W_{s0}^L &= \oint d\mathbf{s}'\cdot\mathbf{A}' = \oint \mathscr{J}' e_{ijk}dx_i'dx_j'A_k' \\ &= \oint\left(1+\delta^{(1)}\right)e_{ijk}\left(dx_i+\frac{\partial\xi_i}{\partial x_l}dx_l\right)\left(dx_j+\frac{\partial\xi_j}{\partial x_m}dx_m\right)\left(A_k^{(0)}+A_k^{(1)}\right) \\ &= \oint d\mathbf{s}\cdot\mathbf{A}+\oint e_{ijk}dx_idx_j\left(A_k^{(0)}\delta^{(1)}+A_k^{(1)}\right) \\ &\quad+\oint dx_idx_me_{ijk}\frac{\partial\xi_j}{\partial x_m}A_k^{(0)}+\oint dx_jdx_le_{ijk}\frac{\partial\xi_i}{\partial x_l}A_k^{(0)}. \end{aligned} \tag{3.71}$$

Here, δ_{ij} is the Kronecker delta, and e_{ijk} is the Levi-Civita symbol or the alternating tensor.

With the fundamentals for Eulerian and Lagrangian descriptions specified, we go on to describe the Euler and Lagrange variational descriptions of ideal MHD in the next sections, sections 3.4.2 and 3.4.3.

3.4.2 EULERIAN DESCRIPTION OF THE LINEAR IDEAL MHD

In this subsection, we review the Eulerian formulation of linear ideal MHD with the general MHD energy conservation law in equation (3.51) used. It will be compared with the Lagrangian theory in the next subsection 3.4.3. The Eulerian specifications of MHD flow field described in subsection 3.4.1 will be used to describe the linear ideal MHD theory in the Eulerian description in this subsection. Both the equilibrium and stability analyses are given, leading to the recoveries of the MHD equilibrium equations and energy principle.

3.4.2.1 Variations of magnetic field and thermal parameters

In this subsection, we describe the variations of magnetic field and thermal parameters in the Eulerian description.

We first treat the magnetic field. Faraday's law and the ideal MHD Ohm's law, one obtains

$$\frac{d\mathbf{B}}{dt} = \frac{\partial \mathbf{B}}{\partial t} = -\nabla \times \mathbf{E} = \nabla \times (\mathbf{v} \times \mathbf{B}) \tag{3.72}$$

and

$$\begin{aligned}\frac{d^2\mathbf{B}}{dt^2} &= \frac{\partial^2 \mathbf{B}}{\partial t^2} = \nabla \times \left(\frac{\partial \mathbf{v}}{\partial t} \times \mathbf{B} + \mathbf{v} \times \frac{\partial \mathbf{B}}{\partial t}\right) \\ &= \nabla \times \{-(\mathbf{v} \cdot \nabla \mathbf{v}) \times \mathbf{B} + \mathbf{v} \times [\nabla \times (\mathbf{v} \times \mathbf{B})]\}. \end{aligned} \tag{3.73}$$

Here, equation (3.53) has been used. Introducing the displacement in equation (3.57), equations (3.72) and (3.73) become

$$\mathbf{B}^{E(1)} \equiv \mathbf{Q} = \nabla \times (\boldsymbol{\xi} \times \mathbf{B}), \tag{3.74}$$

$$2\mathbf{B}^{E(2)} = \nabla \times \{-[(\boldsymbol{\xi} \cdot \nabla \boldsymbol{\xi})] \times \mathbf{B} + \boldsymbol{\xi} \times \mathbf{Q}\}. \tag{3.75}$$

We next evaluate the perturbed pressure in the Eulerian description. In the first order, one has

$$\begin{aligned}\frac{dp}{dt} &\rightarrow \frac{\partial p}{\partial t} = -\mathbf{v} \cdot \nabla p - \Gamma p \nabla \cdot \mathbf{v} \\ &= (\Gamma - 1)\mathbf{v} \cdot \nabla p - \nabla \cdot (\Gamma p \mathbf{v}).\end{aligned}$$

Therefore, one obtains

$$2p^{E(1)} = (\Gamma - 1)\boldsymbol{\xi} \cdot \nabla p - \Gamma \nabla \cdot (p\boldsymbol{\xi}). \tag{3.76}$$

In the second order, one has

$$\begin{aligned}\frac{d^2p}{dt^2} &\rightarrow \frac{\partial^2 p}{\partial t^2} = -\frac{\partial \mathbf{v}}{\partial t} \cdot \nabla p - \mathbf{v} \cdot \nabla \frac{\partial p}{\partial t} - \Gamma \frac{\partial p}{\partial t} \nabla \cdot \mathbf{v} - \Gamma p \nabla \cdot \frac{\partial \mathbf{v}}{\partial t} \\ &= (\mathbf{v} \cdot \nabla \mathbf{v}) \cdot \nabla p + \mathbf{v} \cdot \nabla (\mathbf{v} \cdot \nabla p + \Gamma p \nabla \cdot \mathbf{v}) \\ &\quad + \Gamma (\mathbf{v} \cdot \nabla p + \Gamma p \nabla \cdot \mathbf{v}) \nabla \cdot \mathbf{v} + \Gamma p \nabla \cdot (\mathbf{v} \cdot \nabla \mathbf{v}) \\ &= (1 - \Gamma)(\mathbf{v} \cdot \nabla \mathbf{v}) \cdot \nabla p + \nabla \cdot [\Gamma p (\mathbf{v} \cdot \nabla \mathbf{v})] \\ &\quad + (\Gamma - 1)(\mathbf{v} \cdot \nabla p + \Gamma p \nabla \cdot \mathbf{v}) \nabla \cdot \mathbf{v} + \nabla \cdot [\mathbf{v}(\mathbf{v} \cdot \nabla p + \Gamma p \nabla \cdot \mathbf{v})].\end{aligned}$$

Therefore, one obtains

$$\begin{aligned} 2p^{E(2)} &= (1-\Gamma)(\boldsymbol{\xi}\cdot\nabla\boldsymbol{\xi})\cdot\nabla p+(\Gamma-1)(\mathbf{v}\cdot\nabla p+\Gamma p\nabla\cdot\boldsymbol{\xi})\nabla\cdot\boldsymbol{\xi} \\ &\quad +\nabla\cdot[\Gamma p(\boldsymbol{\xi}\cdot\nabla\boldsymbol{\xi})+\xi(\boldsymbol{\xi}\cdot\nabla p+\Gamma p\nabla\cdot\boldsymbol{\xi})]. \end{aligned} \tag{3.77}$$

Furthermore, the Poynting vector and $p\mathbf{v}$ in the last two divergence terms in the energy conservation law in equation (3.51) also need to be reduced. They will be used later in the variational analyses in subsection 3.4.1. To analyze these two divergence terms, $\int dt$ needs to be applied. We first treat the Poynting vector. In the first order, it yields

$$\left[\int dt(\mathbf{v}\times\mathbf{B})\times\mathbf{B}\right]^{E(1)} = (\boldsymbol{\xi}\times\mathbf{B})\times\mathbf{B}. \tag{3.78}$$

In the second order, noting that

$$\begin{aligned} \frac{d}{dt}(\mathbf{E}\times\mathbf{B}) &= -\frac{\partial}{\partial t}[\mathbf{v}\times\mathbf{B})\times\mathbf{B}] \\ &= [(\mathbf{v}\cdot\nabla\mathbf{v})\times\mathbf{B}]\times\mathbf{B}-\left(\mathbf{v}\times\frac{\partial\mathbf{B}}{\partial t}\right)\times\mathbf{B}-(\mathbf{v}\times\mathbf{B})\times\frac{\partial\mathbf{B}}{\partial t}, \end{aligned}$$

one has

$$\begin{aligned} & 2\left[\int dt(\mathbf{v}\times\mathbf{B})\times\mathbf{B}\right]^{E(2)} \\ = \; & [(\boldsymbol{\xi}\cdot\nabla\boldsymbol{\xi})\times\mathbf{B}]\times\mathbf{B}-(\boldsymbol{\xi}\times\mathbf{Q})\times\mathbf{B}-(\boldsymbol{\xi}\times\mathbf{B})\times\mathbf{Q}. \end{aligned} \tag{3.79}$$

For the $p\mathbf{v}$ term, one has in the first-order

$$\left[\int dt p\mathbf{v}\right]^{E(1)} = p\boldsymbol{\xi}. \tag{3.80}$$

In the second order, noting that

$$\begin{aligned} \frac{d(p\mathbf{v})}{dt} &= \frac{\partial p}{\partial t}\mathbf{v}+p\frac{\partial\mathbf{v}}{\partial t} \\ &= (-\mathbf{v}\cdot\nabla p-\Gamma p\nabla\cdot\mathbf{v})\mathbf{v}-p\mathbf{v}\cdot\nabla\mathbf{v}, \end{aligned}$$

one obtains

$$\left[\int dt p\mathbf{v}\right]^{E(2)} = (-\boldsymbol{\xi}\cdot\nabla p-\Gamma p\nabla\cdot\boldsymbol{\xi})\boldsymbol{\xi}-p\boldsymbol{\xi}\cdot\nabla\boldsymbol{\xi}. \tag{3.81}$$

3.4.2.2 First-order analyses: equilibrium

In this subsection, the energy integral in equation (3.58) in the first order is used to derive the MHD equilibrium equation in the Eulerian description. Using equations

(3.74) and (3.76) in subsection 3.4.2.1 for $\mathbf{Q}$ and p_1, and equations (3.78) and (3.80) for the two surface terms, the energy integral in equation (3.60) in the first order reduces to

$$\begin{aligned}
\delta W_v^{E(1)} &= \int d\mathbf{r}\left\{\frac{1}{\Gamma-1}[(\Gamma-1)\boldsymbol{\xi}\cdot\nabla p-\Gamma\nabla\cdot(p\boldsymbol{\xi})]+\frac{2\mathbf{B}\cdot\mathbf{Q}}{2\mu_0}\right\} \\
&= \int d\mathbf{r}\left[\boldsymbol{\xi}\cdot\nabla p+\frac{1}{\mu_0}\mathbf{B}\cdot\nabla\times(\boldsymbol{\xi}\times\mathbf{B})\right]-\frac{\Gamma}{\Gamma-1}\int d\mathbf{s}\cdot(p\boldsymbol{\xi}) \\
&= \int d\mathbf{r}\boldsymbol{\xi}\cdot(\nabla p-\mathbf{j}\times\mathbf{B})+\frac{1}{\mu_0}\int d\mathbf{s}\cdot[(\boldsymbol{\xi}\times\mathbf{B})\times\mathbf{B}] \\
&\quad -\frac{\Gamma}{\Gamma-1}\int d\mathbf{s}\cdot(p\boldsymbol{\xi}).
\end{aligned} \tag{3.82}$$

The two surface integration terms, which are related to $\delta W_{sv}^{E(1)}$, cancel with the surface integration terms exactly in equation (3.61), *i.e.*, $\delta W_{s0}^{E(1)}$. One finally obtains

$$\delta W^{E(1)} = \int d\mathbf{r}\boldsymbol{\xi}\cdot(\nabla p-\mathbf{j}\times\mathbf{B}).$$

Its variation with respect to $\boldsymbol{\xi}$ yields the equilibrium equation $\mathbf{j}\times\mathbf{B}=\nabla p$.

3.4.2.3 Second-order analyses: stability

In this subsection, we continue to the second order to derive the ideal MHD energy principle from the energy integral in equation (3.58) by using equations (3.75) and (3.77) in subsection 3.4.2.1 for $\mathbf{B}_2$ and p_2, and equations (3.78) and (3.80) for the two surface terms.

The total bulk energy is decomposed into the magnetic and plasma thermal energy: $\delta W_v^{E(2)} = \delta W_m^{E(2)} + \delta W_{pl}^{E(2)}$. By direct expansion, the magnetic energy in the second order is reduced to

$$\begin{aligned}
\delta W_m^{E(2)} &= \frac{1}{2\mu_0}\int d\mathbf{r}\delta B^2\Big|_2 = \frac{1}{\mu_0}\int d\mathbf{r}\,(\mathbf{B}\cdot\mathbf{B}_{2E}+\mathbf{B}_{1E}\cdot\mathbf{B}_{1E}) \\
&= \frac{1}{2\mu_0}\int d\mathbf{r}\Big\{\mathbf{Q}^2+[(\nabla\times\mathbf{B})\times\mathbf{B}]\cdot(\boldsymbol{\xi}\cdot\nabla\boldsymbol{\xi})+(\nabla\times\mathbf{B})\cdot(\boldsymbol{\xi}\times\mathbf{Q}) \\
&\quad +\nabla\cdot\{\mathbf{B}\times[(\boldsymbol{\xi}\cdot\nabla\boldsymbol{\xi})\times\mathbf{B}]-\mathbf{B}\times(\boldsymbol{\xi}\times\mathbf{Q})\}\Big\} \\
&= \frac{1}{2\mu_0}\int d\mathbf{r}\Big\{\mathbf{Q}^2+\underbrace{(\boldsymbol{\xi}\cdot\nabla\boldsymbol{\xi})\cdot\nabla p}_{x1}+(\boldsymbol{\xi}\times\mathbf{Q})\cdot(\nabla\times\mathbf{B})\Big\} \\
&\quad +\frac{1}{2\mu_0}\oint d\mathbf{s}\cdot\left[\underbrace{B^2(\boldsymbol{\xi}\cdot\nabla\boldsymbol{\xi})}_{x2}\underbrace{-\mathbf{B}\times(\boldsymbol{\xi}\times\mathbf{Q})}_{x3}\right].
\end{aligned} \tag{3.83}$$

Similarly, the plasma thermal energy in the second order is reduced as follows:

$$
\begin{aligned}
\delta W_{pl}^{E(2)} &= \frac{1}{2}\int d\mathbf{r}\frac{1}{\Gamma-1}\Big\{(1-\Gamma)(\boldsymbol{\xi}\cdot\nabla\boldsymbol{\xi})\cdot\nabla p+(\Gamma-1)(\mathbf{v}\cdot\nabla p+\Gamma p\nabla\cdot\boldsymbol{\xi})\nabla\cdot\boldsymbol{\xi} \\
&\quad +\nabla\cdot[\Gamma p(\boldsymbol{\xi}\cdot\nabla\boldsymbol{\xi})+\xi(\boldsymbol{\xi}\cdot\nabla p+\Gamma p\nabla\cdot\boldsymbol{\xi})]\Big\} \\
&= \frac{1}{2}\int d\mathbf{r}\Big\{\underbrace{-(\boldsymbol{\xi}\cdot\nabla\boldsymbol{\xi})\cdot\nabla p}_{x1}+(\mathbf{v}\cdot\nabla p+\Gamma p\nabla\cdot\boldsymbol{\xi})\nabla\cdot\boldsymbol{\xi}\Big\} \\
&\quad +\frac{1}{2}\oint d\mathbf{s}\cdot\left[\underbrace{\frac{1}{\Gamma-1}\Gamma p(\boldsymbol{\xi}\cdot\nabla\boldsymbol{\xi})}_{x4}+\underbrace{\frac{1}{\Gamma-1}\xi(\boldsymbol{\xi}\cdot\nabla p+\Gamma p\nabla\cdot\boldsymbol{\xi})}_{o1}\right]. \qquad (3.84)
\end{aligned}
$$

We also need to obtain the boundary contribution from the Poynting vector

$$
\begin{aligned}
&\frac{1}{\mu_0}\left[\oint d\mathbf{s}\cdot[(\boldsymbol{\xi}\times\mathbf{B})\times\mathbf{B}]\right]_2 \\
&= \frac{1}{2\mu_0}\oint d\mathbf{s}\cdot\Big\{[(\boldsymbol{\xi}\cdot\nabla\boldsymbol{\xi})\times\mathbf{B}]\times\mathbf{B}-(\boldsymbol{\xi}\times\mathbf{Q})\times\mathbf{B}-(\boldsymbol{\xi}\times\mathbf{B})\times\mathbf{Q}\Big\} \\
&= \frac{1}{2\mu_0}\oint d\mathbf{s}\cdot\Big\{\underbrace{-B^2(\boldsymbol{\xi}\cdot\nabla\boldsymbol{\xi})}_{x2}\underbrace{-(\boldsymbol{\xi}\times\mathbf{Q})\times\mathbf{B}}_{x3}+\boldsymbol{\xi}(\mathbf{B}\cdot\mathbf{Q})\Big\}. \qquad (3.85)
\end{aligned}
$$

Similarly, another surface integration in equation (3.61) can also be reduced

$$
\begin{aligned}
&\left[\oint d\mathbf{s}\cdot\frac{\Gamma p\boldsymbol{\xi}}{\Gamma-1}\right]_2 \\
&= \frac{1}{2}\frac{\Gamma}{\Gamma-1}\oint d\mathbf{s}\cdot\Big\{\underbrace{(-\boldsymbol{\xi}\cdot\nabla p-\Gamma p\nabla\cdot\boldsymbol{\xi})\boldsymbol{\xi}}_{o1}\underbrace{-p(\boldsymbol{\xi}\cdot\nabla\boldsymbol{\xi})}_{x4}\Big\}. \qquad (3.86)
\end{aligned}
$$

By combining equations (3.83)–(3.86) and noting the cancellations with the same underbrace indices "*xi*" and the combination with the same underbrace indices "*oi*", one finally obtains the total energy

$$
\delta W^E = \delta W_p^E + \delta W_s^E, \qquad (3.87)
$$

where the ordering indices have been dropped and

$$
\begin{aligned}
\delta W_p^E &= \frac{1}{2}\int d\mathbf{r}\left\{\frac{\mathbf{Q}^2}{\mu_0}+\mathbf{j}\cdot(\boldsymbol{\xi}\times\mathbf{Q})+(\boldsymbol{\xi}\cdot\nabla p)\nabla\cdot\boldsymbol{\xi}+\Gamma p(\nabla\cdot\boldsymbol{\xi})^2\right\}, \qquad (3.88) \\
\delta W_s^E &= \frac{1}{2}\oint d\mathbf{s}\cdot\boldsymbol{\xi}\left[(-\boldsymbol{\xi}\cdot\nabla p-\Gamma p\nabla\cdot\boldsymbol{\xi})+\frac{1}{\mu_0}\mathbf{B}\cdot\mathbf{Q}\right]. \qquad (3.89)
\end{aligned}
$$

Using equations (3.87)–(3.89), one can obtain the MHD energy principle as given in reference [1]

$$
\delta W_k^E = \delta W(\boldsymbol{\xi},\boldsymbol{\xi}),
$$

where

$$\begin{aligned}\delta W_k &= \frac{1}{2}\int \rho_m \left(\frac{\partial \boldsymbol{\xi}}{\partial t}\right)^2 d\mathbf{r}, = -\frac{\omega^2}{2}\int \rho_m |\boldsymbol{\xi}|^2 d\mathbf{r},\\ \delta W &= \delta W_p + \delta W_{surf} + \delta W_{vac}.\end{aligned}$$

Here, the plasma energy is given by

$$\delta W_p = \frac{1}{2}\int d\mathbf{r}\left\{\frac{\mathbf{Q}^2}{\mu_0} + \mathbf{j}\cdot(\boldsymbol{\xi}\times\mathbf{Q}) + (\boldsymbol{\xi}\cdot\nabla p)\nabla\cdot\boldsymbol{\xi} + \Gamma p(\nabla\cdot\boldsymbol{\xi})^2\right\},$$

which is the same as equation (3.40), and the surface and vacuum energies are respectively given as follows:

$$\begin{aligned}\delta W_{surf} &= -\frac{1}{2}\oint ds\xi_n^2 \frac{\partial}{\partial n}\left(p_0 + \frac{B_{0i}^2}{2\mu_0} - \frac{B_{0i}^2}{2\mu_0}\right),\\ \delta W_{vac} &= \frac{1}{2\mu_0}\int_V (\nabla\times\mathbf{A})\cdot(\nabla\times\mathbf{A})d\mathbf{r} = \frac{1}{2\mu_0}\int_V B_1^2 d\mathbf{r}.\end{aligned}$$

Here, $\mathbf{A}$ is the vector potential, n is used to denote the normal direction to the plasma-vacuum interface, and the subscripts i and o denote inside and outside the interface.

The results reviewed in this section will be compared with the Lagrangian formulation in the next Section 3.4.3.

3.4.3 LAGRANGIAN DESCRIPTION OF THE LINEAR IDEAL MHD

In this subsection, the variations of the magnetic and plasma energies in the Lagrangian description are described. Both the first- and second-order theories are presented. As pointed out earlier, reference [6] worked out the energy principle from the Lagrangian approach, and reference [10] derived the surface contributions.

3.4.3.1 Variations of magnetic field and thermal quantities

In this subsection, the variations of magnetic field and thermal quantities in the Lagrangian description are described, which is more complicated than the Eulerian description. Noting that the energy integral in equation (3.61) is given in the Eulerian description, the modifications of the volume and surface flux contributions are required in the Lagrangian description.

Noting that the magnetic ($B^2/2\mu_0$) and plasma ($p/(\Gamma-1)$) energies in the general conservation law in equation (3.51) are scalars, we will derive a general scalar variation rule. We attempt to relate the Lagrangian variations to the Eulerian ones. The results in the Eulerian formulation in subsection 3.4.2 can be used. Suppose that the scalar S denotes either the magnetic, plasma energy, or the total energy. One then has

$$\begin{aligned}
\frac{dS}{dt} &= \frac{\partial S}{\partial t} + \mathbf{v}\cdot\nabla S. &(3.90)\\
\frac{d^2S}{dt^2} &= \frac{\partial^2 S}{\partial t^2} + \frac{\partial \mathbf{v}}{\partial t}\cdot\nabla S + \mathbf{v}\cdot\nabla\frac{\partial S}{\partial t} + \mathbf{v}\cdot\nabla\left(\frac{\partial S}{\partial t} + \mathbf{v}\cdot\nabla S\right)\\
&= \frac{\partial^2 S}{\partial t^2} - (\mathbf{v}\cdot\nabla\mathbf{v})\cdot\nabla S - (\nabla\cdot\mathbf{v})\frac{\partial S}{\partial t} - (\nabla\cdot\mathbf{v})\left(\frac{\partial S}{\partial t} + \mathbf{v}\cdot\nabla S\right)\\
&\quad +\nabla\cdot\left(\mathbf{v}\frac{\partial S}{\partial t}\right) + \nabla\cdot\left[\mathbf{v}\left(\frac{\partial S}{\partial t} + \mathbf{v}\cdot\nabla S\right)\right]\\
&= \frac{\partial^2 S}{\partial t^2} + S\nabla\cdot(\mathbf{v}\cdot\nabla\mathbf{v}) - (\nabla\cdot\mathbf{v})\frac{\partial S}{\partial t} - (\nabla\cdot\mathbf{v})\left(\frac{\partial S}{\partial t} + \mathbf{v}\cdot\nabla S\right)\\
&\quad -\nabla\cdot(S\mathbf{v}\cdot\nabla\mathbf{v}) + \nabla\cdot\left(\mathbf{v}\frac{\partial S}{\partial t}\right) + \nabla\cdot\left[\mathbf{v}\left(\frac{\partial S}{\partial t} + \mathbf{v}\cdot\nabla S\right)\right]\\
&= \frac{\partial^2 S}{\partial t^2} + S\nabla\mathbf{v}:\nabla\mathbf{v} + S\mathbf{v}\cdot\nabla(\nabla\cdot\mathbf{v}) - (\nabla\cdot\mathbf{v})\frac{\partial S}{\partial t} - (\nabla\cdot\mathbf{v})\left(\frac{\partial S}{\partial t} + \mathbf{v}\cdot\nabla S\right)\\
&\quad -\nabla\cdot(p\mathbf{v}\cdot\nabla\mathbf{v}) + \nabla\cdot\left(\mathbf{v}\frac{\partial S}{\partial t}\right) + \nabla\cdot\left[\mathbf{v}\left(\frac{\partial S}{\partial t} + \mathbf{v}\cdot\nabla S\right)\right]\\
&= \frac{\partial^2 S}{\partial t^2} + S\nabla\mathbf{v}:\nabla\mathbf{v} - S(\nabla\cdot\mathbf{v})^2 - (\nabla\cdot\mathbf{v})\mathbf{v}\cdot\nabla S - (\nabla\cdot\mathbf{v})\frac{\partial S}{\partial t}\\
&\quad -(\nabla\cdot\mathbf{v})\left(\frac{\partial S}{\partial t} + \mathbf{v}\cdot\nabla S\right) + \nabla\cdot[S\mathbf{v}(\nabla\cdot\mathbf{v})]\\
&\quad -\nabla\cdot(S\mathbf{v}\cdot\nabla\mathbf{v}) + \nabla\cdot\left(\mathbf{v}\frac{\partial S}{\partial t}\right) + \nabla\cdot\left[\mathbf{v}\left(\frac{\partial S}{\partial t} + \mathbf{v}\cdot\nabla S\right)\right]\\
&= \frac{\partial^2 S}{\partial t^2} + S\nabla\mathbf{v}:\nabla\mathbf{v} - S(\nabla\cdot\mathbf{v})^2 - 2(\nabla\cdot\mathbf{v})\left(\frac{\partial S}{\partial t} + \mathbf{v}\cdot\nabla S\right)\\
&\quad +\nabla\cdot[S\mathbf{v}(\nabla\cdot\mathbf{v})] - \nabla\cdot(S\mathbf{v}\cdot\nabla\mathbf{v}) + \nabla\cdot\left(\mathbf{v}\frac{\partial S}{\partial t}\right)\\
&\quad +\nabla\cdot\left[\mathbf{v}\left(\frac{\partial S}{\partial t} + \mathbf{v}\cdot\nabla S\right)\right]. &(3.91)
\end{aligned}$$

Here, we have used equation (3.54) for reduction. Note further that $\partial S/\partial t$ and $\partial^2 S/\partial t^2$ correspond to the Eulerian variations in the first and second orders. Introducing $\boldsymbol{\xi}$, equations (3.90) and (3.91) become, respectively,

$$\begin{aligned}
S^{L(1)} &= S^{E(1)} + \boldsymbol{\xi}\cdot\nabla S\\
&= S^{E(1)} - S\nabla\cdot\boldsymbol{\xi} + \nabla\cdot(\boldsymbol{\xi}S). &(3.92)\\
2S^{L(2)} &= 2S^{E(2)} + S\nabla\boldsymbol{\xi}:\nabla\boldsymbol{\xi} - S(\nabla\cdot\boldsymbol{\xi})^2 - 2(\nabla\cdot\boldsymbol{\xi})S^{L(1)}\\
&\quad +\nabla\cdot[S\boldsymbol{\xi}(\nabla\cdot\boldsymbol{\xi})] - \nabla\cdot(S\boldsymbol{\xi}\cdot\nabla\boldsymbol{\xi}) + \nabla\cdot\left(\boldsymbol{\xi}S^{E(1)}\right)\\
&\quad +\nabla\cdot\left[\boldsymbol{\xi}\left(S^{E(1)} + \boldsymbol{\xi}\cdot\nabla S\right)\right]. &(3.93)
\end{aligned}$$

Here, S^E represents the Eulerian variations of S, which have been derived in subsection 3.4.2.1.

3.4.3.2 First-order analyses: equilibrium

In this subsetion, we derive the MHD equilibrium equation from the Lagrangian analyses in the first order. Equations (3.66) and (3.92) are used to represent the variations of the volume element and the energy integrand in the Lagrangian description. For brevity, we let

$$S = \frac{p}{\Gamma - 1} + \frac{B^2}{2\mu_0}. \tag{3.94}$$

In the first order, the bulk energy integral in equation (3.60) is reduced to

$$\begin{aligned}\delta W_v^{L(1)} &= \left[\int d\mathbf{r}(1+\nabla\cdot\boldsymbol{\xi})\left(S+S^{E(1)}-S\nabla\cdot\boldsymbol{\xi}+\nabla\cdot(\boldsymbol{\xi}S)\right)\right]^{(1)} \\ &= \int d\mathbf{r}\left[S^{E(1)}+S\nabla\cdot\boldsymbol{\xi}-S\nabla\cdot\boldsymbol{\xi}+\nabla\cdot(\boldsymbol{\xi}S)\right] \\ &= \int d\mathbf{r}S^{E(1)}+\oint d\mathbf{s}\cdot\boldsymbol{\xi}S. \end{aligned} \tag{3.95}$$

Note that $\int d\mathbf{r}S^{E(1)}$ is $\delta W_v^{E(1)}$ in the Eulerian description in equation (3.82) and $\delta W_{s0}^{L(1)} = \delta W_{s0}^{E(1)}$ since there is no first-order modification in the Lagrangian description. Therefore, one has

$$\begin{aligned}\delta W^{L(1)} &= \delta W_v^{E(1)}+\oint d\mathbf{s}\cdot\boldsymbol{\xi}S+\delta W_{s0}^{L(1)} \\ &= \delta W_v^{E(1)}+\delta W_{s0}^{E(1)}+\oint d\mathbf{s}\cdot\boldsymbol{\xi}S \\ &= \delta W^{E(1)}+\oint d\mathbf{s}\cdot\boldsymbol{\xi}S.\end{aligned}$$

This indicates that the Lagrangian description gives rise to an extra surface flux contribution $\delta W_{sx}^{L(1)} = \oint d\mathbf{s}\cdot\boldsymbol{\xi}S$, as compared with the results in the Eulerian description in subsection 3.4.2.2. However, only a formal extra term is obtained. In the MHD equilibrium, $\boldsymbol{\xi}$ at the boundary should vanish. Therefore, in the Lagrangian description, the first-order analyses still give rise to the MHD equilibrium equation $\mathbf{j}\times\mathbf{B} = \nabla p$. These constitute the first-order analyses of linear ideal MHD in the Lagrangian description.

3.4.3.3 Second-order analyses: stability

In this subsection, the second-order variational analyses in the Lagrangian description are described [6, 10]. For this, equations (3.66), (3.67), (3.92), and (3.93) are used to represent the variations of the energy integrand and volume element in the

Lagrangian description. For brevity, the definition of S in equation (3.94) is used. In the second order, the bulk energy integral in equation (3.60) becomes

$$\begin{aligned}
\delta W_v^{L(2)} &= \Bigg[\int d\mathbf{r}\left\{1+\nabla\cdot\boldsymbol{\xi}+\frac{1}{2}\left[(\nabla\cdot\boldsymbol{\xi})^2-\nabla\boldsymbol{\xi}:\nabla\boldsymbol{\xi}\right]\right\} \\
&\quad\times\Bigg\{S+\underbrace{S^{E(1)}-S\nabla\cdot\boldsymbol{\xi}+\nabla\cdot(\boldsymbol{\xi}S)}_{S_L^{(1)}}+S^{E(2)} \\
&\quad+\frac{1}{2}S\left[\nabla\boldsymbol{\xi}:\nabla\boldsymbol{\xi}-(\nabla\cdot\boldsymbol{\xi})^2\right]-(\nabla\cdot\boldsymbol{\xi})S^{L(1)} \\
&\quad+\frac{1}{2}\Bigg[\nabla\cdot\left[S\boldsymbol{\xi}(\nabla\cdot\boldsymbol{\xi})\right]-\nabla\cdot(S\boldsymbol{\xi}\cdot\nabla\boldsymbol{\xi})+\nabla\cdot\left(\boldsymbol{\xi}S^{E(1)}\right) \\
&\quad+\nabla\cdot\left[\boldsymbol{\xi}\left(S^{E(1)}+\boldsymbol{\xi}\cdot\nabla S\right)\right]\Bigg]\Bigg\}\Bigg]^{(2)} \\
&= \int d\mathbf{r}\Bigg\{S^{E(2)}+\frac{1}{2}\Bigg[\nabla\cdot\left[S\boldsymbol{\xi}(\nabla\cdot\boldsymbol{\xi})\right]-\nabla\cdot(S\boldsymbol{\xi}\cdot\nabla\boldsymbol{\xi})+\nabla\cdot\left(\boldsymbol{\xi}S^{E(1)}\right) \\
&\quad+\nabla\cdot\left[\boldsymbol{\xi}\left(S^{E(1)}+\boldsymbol{\xi}\cdot\nabla S\right)\right]\Bigg]\Bigg\} \\
&= \int d\mathbf{r}S^{E(2)}+\delta W_{sxv}^{L(2)}.
\end{aligned} \tag{3.96}$$

where $\delta W_v^{E(2)}=\int d\mathbf{r}S_E^{(2)}$ in equation (3.84) and

$$\delta W_{sxv}^{L(2)}=\frac{1}{2}\oint d\mathbf{s}\cdot\left[\boldsymbol{\xi}S(\nabla\cdot\boldsymbol{\xi})-S\boldsymbol{\xi}\cdot\nabla\boldsymbol{\xi}+\boldsymbol{\xi}\left(2S^{E(1)}+\boldsymbol{\xi}\cdot\nabla S\right)\right]. \tag{3.97}$$

There are additional contributions from the surface flux contributions according to equation (3.71). Noting that the sum of the surface flux and work contributions is

$$\mathbf{A} = S\boldsymbol{\xi}+\left(p+\frac{B^2}{2\mu_0}\right)\boldsymbol{\xi},$$

one has

$$\delta W_{s0}^{L(2)} = \delta W_s^{E(2)}+\delta W_{sx0}^{L(2)}, \tag{3.98}$$

where

$$\begin{aligned}
\delta W_{sx0}^{L(2)} &= \oint e_{ijk}dx_i dx_j\left(A_k^{(0)}\delta^{(1)}+A_k^{(1)}\right) \\
&\quad+\oint dx_i dx_m e_{ijk}\frac{\partial\xi_j}{\partial x_m}A_k^{(0)}+\oint dx_j dx_l e_{ijk}\frac{\partial\xi_i}{\partial x_l}A_k^{(0)}.
\end{aligned}$$

Combining equations (3.96) and (3.98) yields

$$\begin{aligned}
\delta W^L &= \delta W_v^{E(2)}+\delta W_{sxv}^{L(2)}+\delta W_s^{E(2)}+\delta W_{sx0}^{L(2)} \\
&= \delta W^E+\delta W_{sx}^L,
\end{aligned} \tag{3.99}$$

where

$$\delta W_{sx}^{L} = \delta W_{sxv}^{L(2)} + \delta W_{sx0}^{L(2)}, \tag{3.100}$$

Here, δW^E is the energy variation in the Eulerian description given in equation (3.87). For brevity, some ordering indices are dropped. The results in equation (3.99) indicate that the Lagrangian description gives rise to an extra surface flux contribution δW_{sx}^{L} in equation (3.100), as compared with the results in the Eulerian description in subsection 3.4.2.3. This completes the second-order analyses of linear ideal MHD in the Lagrangian description.

The Lagrange and Euler variational formulations are often used in fluid mechanics. The formulation described in this section extends it to the ideal MHD case. It is useful both for the analytical and simulation studies of MHD stabilities, especially for the free boundary modes.

3.5 PROPERTIES OF LINEAR MHD

With the fundamental theoretical framework for MHD laid out in the previous sections, in this section, the properties of linearized MHD are described. The description is in a somewhat advanced manner and nevertheless, the connection to the basic pictures is clarified. From the general decomposition of MHD equations, the three fundamental MHD waves: compressional Alfvén, shear Alfvén, and acoustic waves are presented. The alternative expression of the ideal MHD energy principle by J. Greene and J. Johnson is presented [12]. It is used to discuss the two fundamental types of instabilities, the kink and interchange modes.

3.5.1 THREE FUNDAMENTAL MHD WAVES

In this subsection, we describe the decomposition of MHD equations. Using the general decomposition, the three fundamental MHD waves: compressional Alfvén, shear Alfvén, and acoustic waves are described. One also uses the equations to discuss the instabilities. Nevertheless, we describe them instead by using the energy principle in the next subsection.

The linearized MHD equation is given in equation (3.25)

$$-\rho_m \omega^2 \boldsymbol{\xi} = \mathbf{F}(\boldsymbol{\xi}), \tag{3.101}$$

where

$$\begin{aligned} \mathbf{F}(\boldsymbol{\xi}) &= \frac{1}{\mu_0}(\nabla \times \mathbf{Q}) \times \mathbf{B} + \frac{1}{\mu_0}(\nabla \times \mathbf{B}) \times \mathbf{Q} + \nabla(\boldsymbol{\xi} \cdot \nabla p) + \Gamma \nabla(p \nabla \cdot \boldsymbol{\xi}) \\ &= -\nabla\left(\frac{\mathbf{B} \cdot \mathbf{Q}}{\mu_0} + p_1\right) + \frac{1}{\mu_0}(\mathbf{B} \cdot \nabla \mathbf{Q} + \mathbf{Q} \cdot \nabla \mathbf{B}). \end{aligned} \tag{3.102}$$

We have introduced the Fourier representation for time: $\boldsymbol{\xi} = \boldsymbol{\xi} \exp\{-i\omega t\}$. Without causing ambiguity, the same notation has been used for the amplitude of $\boldsymbol{\xi}$.

Equation (3.101) has three components: two perpendicular and one parallel to the magnetic field. We introduce three unit vectors: $\mathbf{e}_b = \mathbf{B}/B$, $\mathbf{e}_1 = \nabla\psi/|\nabla\psi|$, and $\mathbf{e}_2 = \mathbf{e}_b \times \mathbf{e}_1$ for the projections. Here, we choose $\nabla\psi$ to be perpendicular both to the equilibrium current density and magnetic field. This can be realized by noting the equilibrium equation (3.3). It will become clear later that ψ labels the magnetic surface.

The $\mathbf{e}_2$ and $\mathbf{e}_1$ projections of the MHD equation (3.101) give respectively

$$\underbrace{\frac{1}{\mu_0}\mathbf{e}_1 \cdot \nabla \times \delta\mathbf{B}}_{\omega_{A\perp}^2} - \underbrace{\frac{\rho_m \omega^2}{B}\mathbf{e}_2 \cdot \boldsymbol{\xi}}_{\omega^2} = B^2\sigma\mathbf{e}_1 \cdot \delta\mathbf{B} + \frac{1}{B}\mathbf{e}_2 \cdot \nabla\left(|\nabla P|\mathbf{e}_1 \cdot \boldsymbol{\xi}\right) + \frac{\Gamma}{B}\mathbf{e}_2 \cdot \nabla\left(P\nabla \cdot \boldsymbol{\xi}\right), \tag{3.103}$$

$$\begin{aligned}\underbrace{\frac{1}{\mu_0}\mathbf{e}_2 \cdot \nabla \times \delta\mathbf{B}}_{\omega_{A\perp}^2} + \underbrace{\frac{\rho_m \omega^2}{B}\mathbf{e}_1 \cdot \boldsymbol{\xi}}_{\omega^2} &= B^2\sigma\mathbf{e}_2 \cdot \delta\mathbf{B} - \frac{|\nabla P|}{B^2}\mathbf{e}_b \cdot \delta\mathbf{B} - \frac{1}{B}\mathbf{e}_1 \cdot \nabla\left(|\nabla P|\mathbf{e}_1 \cdot \boldsymbol{\xi}\right) \\ &\quad - \frac{\Gamma}{B}\mathbf{e}_1 \cdot \nabla\left(P\nabla \cdot \boldsymbol{\xi}\right),\end{aligned} \tag{3.104}$$

where $\sigma = \mathbf{j} \cdot \mathbf{B}/B^2$. Here, the orderings for the terms on the left-hand sides have been added, which will be discussed later on in this section with $\omega_{A\perp} = k_\perp v_A$ and $k_\perp$ the perpendicular wave vector and the Alfvén speed $v_A = \sqrt{B^2/(\mu_0\rho_m)}$. Alternatively, using the second formulation of operator $\mathbf{F}$ in equations (3.102), (3.103), and (3.104) can be expressed as

$$\underbrace{\mathbf{e}_\perp \cdot \nabla\left(\frac{\mathbf{B} \cdot \mathbf{Q}}{\mu_0} + p_1\right)}_{\omega_{A\perp}^2} - \underbrace{\rho\omega^2\mathbf{e}_\perp \cdot \boldsymbol{\xi}}_{\omega^2} = \frac{1}{\mu_0}\mathbf{e}_\perp \cdot \left(\mathbf{B} \cdot \nabla\mathbf{Q} + \mathbf{Q} \cdot \nabla\mathbf{B}\right). \tag{3.105}$$

Here, $\mathbf{e}_\perp$ denotes the unit vector perpendicular to the magnetic field.

The $\mathbf{e}_b$ projection of MHD equation (3.101) can be obtained by applying the operator $\nabla \cdot [\mathbf{e}_b\mathbf{e}_b \cdot (\cdots)]$ on equation (3.101)

$$\underbrace{\Gamma P\mathbf{B} \cdot \nabla\left(\frac{1}{B^2}\mathbf{B} \cdot \nabla\nabla \cdot \boldsymbol{\xi}\right)}_{\omega_s^2} + \underbrace{\rho_m\omega^2\nabla \cdot \boldsymbol{\xi}}_{\omega^2} = \rho_m\omega^2\nabla \cdot \boldsymbol{\xi}_\perp. \tag{3.106}$$

Here, $\nabla \cdot \boldsymbol{\xi}$ is used as an independent unknown [13], This is appropriate, noting that $\delta\rho_m \propto -\rho_m\nabla \cdot \boldsymbol{\xi}$ and the left-hand side of this equation describes the sound wave along the field line.

Noting that $\delta\mathbf{J}$ and $\delta\mathbf{B}$ are determined completely by $\boldsymbol{\xi}_\perp$, one can see that the set of equations (3.103), (3.104) (or (3.105)), and (3.106) is complete to determine the two components of $\boldsymbol{\xi}_\perp$ and the scalar unknown $\nabla\cdot\boldsymbol{\xi}$.

Two perpendicular equations of motion, equations (3.103) and (3.104), result from perpendicular projections of MHD equation (3.101) and therefore contain the compression/restoration force from the excitation of compressional Alfvén wave. Note that compressional Alfvén wave results from the term $\delta\mathbf{J}\times\mathbf{B}+\mathbf{J}\times\delta\mathbf{B}+\nabla\delta P\rightarrow\nabla(\mathbf{B}\cdot\delta\mathbf{B}+\delta P)$ as shown in equation (3.105). Therefore, a curl operation on equation (3.101) can suppress the compressional Alfvén wave. Applying the operator $\nabla\cdot(\mathbf{B}/B^2)\times(\cdots)$ on equation (3.101), one obtains

$$\underbrace{-\mathbf{B}\cdot\nabla\frac{\mathbf{B}\cdot\delta\mathbf{J}}{B^2}}_{\omega_A^2}+\underbrace{\nabla\cdot\frac{\mathbf{B}}{B^2}\times\rho_m\omega^2\boldsymbol{\xi}}_{\omega^2}$$

$$=\delta\mathbf{B}\cdot\nabla\sigma-\mathbf{J}\cdot\nabla\frac{\mathbf{B}\cdot\delta\mathbf{B}}{B^2}+\nabla\times\frac{\mathbf{B}}{B^2}\cdot\nabla\delta P, \tag{3.107}$$

Equation (3.107) is often referred to as the shear Alfvén law or vorticity equation. One can use equation (3.107) to replace one of the perpendicular momentum equations to form the complete set of equations for linear MHD analyses.

Note that the MHD eigenvalues ω^2 are real as shown in subsection 3.3.2. When $\omega^2>0$, this set of equations characterizes three fundamental MHD waves, respectively: the shear Alfvén, compressional Alfvén, and parallel acoustic waves, mainly by their left-hand sides and their couplings by their right-hand sides. When $\omega^2<0$, one has instabilities. In the following sections, we will first describe the global MHD spectrum in tokamaks; Subsequently, we will study one by one the interchange modes, ballooning modes, peeling and free boundary ballooning modes, toroidal Alfvén eigen modes, and kinetically driven MHD modes.

Let us discuss the underlying physics for the three components of the MHD equations. For simplicity, it is assumed that the equilibrium pressure P and magnetic field $\mathbf{B}$ are constant. We first consider equation (3.103). Introduce the wave vector $\mathbf{k}$. In the case with both $\mathbf{k}$ and $\boldsymbol{\xi}$ being in the $\mathbf{e}_2$ direction, one has $e_1\cdot\nabla\times\delta\mathbf{B}=-\mathbf{e}_1\cdot\mathbf{k}_2\times[\mathbf{k}_2\times(\boldsymbol{\xi}\times\mathbf{B})]=Bk_2^2\xi_2$. Therefore, the left-hand side of equation (3.103) yields

$$\left[k_2^2\left(v_A^2+v_s^2\right)-\omega^2\right]\xi_2=0. \tag{3.108}$$

where $v_s=\sqrt{\Gamma P/\rho_m}$ is the sound speed. Similarly, in the case of both $\mathbf{k}$ and $\boldsymbol{\xi}$ being in the $\mathbf{e}_1$ direction, one obtains from equation (3.104)

$$\left[k_1^2\left(v_A^2+v_s^2\right)-\omega^2\right]\xi_1=0. \tag{3.109}$$

Equations (3.108) and (3.109) describe the same mode feature: the magnetic field and plasma are compressed in the direction perpendicular to the magnetic field as shown in figure 3.2. The field oscillation is referred to as the compressional Alfvén wave or magneto-sound wave if the plasma compression is considered as well.

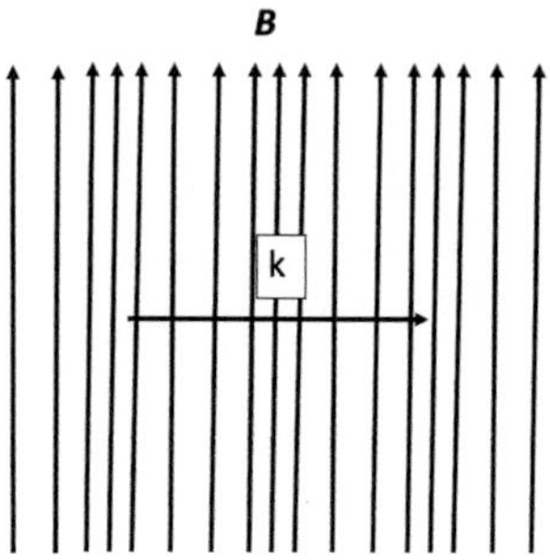

Figure 3.2 Compressional Alfvén wave.

Next, we consider the vorticity equation (3.107). Using the frozen-in law in equation (3.16)

$$-i\omega\boldsymbol{\xi} = \frac{\mathbf{E}\times\mathbf{B}}{B^2} = \frac{-\nabla\delta\varphi\times\mathbf{B}}{B^2}$$

where $\delta\varphi$ represents the electric potential or so-called stream function in the applications to be described later in this book. In this case, one has

$$\begin{aligned} \mu_0\delta\mathbf{J} &= \nabla\times(\nabla\times\boldsymbol{\xi}\times\mathbf{B}) \\ &= -\nabla\times\left[\nabla\times(\mathbf{B}\mathbf{B}\cdot\nabla\delta\varphi/B^2)\right] \\ &= \nabla^2(\mathbf{B}\mathbf{B}\cdot\nabla\delta\varphi/B^2)] - \nabla\left[\mathbf{B}\cdot\nabla\left(\mathbf{B}\cdot\nabla\delta\varphi/B^2\right)\right]. \end{aligned} \tag{3.110}$$

In the case with $k_\perp \gg k_\parallel$, the first term on the right is much larger than the second in equation (3.110). Therefore, one has in the WKB expression

$$\mathbf{B}\cdot\nabla\frac{\mathbf{B}\cdot\delta\mathbf{J}}{B^2} = \frac{1}{\mu_0}k^2k_\parallel^2\delta\varphi \tag{3.111}$$

and the vorticity equation becomes

$$\left(k_\parallel^2 v_A^2 - \omega^2\right)\delta\varphi = 0. \tag{3.112}$$

This describes the so-called shear Alfvén wave propagating along the magnetic field as shown in figure 3.3.

Let us now consider the sound wave equation (3.106). Note that $\nabla\cdot\boldsymbol{\xi}$ and $\nabla\cdot\boldsymbol{\xi}_\perp$ are independent. The homogeneous part of equation (3.106) gives rise to

$$\left(k_\parallel^2 v_s^2 - \omega^2\right)\delta\varphi = 0. \tag{3.113}$$

This characterizes the sound waves propagating along the magnetic field line as shown in figure 3.4.

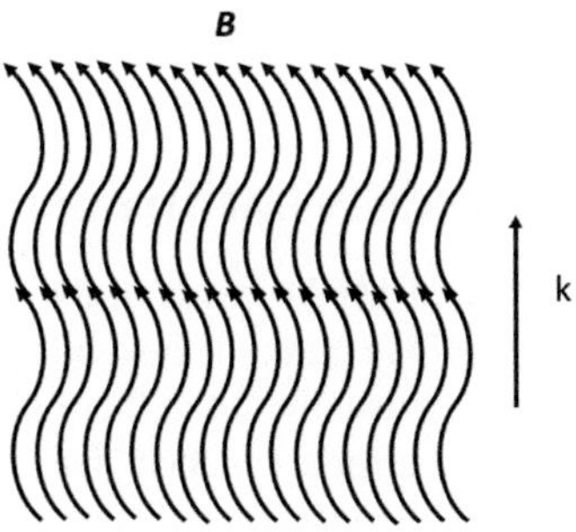

Figure 3.3 Shear Alfvén wave.

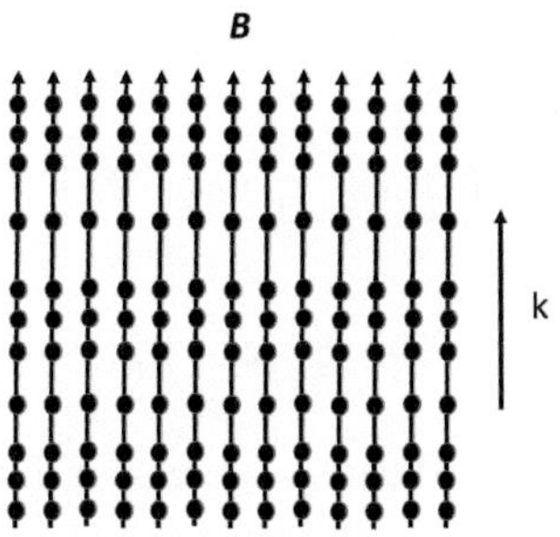

Figure 3.4 Sound wave along the magnetic field.

In equations (3.108) (or (3.109)), (3.112), and (3.113), we show that there are three fundamental MHD waves: the compressional Alfvén waves propagating in the direction perpendicular to the magnetic field, the shear Alfvén waves propagating in the direction parallel to the magnetic field, and the sound waves propagating also in the direction parallel to the magnetic field without the magnetic field being perturbed. For the compressional Alfvén waves, the wave vector **k** and the field displacement $\boldsymbol{\xi}$ are parallel and the shear Alfvén waves; however, the wave vector **k** and the field displacement $\boldsymbol{\xi}$ are perpendicular to each other.

Note that $k_\perp$ is usually much larger than $k_\parallel$. we then have the following ordering sequence for the three MHD waves

$$\omega_{A\perp} \gg \omega_A \gg \omega_s. \tag{3.114}$$

The orderings in equation (3.114) are important for MHD stability analyses. For example, when considering the modes with frequency much lower than the compressional Alfvén frequency, one can obtain from equation (3.105) that

$$\frac{\mathbf{B} \cdot \mathbf{Q}}{\mu_0} + p_1 = 0.$$

We have indicated the orderings for three fundamental waves in equations (3.103) - (3.107).

3.5.2 KINK AND INTERCHANGE INSTABILITY DRIVES

We have shown three fundamental MHD waves in the previous subsection. In this subsection, we show another MHD feature: the instability drives. This is because the magnetically confined plasmas are usually not in thermal equilibrium. The so-called MHD equilibrium is actually just a force balance. Therefore, there is an instability issue. This is just like putting a ball on a hilltop. One can establish the force balance. But the status is unstable. A small perturbation can cause the ball to roll down the hill. The energy principle, in particular, given in reference [12] will be used for identifying the instability drives: the kink and interchange modes.

The plasma energy has been derived in equation (3.40) in Section 3.3. We now transform it to that given by J. Greene and J. Johnson in reference [12] as follows

$$\begin{aligned} \delta W_p &= \frac{1}{2}\int \left\{ \frac{1}{\mu_0}\left| \mathbf{Q} - \mathbf{B}\frac{\mu_0 \boldsymbol{\xi}\cdot\nabla p}{B^2}\right|^2 - \frac{J_\parallel}{B}\boldsymbol{\xi}\times\mathbf{B}\cdot\mathbf{Q} \right. \\ & \left. -2(\boldsymbol{\xi}\cdot\nabla p)(\boldsymbol{\xi}\cdot\boldsymbol{\kappa}) + \Gamma p(\nabla\cdot\boldsymbol{\xi})^2 \right\} d\mathbf{r}, \end{aligned} \tag{3.115}$$

To prove equation (3.115), we first derive the relationship between the field-line curvature and the magnetic and plasma pressures, noting that

$$\begin{aligned} \mathbf{B}\cdot\nabla\mathbf{B} &= \frac{1}{2}\nabla B^2 + (\nabla\times\mathbf{B})\times\mathbf{B} \\ &= \frac{1}{2}\nabla B^2 + \mu_0\mathbf{J}\times\mathbf{B} = \frac{1}{2}\nabla(2\mu_0 p + B^2) \end{aligned}$$

and

$$\begin{aligned} \boldsymbol{\kappa} &= \mathbf{b}\cdot\nabla\mathbf{b} = \frac{1}{B^2}\left(\mathbf{B}\cdot\nabla\mathbf{B} - \frac{\mathbf{B}}{B}\mathbf{B}\cdot\nabla B\right) \\ &= \frac{(\mathbf{B}\times\mathbf{B}\cdot\nabla\mathbf{B})\times\mathbf{B}}{B^4}, \end{aligned}$$

one obtains

$$\boldsymbol{\kappa} = \frac{\left[\mathbf{B}\times\nabla(2\mu_0 p + B^2)\right]\times\mathbf{B}}{2B^4}. \tag{3.116}$$

One needs another useful relationship for the proof. Note that

$$\begin{aligned} &\nabla\cdot\boldsymbol{\xi} - \mathbf{B}\cdot\nabla\left(\frac{\boldsymbol{\xi}\cdot\mathbf{B}}{B^2}\right) = \nabla\cdot\frac{\mathbf{B}\times(\boldsymbol{\xi}\times\mathbf{B})}{B^2} \\ &= \frac{\boldsymbol{\xi}\times\mathbf{B}}{B^2}\cdot\nabla\times\mathbf{B} - \frac{\mathbf{B}}{B^2}\cdot\nabla\times(\boldsymbol{\xi}\times\mathbf{B}) + \mathbf{B}\times(\boldsymbol{\xi}\times\mathbf{B})\cdot\nabla\frac{1}{B^2} \\ &= -\frac{\mu_0\boldsymbol{\xi}\cdot\nabla p}{B^2} - \frac{\mathbf{B}\cdot\delta\mathbf{B}}{B^2} - \frac{1}{B^2}\boldsymbol{\xi}\cdot\nabla B^2 + \frac{\boldsymbol{\xi}\cdot\mathbf{B}}{B^4}\mathbf{B}\cdot\nabla B^2. \end{aligned}$$

Using equation (3.116), one obtains

$$2\boldsymbol{\xi}\cdot\boldsymbol{\kappa} = \frac{\boldsymbol{\xi}\cdot\nabla(2\mu_0 P + B^2)}{B^2} - \frac{\boldsymbol{\xi}\cdot\mathbf{B}}{B^4}\mathbf{B}\cdot\nabla B^2. \tag{3.117}$$

Combining these two equations, one obtains

$$\frac{1}{B^2}\left(\mathbf{Q} - \mathbf{B}\frac{\mu_0\boldsymbol{\xi}\cdot\nabla p}{B^2}\right)\cdot\mathbf{B} = -\nabla\cdot\boldsymbol{\xi} + \mathbf{B}\cdot\nabla\left(\frac{\boldsymbol{\xi}\cdot\mathbf{B}}{B^2}\right) - 2\boldsymbol{\xi}\cdot\boldsymbol{\kappa}, \tag{3.118}$$

Let us now prove the energy principle in equation (3.115). First, we note that the first term in the integrand can be reduced as follows

$$\frac{1}{\mu_0}\left|\mathbf{Q} - \mathbf{B}\frac{\mu_0\boldsymbol{\xi}\cdot\nabla p}{B^2}\right|^2 = \frac{1}{\mu_0}\mathbf{Q}^2 \underbrace{-2\mathbf{B}\cdot\mathbf{Q}\frac{\boldsymbol{\xi}\cdot\nabla p}{B^2}}_{o1} + \underbrace{\frac{\mu_0(\boldsymbol{\xi}\cdot\nabla p)^2}{B^2}}_{x1}. \tag{3.119}$$

Using equation (3.118), one can find that the third term in the integrand can be reduced as follows:

$$\begin{aligned}-2(\boldsymbol{\xi}\cdot\nabla p)(\boldsymbol{\xi}\cdot\boldsymbol{\kappa}) &= (\boldsymbol{\xi}\cdot\nabla p)\nabla\cdot\boldsymbol{\xi} \underbrace{-(\boldsymbol{\xi}\cdot\nabla p)\mathbf{B}\cdot\nabla\left(\frac{\boldsymbol{\xi}\cdot\mathbf{B}}{B^2}\right)}_{o2} \\ &\quad \underbrace{+\frac{1}{B^2}(\boldsymbol{\xi}\cdot\nabla p)\mathbf{B}\cdot\mathbf{Q}}_{o1}\underbrace{-\frac{\mu_0(\boldsymbol{\xi}\cdot\nabla p)^2}{B^2}}_{x1}.\end{aligned} \tag{3.120}$$

In equations (3.119) and (3.120), the terms marked with "xi" cancel each other; the terms marked with "oi" can be combined to yield the following term

$$\begin{aligned}&(\mathbf{j}_\perp\times\boldsymbol{\xi})\cdot\nabla\times\boldsymbol{\xi}\times\mathbf{B} \\ =\ & \frac{\mathbf{B}\times\nabla p}{B^2}\times\boldsymbol{\xi}\cdot\nabla\times\boldsymbol{\xi}\times\mathbf{B} \\ =\ & \frac{1}{B^2}[(\mathbf{B}\cdot\boldsymbol{\xi})(\mathbf{Q}\cdot\nabla p) - (\mathbf{B}\cdot\mathbf{Q})(\boldsymbol{\xi}\cdot\nabla p)] \\ =\ & \frac{1}{B^2}[(\mathbf{B}\cdot\boldsymbol{\xi})(\mathbf{B}\cdot\nabla(\boldsymbol{\xi}\cdot\nabla p)) - (\mathbf{B}\cdot\mathbf{Q})(\boldsymbol{\xi}\cdot\nabla p)] \\ =\ & -\mathbf{B}\cdot\nabla\left(\frac{\boldsymbol{\xi}\cdot\mathbf{B}}{B^2}\right)(\boldsymbol{\xi}\cdot\nabla p) - \frac{1}{B^2}(\mathbf{B}\cdot\mathbf{Q})(\boldsymbol{\xi}\cdot\nabla p) + \mathbf{B}\cdot\nabla\left[\left(\frac{\boldsymbol{\xi}\cdot\mathbf{B}}{B^2}\right)(\boldsymbol{\xi}\cdot\nabla p)\right] \\ =\ & -\mathbf{B}\cdot\nabla\left(\frac{\boldsymbol{\xi}\cdot\mathbf{B}}{B^2}\right)(\boldsymbol{\xi}\cdot\nabla p) - \frac{1}{B^2}(\mathbf{B}\cdot\mathbf{Q})(\boldsymbol{\xi}\cdot\nabla p) + \nabla\cdot\left[\mathbf{B}\left(\frac{\boldsymbol{\xi}\cdot\mathbf{B}}{B^2}\right)(\boldsymbol{\xi}\cdot\nabla p)\right] \\ =\ & \underbrace{-\mathbf{B}\cdot\nabla\left(\frac{\boldsymbol{\xi}\cdot\mathbf{B}}{B^2}\right)(\boldsymbol{\xi}\cdot\nabla p)}_{o2}\underbrace{-\frac{1}{B^2}(\mathbf{B}\cdot\mathbf{Q})(\boldsymbol{\xi}\cdot\nabla p)}_{o1}.\end{aligned} \tag{3.121}$$

Here, in the last step, $d\mathbf{s} \cdot \mathbf{B} = 0$ on the boundary has been used. Using equations (3.119)–(3.121), one can prove that the energy principle in equation (3.115) is equivalent to that in equation (3.40).

In the expression of the energy principle in equation (3.115), the first term in the integrand corresponds to the stabilizing magnetic energies from the magnetic field compression and the field line bending. They are related to the excitation of the compressional and shear Alfvén modes discussed in Section 3.5.1. The last term reflects the stabilizing effects of the plasma compression energy. It is related to the sound wave discussed in Section 3.5.1.

Using the energy principle in equation (3.115), one can discuss two types of instability drives in the magnetically confined plasmas: the kink and pressure-driven instabilities. Note that the second term in the integrand contains the local helicity $\boldsymbol{\xi} \times \mathbf{B} \cdot \delta\mathbf{B} = \delta\mathbf{A} \cdot \nabla \times \delta\mathbf{A}$. It reflects the sausage-type deformation of the field line bundle or the twist of the magnetic field line bundle due to the perturbation. Because there is the parallel current—the coefficient tied to the helicity, the sausage and twist can further develop due to the magnetic pressure induced by the parallel current. In the sausage case, a plasma column contraction can cause the local poloidal field to become stronger so that the sausage type of instability can develop further. Also, in the twisting case, the magnetic field becomes stronger on the concave side so that the twist can further develop. They are shown in figure 3.5. This type of instability is referred to as the kink modes or the current-driven modes.

The third term in the integrand of equation (3.115) can cause another type of instability. Note that the sign of this term depends on the relative directions of the pressure gradient and magnetic field line curvature. It is referred to as the bad curvature if their directions are the same as shown in figure 3.6; otherwise, the good curvature. From the energy principle expressed in equation (3.115), one can see that, in the bad curvature case, the third term is destabilizing. This is similar to the Rayleigh-Taylor instability in fluid theory. The instabilities driven by the bad curvature mechanism, such as the interchange and ballooning modes, are referred to as the pressure-driven modes.

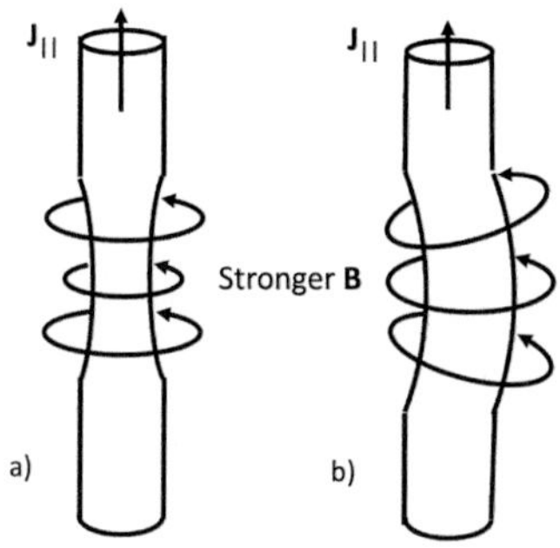

Figure 3.5 Current driven modes: a) sausage type; b) kink type.

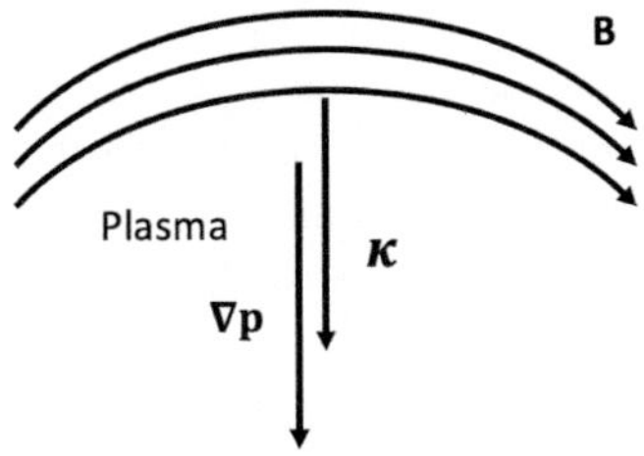

Figure 3.6 Pressure driven modes: unstable configuration.

3.6 NONIDEAL MHD EFFECTS

We have mainly discussed the ideal MHD theories in the previous sections. In this section, we will discuss the nonideal MHD effects. This covers the resistivity, two-fluid, and kinetic effects. Here, we only focus on explaining the underlying mathematical and physical thoughts. The details will be addressed in the concrete stability analyses in the later chapters.

3.6.1 RESISTIVE EFFECTS

We first discuss the resistivity effects. Taking into account the resistivity effects, the Ohm's law becomes

$$\mathbf{E}+\mathbf{v}\times\mathbf{B}=\eta\mathbf{j} \tag{3.122}$$

Using the Farady and Ampere laws with the displacement current neglected, the linearization of equation (3.122) becomes

$$i\omega\left[\delta\mathbf{B}-\nabla\times(\boldsymbol{\xi}\times\mathbf{B})\right]=\eta\nabla\times\delta\mathbf{j}. \tag{3.123}$$

Here, it is assumed η is constant. When the resistivity effects are taken into account, two types of nonideal MHD modes appear: the resistive interchange and tearing types. Suppose that there is the magnetic surface ψ such that $\mathbf{j}\cdot\nabla\psi=\mathbf{B}\cdot\nabla\psi=0$. We further introduce the concept of rational surface where $k_{\|}\propto\mathbf{B}\cdot\nabla=0$. Depending on the parity of $\delta\mathbf{B}\cdot\nabla\psi$ across the rational magnetic surface, the modes are, respectively, referred to as odd, or even modes. The interchange modes are odd and the tearing modes are even. Let us consider them one by one.

For resistive interchange-type modes, one needs to consider the vorticity equation (3.107). Operating $\nabla\times(\cdots)$ on equation (3.123) yields

$$\left(1+\frac{\eta}{i\omega\mu_0}\nabla^2\right)\delta\mathbf{j}=\frac{1}{\mu_0}\nabla\times\left[\nabla\times(\boldsymbol{\xi}\times\mathbf{B})\right], \tag{3.124}$$

where $\nabla\cdot\delta\mathbf{j}=0$ has been used. The first term in the vorticity equation contains the derivative of the highest order. From equations (3.110) and (3.111), one can see that $k_{\parallel} = 0$ is the singular point of the equation. Including the resistivity effects, the solution of equation (3.124) allows a finite $\delta \mathbf{j}_{\parallel}$ when $k_{\parallel} = 0$. The resistivity effects then remove the singularity of the vorticity equation. Since the singularity effects cause the continuum damping as will be seen in the later chapters, the resistivity effects give rise to the so-called resistive interchange type of modes even when the ideal MHD shows stablility. The resistivity weakens the stabilization effects from the field bending effects.

For tearing type of modes, one can alternatively work on equation (3.123) to obtain

$$\left(1 + \frac{\eta}{i\omega\mu_0}\nabla^2\right)\delta\mathbf{B} = \nabla \times (\boldsymbol{\xi} \times \mathbf{B}). \tag{3.125}$$

Note that $\nabla\psi \cdot \nabla \times (\boldsymbol{\xi} \times \mathbf{B}) = \nabla \cdot (\boldsymbol{\xi} \times \mathbf{B}) \times \nabla\psi = \mathbf{B} \cdot \nabla(\boldsymbol{\xi} \cdot \nabla\psi)$. In the ideal MHD case, equation (3.125) yields $\delta\mathbf{B} \cdot \nabla\psi = \mathbf{B} \cdot \nabla(\boldsymbol{\xi} \cdot \nabla\psi)$. This indicates that the radial magnetic field $\delta\mathbf{B} \cdot \nabla\psi$ vanishes on the rational surface where $k_{\parallel} \propto \mathbf{B} \cdot \nabla = 0$. In the presence of resistivity, equation (3.125) becomes the differential equation for determining the radial magnetic field $\delta\mathbf{B} \cdot \nabla\psi$, which leads to a finite $\delta\mathbf{B} \cdot \nabla\psi$ at the rational surface. Suppose that the direction of magnetic field $\mathbf{B}$ varies across the magnetic surface. The finite $\delta\mathbf{B} \cdot \nabla\psi$ leads to field line reconnection at the rational surface, resulting in the magnetic island and stochastic field structure.

Since resistive MHD is one of the main topics of this book. It will be introduced in detail in the following chapters.

3.6.2 TWO FLUID EFFECTS

Let us now discuss the two-fluid effects in this subsection. Discussion of two-fluid effects helps to explain the roots of nonideal MHD effects. In the ideal MHD description, the ideal MHD Ohm's law indicates that the electric field in the direction parallel to the magnetic field should vanish. In the two-fluid description, however, the different responses of ion and electron fluids can lead to charge separation so that the parallel electric field can be excited. Besides, the two-fluid description can also take into account the so-called finite Larmor radius (FLR) effects in the second order.

We first discuss the parallel electric effects. Without taking into account the FLR effects, the motions of ions and electrons in the direction perpendicular to the magnetic field basically follow the field line displacement $\boldsymbol{\xi}$, which is different from the fluid displacement. Note that for similar ion and electron temperatures, the electron thermal speed is much faster than the ion's in the direction parallel to the magnetic field. This can cause the difference in ion and electron responses.

We introduce ion and electron transit frequencies: $\omega_{ti,e} = v_{Ti,e}/L$, where v_T denotes the thermal speed, subscripts i and e represent the ion and electron species, respectively, and L is the characteristic scale length in the parallel direction. We define three frequency regimes: the low, comparable, and intermediate frequency regimes, respectively, according to

$$\omega \ll \omega_{ti}, \quad \omega \sim \omega_{ti}, \quad \text{and} \quad \omega_{ti} \ll \omega \ll \omega_{te}, \tag{3.126}$$

One needs not to consider the parallel electric field effects in the low-frequency regime since both ions and electrons have sufficient time to be thermalized. In the comparable and intermediate frequency regimes, the electron and ion responses are different. Kinetic treatment is necessary for the comparable frequency regime due to the wave-ion resonance that occurs. Therefore, we mainly focus on the intermediate frequency regime.

The ideal MHD Ohm's law in equation (3.16) results in the parallel electric field vanishes, $\delta \mathbf{E}_\parallel = 0$. This needs to be modified in the comparable or intermediate frequency regimes even without taking into account the resistivity effects. Introducing the scalar $\delta\varphi$ and vector potentials $\delta\mathbf{A}$, the perturbed electric and magnetic fields can be represented as

$$\delta\mathbf{E} = -\nabla\delta\varphi - \frac{\partial\delta\mathbf{A}}{\partial t}, \quad \delta\mathbf{B} = \nabla\times\delta\mathbf{A}.$$

For MHD-like modes, one can take the gauge $\delta\mathbf{A}_\parallel = 0$ so that

$$\delta\mathbf{A} = \boldsymbol{\xi}\times\mathbf{B}.$$

In the perpendicular direction, the momentum equations (2.13) and (2.12) of ion and electron fluid yield in the lowest order

$$\delta\mathbf{v}_\perp = -i\omega\boldsymbol{\xi}_\perp - \frac{\nabla\delta\varphi\times\mathbf{B}}{B^2}.$$

Because the FLR effects are neglected, the perpendicular velocities for the ion and electron fluids become identical.

$$\begin{aligned} \nabla\cdot\mathbf{v}_\perp &= -i\omega\nabla\cdot\boldsymbol{\xi}_\perp + \nabla\cdot\frac{\mathbf{B}\times\nabla\delta\varphi}{B^2} \\ &= -i\omega\nabla\cdot\boldsymbol{\xi}_\perp + i\frac{e_{i,e}}{T_{i,e}}\omega_{di,e}\delta\varphi, \end{aligned} \tag{3.127}$$

where

$$i\omega_{di,e} = \frac{T_{i,e}}{e_{i,e}}\nabla\times\frac{\mathbf{B}}{B^2}\cdot\nabla.$$

We first consider the electron fluid. The parallel mobility of electron fluid leads it to be isothermal along the field line, i.e., $\mathbf{B}\cdot\nabla T_e = 0$. In the first order, one obtains

$$\delta\mathbf{B}\cdot\nabla T + \mathbf{B}\cdot\nabla\delta T_e = \nabla\times(\boldsymbol{\xi}\times\mathbf{B})\cdot\nabla T_e + \mathbf{B}\cdot\nabla\delta T_e = 0.$$

This yields

$$\delta T_e = -\boldsymbol{\xi}_\perp\cdot\nabla T_e. \tag{3.128}$$

We next consider the electron parallel momentum equation. Since electron inertia is negligible, one has

$$-T_e\mathbf{B}\cdot\nabla\delta n_e - T_e\delta\mathbf{B}\cdot\nabla n_e - n_0 e_e\mathbf{B}\cdot\nabla\delta\varphi = 0. \tag{3.129}$$

This gives

$$\delta n_e = -\boldsymbol{\xi}_\perp \cdot \nabla n_0 - \frac{n_0 e_e \delta\varphi}{T_e}. \tag{3.130}$$

Combining equations (3.128) and (3.130) yields

$$\delta p_e = -\boldsymbol{\xi}_\perp \cdot \nabla p_e - \frac{p_e e_e \delta\varphi}{T_e}. \tag{3.131}$$

For an ion fluid, the continuity equation yields

$$-i\omega\delta n_i - i\omega\boldsymbol{\xi}_\perp \cdot \nabla n_0 - \frac{\nabla\delta\varphi \times \mathbf{B}}{B^2} \cdot \nabla n_0 + n_0 \nabla \cdot \delta\mathbf{v}_i = 0,$$

where it has been considered that $\mathbf{B} \cdot \nabla n_0 = 0$. It can be solved to yield

$$\delta n_i \quad = \quad -\boldsymbol{\xi}_\perp \cdot \nabla n_0 - n_0 \nabla \cdot \boldsymbol{\xi} + \frac{n_0 e_i}{T_i}\frac{\omega_{di}}{\omega}\delta\varphi - \frac{n_0 e_i}{T_i}\frac{\omega_{*n}}{\omega}\delta\varphi, \tag{3.132}$$

where the parallel ion fluid velocity $\delta\mathbf{v}_{\parallel i} = -i\omega\boldsymbol{\xi}_\parallel$ is used and

$$i\omega_{*ni} = \frac{T_i}{n_0 e_i B^2}\mathbf{B} \times \nabla n_0 \cdot \nabla.$$

The adiabatic law for ion fluid gives

$$-i\omega\delta p_i - i\omega\boldsymbol{\xi}_\perp \cdot \nabla p + \frac{\nabla\delta\varphi \times \mathbf{B}}{B^2} \cdot \nabla p_i + \Gamma p_i \nabla \cdot \delta\mathbf{v}_i = 0.$$

One obtains

$$\delta p_i \quad = \quad -\boldsymbol{\xi}_\perp \cdot \nabla p_i - \Gamma p_i \nabla \cdot \boldsymbol{\xi}_i + \frac{\Gamma p_i e_i}{T_i}\frac{\omega_{di}}{\omega}\delta\varphi - \frac{p_i e_i}{T_i}\frac{\omega_{*pi}}{\omega}\delta\varphi, \tag{3.133}$$

where

$$i\omega_{*pi} = \frac{1}{n_0 e_i B^2}\mathbf{B} \times \nabla p_i \cdot \nabla.$$

Because the ion and electron fluid motions in the perpendicular direction are the same, our focus is put on the parallel motion. In the ideal MHD description, one has the single-fluid sound-wave equation (3.106). Since the ion and electron fluids behave differently in the two-fluid picture, we need to treat them separately to include the parallel electric effects using the quasineutrality condition

$$\delta n_e e_e + \delta n_i e_i = 0. \tag{3.134}$$

The ion momentum equation in the parallel direction with the adiabatic ion fluid response yields

$$\begin{aligned}
-\omega^2\rho_m\boldsymbol{\xi}_i\cdot\mathbf{B} &= -\mathbf{B}\cdot\nabla\delta p_i - \delta\mathbf{B}\cdot\nabla p_i - n_0e_i\mathbf{B}\cdot\nabla\delta\varphi \\
&= -\mathbf{B}\cdot\nabla\left(-\boldsymbol{\xi}_\perp\cdot\nabla p_i - \Gamma p_i\nabla\cdot\boldsymbol{\xi}_i + \frac{\Gamma p_ie_i}{T_i}\frac{\omega_{di}}{\omega}\delta\varphi - \frac{p_ie_i}{T_i}\frac{\omega_{*pi}}{\omega}\delta\varphi\right) \\
&\quad -(\nabla\times\boldsymbol{\xi}\times\mathbf{B})\cdot\nabla p_i - n_0e_i\mathbf{B}\cdot\nabla\delta\varphi \\
&= -\mathbf{B}\cdot\nabla\left(-\Gamma p_i\nabla\cdot\boldsymbol{\xi} + \frac{\Gamma p_ie_i}{T_i}\frac{\omega_{di}}{\omega}\delta\varphi - \frac{p_ie_i}{T_i}\frac{\omega_{*pi}}{\omega}\delta\varphi\right) \\
&\quad -n_0e_i\mathbf{B}\cdot\nabla\delta\varphi.
\end{aligned} \tag{3.135}$$

Similar to the derivation of equation (3.106) in the ideal MHD case, applying $\mathbf{B}\cdot\nabla\left[(1/B^2)(\cdots)\right]$ on equation (3.135) yields

$$\begin{aligned}
&\Gamma p_i\mathbf{B}\cdot\nabla\left(\frac{1}{B^2}\mathbf{B}\cdot\nabla\nabla\cdot\boldsymbol{\xi}_i\right) + \rho_m\omega^2\nabla\cdot\boldsymbol{\xi}_i \\
&= \rho_m\omega^2\nabla\cdot\boldsymbol{\xi}_\perp - \mathbf{B}\cdot\nabla\left[\frac{1}{B^2}\mathbf{B}\cdot\nabla\left(n_0e_i\delta\varphi + \frac{\Gamma p_ie_i}{T_i}\frac{\omega_{di}}{\omega}\delta\varphi\right.\right. \\
&\quad \left.\left. -\frac{p_ie_i}{T_i}\frac{\omega_{*pi}}{\omega}\delta\varphi\right)\right].
\end{aligned} \tag{3.136}$$

Note here that $\boldsymbol{\xi}_{\parallel i}$ is related to the parallel velocity of ion fluid, while $\boldsymbol{\xi}_{\perp i} = \boldsymbol{\xi}_{\perp e}$.

Using (3.130) and (3.132), the quasineutrality condition in equation (3.134) yields

$$-\frac{n_0e_e^2\delta\varphi}{T_e} + \frac{n_0e_i^2}{T_i}\frac{\omega_{di}}{\omega}\delta\varphi - \frac{n_0e_i^2}{T_i}\frac{\omega_{*n}}{\omega}\delta\varphi = n_0e_i\nabla\cdot\boldsymbol{\xi}_i. \tag{3.137}$$

where $\nabla\cdot\boldsymbol{\xi}$ can be determined by equation (3.136).

With these basic equations derived, we can discuss the two fluid modifications to the ideal MHD. Because of the coupling of electrostatic modes to the MHD modes, let us briefly discuss the electrostatic modes. For electrostatic modes, $\delta\mathbf{A}$ or $\boldsymbol{\xi}_\perp$ can be neglected. Therefore, equation (3.136) can be solved for $\nabla\cdot\boldsymbol{\xi}_{\parallel i}$ or $\boldsymbol{\xi}_{\parallel i}$ as a function of $\delta\varphi$. Inserting the solution into the quasinetrality condition in equation (3.137), one obtains the eigenmode equation for $\delta\varphi$. Here, it should be pointed out that in the intermediate-frequency regime, the first term in the equation of parallel motion, equation (3.136) can be neglected. The solution of $\nabla\cdot\boldsymbol{\xi}_i$ can be then obtained readily. Therefore, the dispersion relation of electrostatic modes can be derived. In the comparable regime, however, the sound wave resonance occurs on the left-hand side of equation (3.136). The kinetic or perpendicular fluid treatments become necessary in this case.

For electromagnetic modes, one needs to keep $\delta\mathbf{A}$ or $\boldsymbol{\xi}_\perp$. From equations (3.131) and (3.133), one can see that the total perturbed pressure can be expressed as

$$\delta p = \delta p_e + \delta p_i = -\boldsymbol{\xi}\cdot\nabla p - \Delta p. \tag{3.138}$$

In the ideal MHD case, $\delta p = \Gamma \nabla \cdot \boldsymbol{\xi}$ with $\nabla \cdot \boldsymbol{\xi}$ determined by equation (3.136). In the two-fluid description, however, one has

$$\Delta p = -\Gamma p \nabla \cdot \boldsymbol{\xi}_i + \frac{\Gamma p_i e_i}{T_i} \frac{\omega_{di}}{\omega} \delta \varphi - \frac{p_i e_i}{T_i} \frac{\omega_{*pi}}{\omega} \delta \varphi, \tag{3.139}$$

with $\nabla \cdot \boldsymbol{\xi}$ determined by the equation of parallel motion of ion fluid, equation (3.136), and $\delta\varphi$ determined by the quasineutrality condition in equation (3.137). With Δp determined, one can use it to modify the ideal MHD equations as follows.

In the ideal MHD case, the basic set of equations consists of two components of Ampere's law, equations (3.103) and (3.104), and the parallel momentum equation (3.106). There are three unknowns: two components of $\boldsymbol{\xi}_\perp$ and $\nabla \cdot \boldsymbol{\xi}$. In the two-fluid description, the parallel momentum equation, equation (3.106), for $\nabla \cdot \boldsymbol{\xi}$ should be replaced. The two-fluid modifications of ideal MHD formalism can be realized by replacing $-\Gamma p \nabla \cdot \boldsymbol{\xi}$ with Δp in equation (3.139) in the two components of Ampere's law, equations (3.103) and (3.104). In equation (3.139) for Δp, $\nabla \cdot \boldsymbol{\xi}_i$ is determined by the ion parallel momentum equation (3.136) and $\delta\varphi$ by the quasineutrality condition in equation (3.137). Therefore, the basic set of equations consists of two components of Ampere's law equations for perpendicular components of field line displacement $\boldsymbol{\xi}_\perp$ with the couplings to the parallel momentum equation of ion fluid for $\nabla \cdot \boldsymbol{\xi}_i$ and the quasineutrality condition for parallel electric field effects $\delta\varphi$.

For resistive MHD, noting that the parallel current is mostly carried by electrons, one can add the resistivity and electron inertia into the parallel momentum equation of electron fluid, equation (3.129). In passing, we note that the parallel electron fluid velocity can be related to $\mathbf{j}_\parallel = (1/\mu_0)\nabla \times \mathbf{B}$ since the parallel ion fluid velocity has been determined by $\boldsymbol{\xi}_i$. Therefore, introducing the electron inertia does not increase the number of unknowns. The ideal MHD is featured by the frozen-in-law so that $\delta \mathbf{B} = \nabla \times \boldsymbol{\xi} \times \mathbf{B}$. Note that both the resistivity and electron inertia directly affect the frozen-in law. However, $\nabla \delta\varphi$ in general does not since $\nabla \times \nabla \delta\varphi = 0$.

The two-fluid description can also include the low-order FLR effects. First, one can include the diamagnetic velocity:

$$\mathbf{v}_p = \frac{\mathbf{B} \times \nabla p}{n e_i B^2},$$

To include this for electrostatic modes is trivial. Let us discuss the so-called diamagnetic drift cancellation for electromagnetic modes. Note that the electron mass is much smaller than the ion one. The inertia term in the MHD equation (3.101) can be regarded as the ion inertia. Taking into account the equilibrium diamagnetic velocity, the inertia term becomes

$$\rho_m \frac{d^2 \boldsymbol{\xi}_\perp}{dt^2} = \left(\frac{\partial}{\partial t} + \mathbf{v}_{pi} \cdot \nabla \right)^2 \boldsymbol{\xi}_\perp = -(\omega - \omega_{*pi})^2 \boldsymbol{\xi}_\perp.$$

where it is assumed that the perpendicular wavelength $\lambda_\perp$ is much smaller than the equilibrium scale length. For ordering consistency, taking into account the equilibrium diamagnetic effect on the inertia term should be accompanied by including the gyroviscosity as well, i.e., $-\nabla p_i \to -\nabla p_i - \nabla \cdot \pi_{gyro}$, where π_{gyro} is the ion gyroviscous tensor. Note that the gyroviscous tensor can give rise to a term $-\omega_{*pi}(\omega-\omega_{*pi})\boldsymbol{\xi}$. This leads to

$$\rho_m \frac{d^2\boldsymbol{\xi}_\perp}{dt^2} + \nabla\cdot\pi_{gyro} = -\omega(\omega-\omega_{*pi})\boldsymbol{\xi}_\perp + \cdots.$$

This gives the so-called diamagnetic stabilization effects. Note that the ion gyroviscous tensor contributes additional effects. They can only be neglected in the case with rather steep pressure gradient [14].

3.6.3 KINETIC AND PERPENDICULAR FLUID DESCRIPTIONS

In this subsection, we described the effects with the kinetic or perpendicular fluid descriptions. The left-hand side of the parallel momentum equation (3.136), or (3.106), describes the sound wave. The parallel motion in this equation is described by the thermal velocity. Because magnetically confined plasmas usually have low collisionality, charged particles can move rather freely along the magnetic field lines and are not localized in a fluid cell. Therefore, the kinetic or perpendicular fluid descriptions become necessary. This is particularly relevant when the particles are trapped by the equilibrium magnetic field or resonate with the waves. Here, we outline how the fluid description is modified by the kinetic description in the parallel responses.

Let us explain how the kinetic description modifies the fluid one for the trapped particle and resonance effects. The structure of the linearized Vlasov equation in the guiding center approximation is like (see [15, 11] and the references therein)

$$-i\omega\delta f + v_\parallel \mathbf{B}\cdot\nabla\delta f + v_\parallel\delta\mathbf{B}\cdot\nabla F - i\omega\boldsymbol{\xi}\cdot\nabla F = e(\delta\mathbf{E}+\mathbf{v}\times\delta\mathbf{B})\cdot\nabla_v F, \quad (3.140)$$

where F and δf are respectively the equilibrium and perturbed distribution functions, and it has been noted that $\mathbf{v}_\parallel\cdot\nabla\delta f = v_\parallel\mathbf{B}\cdot\nabla\delta f + v_\parallel\delta\mathbf{B}\cdot\nabla F$ in the linearization. When the convective part is extracted by defining

$$\delta f = -\boldsymbol{\xi}\cdot\nabla F + \delta g, \quad (3.141)$$

equation (3.140) becomes

$$-i\omega\delta g + v_\parallel\mathbf{B}\cdot\nabla\delta g = e(\delta\mathbf{E}+\mathbf{v}\times\delta\mathbf{B})\cdot\nabla_v F \equiv \mathscr{R}.$$

Here, it has been noted that $\delta\mathbf{B}\cdot\nabla F = \nabla\times\boldsymbol{\xi}\times\mathbf{B}\cdot\nabla F = \mathbf{B}\cdot\nabla(\boldsymbol{\xi}\cdot\nabla F)$. Comparing equations (3.141) with (3.138), one can see the similar structure for the convective term—the first one on the right. Note that the velocity space integration of the distribution function contains both $v_\parallel > 0$ and $v_\parallel < 0$ parts, δg_+ and δg_-. We introduce the even and odd parts of the nonconvective distribution function

$$\delta g = \delta g_{even} + \delta g_{odd},$$

where

$$\delta g_{even} = \frac{\delta g_+ + \delta g_-}{2} \quad \text{and} \quad \delta g_{odd} = \frac{\delta g_+ - \delta g_-}{2}.$$

The equations for δg_+ and δg_- are as follows

$$\begin{aligned} -i\omega\delta g_+ + |v_\parallel|\mathbf{B}\cdot\nabla\delta g_+ &= \mathscr{R}_+, \\ -i\omega\delta g_- - |v_\parallel|\mathbf{B}\cdot\nabla\delta g_- &= \mathscr{R}_-, \end{aligned}$$

They can be combined to obtain

$$-i\omega\delta g_{even} + |v_\parallel|\mathbf{B}\cdot\nabla\delta g_{odd} = \mathscr{R}_{even}, \tag{3.142}$$

$$-i\omega\delta g_{odd} + |v_\parallel|\mathbf{B}\cdot\nabla\delta g_{even} = \mathscr{R}_{odd}. \tag{3.143}$$

Applying $|\mathbf{v}_\parallel|\mathbf{B}\cdot\nabla$ on equation (3.143) and using equation (3.142) for δg_{odd}, one obtains

$$|v_\parallel|\mathbf{B}\cdot\nabla\left(|v_\parallel|\mathbf{B}\cdot\nabla\delta g_{even}\right) + \omega^2\delta g_{even} = |v_\parallel|\mathbf{B}\cdot\nabla\mathscr{R}_{odd} + i\omega\mathscr{R}_{odd}. \tag{3.144}$$

Comparing with the parallel momentum equation (3.136) or (3.106) in the fluid description, one can see a similar structure. The difference lies in that the thermal velocity in the fluid description is replaced by the particle velocity in the kinetic description. Using the kinetic equation (3.144), one can treat the trapped particle effects by the equilibrium magnetic field and the wave-particle resonance effects. The perpendicular magnetofluid description gives similar modifications [14].

The parallel electric field comes with the kinetic description naturally as soon as the different responses of ions and electrons are taken into account, as in the two-fluid description. In the kinetic description, the FLR effects are recovered by the coordinate transform to the guiding center and the introduction of the gyrophase and adiabatic invariants for velocity space coordinates (see [11] and references therein).

3.7 CONCLUSIONS AND DISCUSSION

In this section, the fundamentals of MHD analyses are described. Besides, the general toptic like the MHD force operator, the boundary conditions, the MHD energy principle, etc., the advanced theories with the Eulerian and Lagrangian variational descriptions are also given. The variational formulation paves the way for abstract theoretical extensions of MHD formalism, for example, to include the plasma flow effects and to extend to the guiding center description.

These fundamental analyses give the theoretical framework for further MHD analyses in this book. More importantly, the underlying physics is explained in the abstract theoretical framework. Three MHD waves, the compressional Alfvén, shear Alfvén, and acoustic waves are fundamental to the plasma wave excitations and heatings. Also, they will be seen to connect to the various continuum dampings in the stability analyses. From the energy principle, we have also shown the stabilizing effects from the compression and bending of magnetic fields and identified two basic kinds of instability drives: the kink and interchange modes.

The nonideal MHD effects are explained afterward. First, the resistivity effects on the odd and even modes, i.e., the resistive interchange and tearing modes, are outlined. The two-fluid modifications in the three frequency regimes: the low, comparable, and intermediate frequency regimes, are explained. How the parallel electric field is excited and how the FLR effects appear are the focuses. They explain the roots of the kinetic description in the fluid formulation. The MHD root of kinetic resonance effects is also particularly explained. There are other fluid closure regimes, such as the finite Larmor-radius closure and drift model. Readers are referred to reference [6].

In conclusion, two-fold efforts have been made in this chapter: first, to lay the theoretical foundation for further analyses in this book, second, to point out the direction for the nonideal MHD extensions with their roots explained.

References

1. Bernstein I B, Frieman E A, Kruskal M D, Kulsrud R M 1958 An energy principle for hydromagnetic stability problems *Proc. R. Soc. A 244*, 17–40. https://doi.org/10.1098/rspa.1958.0023
2. Freidberg J P 1987 Ideal Magnetohydrodynamics (New York: Plenum Press).
3. Miyamoto K Plasma Physics for Nuclear Fusion (New York: The MIT Press).
4. Zheng L 2020 Magnetically Confined Fusion Plasma Physics, Ideal MHD Theory (London: IOP Publishing). DOI: 10.1088/978-1-64327-138-5.
5. Guest G E, Hedrick C L 1972 Elements of the stability theory of the guiding-center plasma, *Oak Ridge National Laboratory Report No. ORNL-TM-3943*.
6. Hazeltine R D, Meiss J D 1992 Plasma Confinement (New York: Addison-Wesley).
7. Andreussi T, Morrison P J, Pegoraro F 2016 Hamiltonian magnetohydrodynamics: Lagrangian, Eulerian, and dynamically accessible stability - Examples with translation symmetry *Phys. Plasmas 23*, 102112. https://doi.org/10.1063/1.4964900
8. Rosswog S, Price D 2007 MAGMA: a three-dimensional, Lagrangian magnetohydrodynamics code for merger applications *Mon. Not. R. Astron. Soc. 379*, 915-931. https://doi.org/10.1111/j.1365-2966.2007.11984.x
9. Hammett G W, Mandell N R, Bernard T, *et al.* 2025 Sheath Resistivity Effects on SOL Instabilities and Gkeyll Simulations of Edge Turbulence, *The 2025 Sherwood Fusion Theory Conference*, P2.37.
10. Linjin Zheng 2025 On the Lagrange and Euler variational formulations of linear ideal magnetohydrodynamics *Phys. Plasmas 32*, 102113. https://doi.org/10.1063/5.0290171
11. Zheng L 2022 Magnetically Confined Fusion Plasma Physics, Volume 3, Kinetic theory (London: IOP Publishing). DOI: 10.1088/978-0-7503-3779-3
12. Greene J M, Johnson J L 1962 Stability criterion for arbitrary hydromagnetic equilibria. *Phys. Fluids 5* 510-7. DOI:10.1063/1.1706651
13. Correa-Restrepo D 1982 Resistive ballooning modes in three-dimensional configurations *Z. Naturforsch. A 37* 848. https://doi.org/10.1515/zna-1982-0818.
14. Zheng L 2020 Magnetically Confined Fusion Plasma Physics, Volume 2, Multifluid theory (London: IOP Publishing). DOI: 10.1088/978-0-7503- 3575-1.
15. Zheng L 2015 IOP Concise Physics: Advanced Tokamak Stability Theory (San Rafael, CA: Morgan & Claypool and London: IOP Publishing).

4 MHD equilibrium

After introducing the MHD equations from the kinetic theory and discussing the fundamentals of MHD theory, we move to describe the MHD equilibrium theory in this chapter. We first point out that an MHD equilibrium is not a thermal equilibrium but just a force balance. In relaxing to the thermal equilibrium, the charged particles may still escape the system due to various transport processes. Also, there is a stability issue for the force balance. The equilibrium studies are fundamental to stability and transport studies. Since the equilibrium is just a bottom line, the possible further evolution defines the quality of the equilibrium.

While we cannot know in advance which type of equilibrium is better, one does foresee the importance of simplicity and beauty. This leads the magnetic fusion field to consider the symmetric systems, such as tokamak, mirror, and quasisymmetric stellarators. Symmetry assures the existence of the magnetic surfaces for equilibrium, improves particle confinement due to the motion invariances, and even the system stability. This leads us to focus on the symmetric or quasisymmetric configurations in the description of MHD equilibrium theories.

Since proving the existence of magnetic surfaces and particle motion invariances requires more demanding Hamilton and Lagrange theory, we start with the conventional topics about the magnetic coordinates and the Grad-Shafranov equation in the symmetric system, followed by the discussion of free boundary equilibrium, how to determine the coil structure from the fixed boundary equilibrium, and the feedback control of the hot plasmas. After having a basic picture of MHD equilibrium in the symmetric system, the advanced topics such as the quasi-symmetric stellarator concept and the Hamilton and Lagrange theories of magnetic-field-line flow will be described. Besides the description of general equilibrium theories, the special topics on X-point physics are described afterward. The application in the large aspect ratio case will be described in the end as an intuitive example.

In Section 3.1, we have obtained the MHD equilibrium equations:

$$\mathbf{j} \times \mathbf{B} = \nabla P, \tag{4.1}$$

$$\nabla \times \mathbf{B} = \mu_0 \mathbf{j}, \tag{4.2}$$

$$\nabla \cdot \mathbf{B} = 0. \tag{4.3}$$

Here, we have assumed the equilibrium flow velocity is negligible and omitted the subscript "0" for brevity. Equations (4.1)–(4.3) will be used as the basic set of equations for equilibrium studies in this chapter.

From equation (4.1), one expresses the current density

$$\mathbf{j} = \sigma \mathbf{B} + \frac{\mathbf{B} \times \nabla P}{B^2}, \tag{4.4}$$

DOI: 10.1201/9781003522133-4

where $\sigma = \mathbf{j} \cdot \mathbf{B}/B^2$. The second term on the right-hand side is referred to as the diamagnetic current. The first term on the right is related to the so-called Pfirsch-Schlüter current. Noting $\nabla \cdot \mathbf{j} = 0$, one obtains

$$\sigma = \int_{l_0}^{l} \nabla \times \frac{\mathbf{B}}{B^2} \cdot \nabla P \frac{dl}{B} + \sigma_0. \tag{4.5}$$

Here, the integration is along the magnetic field line with l the arclength and σ_0 is the integration constant, which can be determined by Ohm's law to relate it to the loop voltage. in the tokamak discharge [1].

4.1 THE MAGNETIC COORDINATES

We start the MHD equilibrium theory with the introduction of the magnetic coordinates. This is related to the magnetic flux surface. A flux surface is a surface on which magnetic field lines lie. A divergence-free magnetic field requires that such a surface be either a torus or a knot. We consider the toroidal shapes as appearing in tokamaks and stellarators. We start with the general representation of magnetic field to explain the concept of magnetic coordinates to explain the symmetry property of the existence of flux surfaces and what is a magnetic coordinate system. Then the magnetic coordinate system in the symmetric system is discussed, although the formalism applies to the non-symmetric system as soon as there exist perfect magnetic surfaces. The current density representation is described next. The particle confinement issue is then discussed together with the so-called canonical flux coordinates.

4.1.1 REPRESENTATION OF MAGNETIC FIELD

In this subsection, we first describe the general representation of magnetic field to explain the concept of magnetic coordinates or straight field line coordinates. Because the magnetic field is divergence-free, as shown in equation (4.3), one can introduce the vector potential $\mathbf{A}$ such that $\mathbf{B} = \nabla \times \mathbf{A}$. Let us consider a torus and introduce the coordinate system (r, θ, ζ) to describe it, where r is the minor radius from the toroidal axis, θ is the poloidal angle, and ζ is the toroidal angle. The contravariant representation of the vector potential can be given as follows:

$$\mathbf{A} = A_r \nabla r + A_\theta \nabla \theta + A_\zeta \nabla \zeta.$$

Because of gauge invariance, one can add a gradient ∇F to $\mathbf{A}$, so that

$$\begin{aligned} \mathbf{A} &= \nabla F + A_r \nabla r + A_\theta \nabla \theta + A_\zeta \nabla \zeta \\ &= (\partial_r F + A_r)\nabla r + (\partial_\theta F + A_\theta)\nabla \theta + (\partial_\zeta F + A_\zeta)\nabla \zeta. \end{aligned}$$

We can let $\partial_r F + A_r$ be a function of r and define $\psi_t = \partial_\theta F + A_\theta$ and $-\psi_p = \partial_\zeta F + A_\zeta$, yielding [2]

$$\mathbf{A} = \psi_t \nabla \theta + \psi_p \nabla \zeta + \nabla F.$$

Therefore, the covariant-like representation of the magnetic field becomes

$$\mathbf{B} = \nabla\zeta \times \nabla\psi_p + \nabla\psi \times \nabla\theta. \tag{4.6}$$

Equation (4.6) is just a contravariant representation of the magnetic field. To be a magnetic coordinate system, the first requirement is the existence of flux surfaces, and the other is a proper choice of θ and ζ as explained in the next section in equation (4.16).

4.1.2 MAGNETIC COORDINATES

In this subsection, we describe the magnetic coordinates or straight field line coordinates. The precondition for constructing the magnetic coordinates lies in the existence of magnetic flux surfaces. We therefore proceed with the description by assuming that there is a symmetric direction. The proof of the existence of magnetic flux surfaces in a symmetric system from the Euler-Lagrange formalism will be described in Section 4.5. The quasi-symmetric system will be described in Section 4.4. Nevertheless, beyond the assumption of the existence of magnetic flux surfaces, the theory is applicable for general toroidal systems, such as stellarators.

In the axisymmetric system, we can introduce the cylinder coordinate system (R, Z, ϕ) where R is the radius, Z is the height, and ϕ is the azimuth. In a particular $\phi = const$ plane, X is used to represent R. Introducing the vector potential $\mathbf{A}$ and noting that ϕ is an ignorable coordinate, one has $\partial A_R/\partial\phi = \partial A_Z/\partial\phi = 0$. Using the expression for the curl operator in the cylinder coordinates, it can be proved that the two components of the vector potential on the $\phi = const.$ plane (A_R, A_Z) can be expressed by a single toroidal component: $\mathbf{A}_\phi$. Without losing generality, it can be denoted as $\mathbf{A} = -\chi\nabla\phi$. The equilibrium magnetic field can then be expressed as

$$\mathbf{B} = \nabla \times \mathbf{A}_\phi + RB_\phi\nabla\phi = \nabla\phi \times \nabla\chi + g\nabla\phi, \tag{4.7}$$

where the toroidal component $RB_\phi\nabla\phi$ has been added with $g = RB_\phi$. From equation (4.7) it can be proved that $\mathbf{B}\cdot\nabla\chi = 0$ and therefore $\chi = const.$ labels magnetic flux surfaces. Inserting equation (4.7) into Ampere's law in equation (4.2), as shown in detail in references [3, 4], one can express equilibrium current density as follows:

$$\mu_0\mathbf{j} = \nabla g \times \nabla\phi + R^2\nabla\cdot\left(\frac{\nabla\chi}{R^2}\right)\nabla\phi. \tag{4.8}$$

Because $\mathbf{j}\cdot\nabla\chi = 0$, from equation (4.8), one can conclude that $g(\chi)$ is a surface function.

The magnetic field is described in equation (4.7) with χ and ϕ. One can introduce another coordinate θ_{eq} to describe the poloidal dependence so that

$$\mathbf{B} = \nabla\phi \times \nabla\chi + q_{local}(\chi,\theta)\nabla\chi \times \nabla\theta, \tag{4.9}$$

Equating equations (4.9) and (4.7) yields

$$q_{local} = \frac{g\mathcal{J}_{eq}}{X^2}, \tag{4.10}$$

where $\mathcal{J}_{eq} = 1/\nabla\chi \times \nabla\theta_{eq} \cdot \nabla\phi$ is the Jacobian of the coordinate system. The representation is still not in magnetic coordinates or straight field line coordinates since $q_{local}(\chi, \theta)$ is not a flux function.

Next, we construct the magnetic coordinates from equation (4.10). We first describe the so-called PEST coordinates [5]. With q_{local} obtained in equation (4.10), the PEST poloidal angle is defined as

$$\theta_{pest} = \frac{1}{q}\int_0^{\theta_{eq}} d\theta_{eq}\frac{g\mathcal{J}_{eq}}{X^2}, \tag{4.11}$$

where the so-called safety factor is given by

$$q(\chi) = \frac{1}{2\pi}\int_0^{2\pi} d\theta_{eq}\frac{g\mathcal{J}_{eq}}{X^2},$$

Transforming θ_{eq} to θ_{pest}, q_{local} becomes independent of θ_{pest} and therefore the coordinate system $(\chi, \theta_{pest}, \phi)$ is a magnetic coordinate system. In the PEST coordinate system, the poloidal flux is also specified. Denote the PEST coordinates as (ψ_{pest}, θ_{pest}, ϕ). where θ_{pest} is given in equation (4.11) and ψ_{pest} is the flux surface label

$$\psi_{pest} = \frac{2\pi X_0}{c_{pest}}\int_0^{\chi} d\chi\frac{q}{g}, \quad c_{pest} = \frac{X_0}{2\pi}\int_v d\tau\frac{1}{X^2}, \tag{4.12}$$

where X_0 is the major radius at magnetic axis and $\int_v d\tau$ denotes volume integration over entire plasma volume. Consequently, the equilibrium magnetic field in the PEST coordinate system can be represented as

$$\mathbf{B} = \chi'(\psi_{pest})(\nabla\phi \times \nabla\psi_{pest} + q\nabla\psi_{pest} \times \nabla\theta_{pest}). \tag{4.13}$$

Here, we have used the prime to denote the derivative with respect to the corresponding flux coordinate, which is ψ_{pest} in this case. Note that $\mathcal{J}_{eq} = 1/\nabla\chi \times \nabla\theta_{eq} \cdot \nabla\phi = \mathcal{J}_{pest}(\partial\theta_{pest}/\partial\theta_{eq})(d\psi_{pest}/d\chi)$ and $\mathcal{J}_{eq}$ can be computed from the equilibrium solution. The Jacobian in the PEST coordinates can be obtained as follows

$$\mathcal{J}_{pest} \equiv \frac{1}{\nabla\psi_{pest} \times \nabla\theta_{pest} \cdot \nabla\phi} = \frac{q\chi' X^2}{g}. \tag{4.14}$$

The magnetic coordinate system is not unique. Apparently, the PEST coordinate system is just one of magnetic coordinate systems. We describe next the construction of general magnetic coordinates. Introducing the following coordinate transform in equation (4.13)

$$\zeta_g = \phi + \nu_g(\psi, \theta), \quad \theta_g = \theta_{pest} + \nu_g(\psi, \theta)/q, \quad \text{and } \psi = \psi(\chi), \tag{4.15}$$

the straight field feature under the contravariant representation is preserved as follows:

$$\mathbf{B} = \chi'(\psi)(\nabla\zeta_g \times \nabla\psi + q\nabla\psi \times \nabla\theta_g). \tag{4.16}$$

Here, θ_g and ζ_g are referred to as generalized poloidal and toroidal angles, respectively. From equation (4.16), one obtains

$$
\begin{aligned}
2\pi\chi &= \int\int \frac{\mathbf{B}\cdot\nabla\theta_g}{\nabla\zeta_g\times\nabla\chi\cdot\nabla\theta_g} d\zeta_g d\chi, \\
2\pi\psi &= \int\int \frac{\mathbf{B}\cdot\nabla\zeta_g}{\nabla\zeta_g\times\nabla\chi\cdot\nabla\theta_g} d\theta_g d\chi,
\end{aligned}
$$

By equating the $\nabla\phi$ components of equations (4.7) and (4.16) one obtains

$$
\left.\frac{\partial \nu}{\partial \theta_{eq}}\right|_{\psi,\phi} \frac{1}{\mathscr{J}_{eq}} + g\frac{1}{X^2} = \chi' q \frac{1}{\mathscr{J}_g}, \tag{4.17}
$$

where $\mathscr{J}_g = 1/\nabla\psi\times\nabla\theta_g\cdot\nabla\zeta_g$. Using the definitions of $\mathscr{J}_{eq}$ and $\mathscr{J}_g$ one can prove that

$$
\left.\frac{\partial \theta_g}{\partial \theta_{eq}}\right|_{\psi,\phi} = \frac{\mathscr{J}_{eq}}{\chi'\mathscr{J}_g}. \tag{4.18}
$$

Using equation (4.18), one can then solve equation (4.17)

$$
\nu_g(\psi,\theta_{eq}) = \int_0^{\theta_{eq}} d\theta_{eq}\mathscr{J}_{eq}\left(q\frac{1}{\mathscr{J}_g} - g\frac{1}{X^2}\right) = q\theta_g - \int_0^{\theta_{eq}} d\theta_{eq}\frac{g\mathscr{J}_{eq}}{X^2}, \tag{4.19}
$$

where the integration is on the path with constant ψ and ϕ.

We can use equations (4.15), (4.18), and (4.19) to construct various types of straight-field-line flux coordinate systems. There are two classes of them, either by specifying Jacobian (e.g., Hamada coordinates [6], and Boozer coordinates [7]) or by directly choosing the generalized poloidal angle (e.g., the equal arc-length coordinate).

In the Hamada coordinate system [6], the volume inside a magnetic surface is used to label magnetic surfaces, i.e., $\psi = V$, and Jacobian $\mathscr{J}_h = 1/\nabla V\cdot\nabla\theta_h\times\nabla\zeta_h$ is set to be unity. With Jacobian $\mathscr{J}_h$ specified, equation (4.18) can be used to solve for θ_h at given (V,ϕ), ν can be determined by equation (4.19), and finally the definition equation (4.15) can be used to specify ζ_h.

In the Boozer coordinate system [7], Jacobian is set to be $\mathscr{J}_B = V'\langle B^2\rangle_s/(4\pi^2B^2)$, where $\langle\cdot\rangle_s$ represents the surface average. Since the Jacobian is given, the procedure for determining Boozer poloidal and toroidal coordinates θ_B and ζ_B is similar to that for Hamada coordinates.

In the equal-arc-length coordinate system, the poloidal angle is given as the equal-arc-length coordinate θ_e on each magnetic surface. In this case, its Jacobian $\mathscr{J}_e$ can magnetic be then determined by equation (4.18) and ν by equation (4.17). Finally, equation (4.15) is used to derive the generalized toroidal coordinate ζ_e.

In passing, it is noted that in reference [2], the various coordinates are alternatively specified by the covariant representation of the magnetic field, which will be reviewed in Section 4.4.

4.1.3 CURRENT DENSITY REPRESENTATION

In this subsection, we derive the current density representation in the magnetic coordinates.

Using Ampere's law, we have derived the current density in equation (4.8)

$$\mu_0\mathbf{j} = \nabla g \times \nabla\phi + R^2\nabla\cdot\left(\frac{\nabla\chi}{R^2}\right)\nabla\phi. \tag{4.20}$$

From this expression, one can obtain

$$\mathbf{J}\cdot\nabla\theta_g = -\frac{1}{\mu_0\mathscr{J}_g}g'_\psi.$$

Using the force balance equation (4.1), one obtains

$$\mathbf{j}\times\mathbf{B}\cdot\nabla\theta_g\times\nabla\zeta_g = p'\nabla\psi\cdot\nabla\theta_g\times\nabla\zeta_g.$$

This gives rise to

$$\mathbf{j}\cdot\nabla\zeta = q\mathbf{j}\cdot\nabla\theta - \frac{p'}{\chi'} = -q\frac{1}{\mu_0\mathscr{J}_g}g'_\psi\cdot - \frac{P'}{\chi'}.$$

Therefore, the current density can be expressed as

$$\mathbf{j} = -\frac{1}{\mu_0}g'_\psi\nabla\zeta_g\times\nabla\psi - \left(\frac{q}{\mu_0}g'_\psi + \frac{p'_\psi}{\chi'_\psi}\mathscr{J}_g\right)\nabla\psi\times\nabla\theta_g. \tag{4.21}$$

Equation (4.21) is given in the general magnetic coordinates. In the PEST coordinates it is reduced to

$$\mathbf{j} = -\frac{1}{\mu_0}g'_\psi\nabla\phi\times\nabla\psi_{pest} - \left(\mu_0R^2p'_\chi + gg'_\chi\right)\nabla\phi, \tag{4.22}$$

where the Jacobian in the PEST coordinates in equation (4.14) has been used. Note here that in the general magnetic coordinates, the current density vector is not a straight line. In the Hamada coordinates, however, due to $\mathscr{J}_h = 1$, equation (4.21) is reduced to

$$\mathbf{j} = J'_V\nabla\zeta\times\nabla V + I'_V\nabla V\times\nabla\theta,$$

where $I(V)$ and $J(V)$ are toroidal and poloidal current fluxes, $I' = -g'_V/\mu_0$, and $J' = -qg'_V/\mu_0 - P'_V/\chi'_V$. This shows that the current density vector is also straight in the Hamada coordinates. Consequently, the force balance equation (4.1) is also simplified in the Hamada coordinates. Using the definitions of I'_V and J'_V one can obtain

$$\mu_0p'_V = J'_V\psi'_V - I'_V\chi'_V. \tag{4.23}$$

In passing, one can express the current density parallel to the magnetic field in Hamada coordinates using equations (4.7) and (4.22),

$$j_\parallel = -\frac{\mu_0gp'_\chi}{B} - g'_\chi B, \tag{4.24}$$

which is referred to as the Pfirsch-Schlüter current as introduced in equation (4.5).

4.1.4 CANONICAL MAGNETIC COORDINATES AND THE GUIDING CENTER LAGRANGIAN

The general theory for magnetic coordinates has been reviewed in subsection 4.1.2. However, to consider the Hamiltonian description of guiding center motion, one needs to consider the so-called canonical coordinates. The coordinate system was first proposed and developed by J. D. Meiss and R. D. Hazeltine in reference [8], in which the Shafranov equilibrium is used to show the coordinates. Later, this coordinate system is numerically implemented for tokamak ITER-type of configurations in reference [9].

In the numerical simulation of a many-charged-particle system, the Hamiltonian description is favored. In the long-time particle pushing, various conservation laws are guardrails. Preserving the Liouville theorem to various invariances or adiabatic invariants is essential. To explain why the canonical coordinates are needed, let us first introduce the so-called Littlejohn phase space Lagrangian for guiding center motion [10]

$$\mathscr{L} = \left(\frac{e}{c}\mathbf{A} + m\frac{v_\parallel}{B}\mathbf{B}\right)\cdot\dot{\mathbf{X}} + \frac{mc}{e}\mu\dot{\alpha} - \mu B - \frac{1}{2}mv_\parallel^2 - e\varphi \tag{4.25}$$

where $\mathbf{X} = \mathbf{x} + \mathbf{v}\times\mathbf{B}/\Omega$ is the guiding center position, μ denotes the magnetic moment, α is the gyrophase, $v_\parallel$ denote the parallel velocity. The complete phase space coordinates $q^\nu = (X^i, \mu, v_\parallel, \alpha)$, where X^i denote 3 contravariant components of $\mathbf{X}$. This is a six-dimensional problem in the phase space. One would expect to have three momenta, which correspond to three coordinates. However, in the Littlejohn guiding center Lagrangian in equation (4.25), there are four variables with time derivative: $(\dot{\mathbf{X}}, \dot{\alpha})$. The corresponding canonical momenta are given by

$$\mathbf{p} = \frac{\partial\mathscr{L}}{\partial\dot{\mathbf{X}}} = \frac{e}{c}\mathbf{A}(\mathbf{X}) + m\frac{v_\parallel}{B}\mathbf{B}(\mathbf{X}),$$
$$p_\alpha = \frac{\partial\mathscr{L}}{\partial\dot{\alpha}} = \frac{mc}{e}\mu.$$

This is overshot. One needs to reduce them to three. Reference [8] comes out with a solution. Suppose both the vector potential and the magnetic field contain only two covariant components

$$\mathbf{A} = A_\theta\nabla\theta + A_\zeta\nabla\zeta, \tag{4.26}$$
$$\mathbf{B} = B_\theta\nabla\theta + B_\zeta\nabla\zeta, \tag{4.27}$$

where θ and ζ are the general poloidal and toroidal angles. Then, the Littlejohn guiding center Lagrangian in equation (4.25) becomes to depend only on three variables with time derivative: $(\dot{\theta}, \dot{\zeta}, \dot{\alpha})$. The three corresponding canonical momenta are

$$p_\theta = \frac{e}{c}A_\theta + m\frac{v_\parallel}{B}B_\theta, \tag{4.28}$$
$$p_\zeta = \frac{e}{c}A_\zeta + m\frac{v_\parallel}{B}B_\zeta, \tag{4.29}$$
$$p_\alpha = \frac{mc}{e}\mu. \tag{4.30}$$

Then, new six-phase space variables become $(\theta, \zeta, \alpha, p_\theta, p_\zeta, \mu)$. They are referred to as the canonical coordinates. Physically, the three momenta in equations (4.28) - (4.30) describe the bounce or circulating motion, the toroidal procession, and gyro motion, respectively.

In general, the canonical coordinates are not the magnetic coordinates or the straight field line coordinates as shown in equation (4.16). Let us describe how to numerically construct the canonical magnetic coordinates [9]. In subsection 4.1.2, it is shown that the magnetic field representation is invariant under the transformation

$$\theta_c = \theta + \nu(\chi, \theta, \zeta), \quad \zeta_c = \zeta + q(\chi)\nu(\chi, \theta, \zeta). \tag{4.31}$$

The requirement of $B_\chi = 0$ as in equation (4.27) becomes

$$\begin{aligned}
0 &= \mathbf{B} \cdot \mathscr{J} \nabla\theta \times \nabla\zeta \\
&= (\nabla\zeta \times \nabla\chi + q(\chi)\nabla\chi \times \nabla\theta) \cdot \nabla\theta_c \times \nabla\zeta_c \\
&= (\nabla\zeta \times \nabla\chi + q(\chi)\nabla\chi \times \nabla\theta) \\
&\quad \cdot \left(\nabla\theta \times \nabla\zeta - \frac{\partial \nu}{\partial \chi}\nabla\zeta \times \nabla\chi + \frac{\partial \nu}{\partial \theta}\nabla\theta \times \nabla\zeta \right. \\
&\quad \left. - q\frac{\partial \nu}{\partial \chi}\nabla\chi \times \nabla\theta + q\frac{\partial \nu}{\partial \zeta}\nabla\theta \times \nabla\zeta - \nu\frac{\partial q}{\partial \chi}\nabla\chi \times \nabla\theta \right).
\end{aligned}$$

This leads to

$$\frac{\partial \nu}{\partial \chi} - C_1\frac{\partial \nu}{\partial \theta} - C_1 q\frac{\partial \nu}{\partial \zeta} - C_2\nu = C_1, \tag{4.32}$$

where

$$\begin{aligned}
C_1 &= \frac{G_{12} + qG_{31}}{G_{22} + 2qG_{23} + q^2 G_{33}}, \\
C_2 &= -\frac{q' G_{33} + qq' G_{23}}{G_{22} + 2qG_{23} + q^2 G_{33}},
\end{aligned}$$

with

$$\begin{aligned}
G_{12} &= \nabla\theta \times \nabla\zeta \cdot \nabla\zeta \times \nabla\chi, \\
G_{22} &= \nabla\zeta \times \nabla\chi \cdot \nabla\zeta \times \nabla\chi, \\
G_{23} &= \nabla\zeta \times \nabla\chi \cdot \nabla\chi \times \nabla\theta, \\
G_{31} &= \nabla\chi \times \nabla\theta \cdot \nabla\theta \times \nabla\zeta, \\
G_{33} &= \nabla\chi \times \nabla\theta \cdot \nabla\chi \times \nabla\theta.
\end{aligned}$$

Equation (4.32) can be solved using the Fourier decomposition in θ and ζ and shooting with respect χ as the AEGIS code [11]. With ν determined, one can use equation (4.31) to construct the canonical magnetic coordinates.

There is a constraint on the existence of the canonical magnetic coordinates. The Jacobian, as given below should be regular.

$$\begin{aligned}\mathcal{J} &= \frac{1}{\nabla\chi\times\nabla\theta_c\cdot\nabla\zeta_c} \\ &= \frac{1}{\nabla\chi\times\nabla\theta\cdot\nabla\zeta\left(1+\frac{\partial v}{\partial\theta}+q\frac{\partial v}{\partial\zeta}\right)}.\end{aligned}\tag{4.33}$$

Note that equation (4.32) is inhomogeneous with the source term C_1 on the right. The orderings for $C_{1,2}$ can be estimated as follows

$$\begin{aligned}C_1 &\sim \frac{G_{12}}{q^2G_{33}}\sim\left(\frac{r}{R}\right)^2\frac{\nabla\chi\cdot\nabla\theta}{|\nabla\chi|^2}, \\ C_2 &\sim \frac{q'G_{33}}{q^2G_{33}}\sim\frac{1}{\chi}.\end{aligned}$$

From these ordering analyses, one can see that $C_2 \gg C_1$. Noting further that equation (4.32) yields $v \sim C_1/C_2$, one can see that v is small. Therefore, the Jacobian in equation (4.33) usually remains finite. The modification from the existing magnetic coordinates to canonical magnetic coordinates is small. The example in the Shafranov equilibrium can be found in reference [8], and the example for ITER-like equilibrium can be found in reference [9].

4.2 GRAD-SHAFRANOV EQUATION

The MHD equilibrium equations have been given in equations (4.1)-(4.3). They are a set of vector equations and therefore can be reduced, especially in the axisymmetric case. In Section 4.1 we show that in the axisymmetric case, the magnetic field can be expressed in the magnetic coordinates. As soon as the poloidal magnetic flux is solved from the given pressure $p(\chi)$ and poloidal current flux $g(\chi)$, the magnetic field is determined. This leads ones to derive the equation for χ, Grad-Shafranov's equation [12, 13].

The derivation of Grad-Shafranov's equation is straightforward using the results in Section 4.1. This is because the set of equilibrium equations (4.1)-(4.3) have been used to express the current density. We just need to equate the toroidal projections of two current density expressions in equations (4.20) and (4.22) in subsection 4.1.3, yielding the so-called Grad-Shafranov's equation

$$\Delta^*\chi = \nabla\cdot\left(\frac{\nabla\chi}{R^2}\right) = -\mu_0 p'_\chi - \frac{gg'_\chi}{R^2}.\tag{4.34}$$

Here, we have noted that p and g are functions of χ only.

One can use equation (4.34) to determine χ with $p(\chi)$ and $g(\chi)$ given. However, complete determination of χ also requires specifying the boundary conditions. Note that plasma cannot be confined without an external coil structure, and between the plasma and coil structure, there is a vacuum region. It is also noted that the stability of plasma confinement by the magnetic field depends on the shape of plasma-vacuum boundary. In most cases, one considers the so-called fixed boundary equilibrium for the plasma region first and then determines the external coil structure afterward. We discuss the plasma region solution here. The vacuum and coil structure will be studied in the next section.

First, we note that the Grad-Shafranov equation can be solved iteratively as follows:

$$\mathscr{L}(\chi^{(n+1)}) = \mathscr{S}(\chi^{(n)}). \tag{4.35}$$

where $\mathscr{L}$ and $\mathscr{S}$ represent the left and right hand sides of the Grad-Shafranov equation (4.34) and superscripts $(n+1)$ and (n) respectively denote the shooting steps. The solution of equation (4.35) for $\chi^{(n+1)}$ contains two parts: the homogeneous and special parts. The homogeneous solutions are used to satisfy the boundary conditions.

In the fixed boundary case, the boundary condition on a given boundary surface is as follows:

$$\chi|_{\Gamma} = 0. \tag{4.36}$$

Since the Grad-Shafranov equation is invariant to adding a constant, it does not lose generality to set the right-hand side of equation (4.36) to zero. There is another condition at the magnetic axis. The solution needs to be small there. This needs to use the asymptotic solution at the magnetic axis. Let us describe the condition using the Fourier decomposition $\chi = \sum \chi_m(r)\exp\{im\theta\}$, where r is the minor radius from the magnetic axis and θ is the poloidal angle. Near the magnetic axis, equation (4.34) becomes

$$\frac{1}{r}\frac{d}{dr}r\frac{d\chi_m}{dr} + m^2\chi_m = 0,$$

The solution is

$$\begin{aligned} \chi_0 &= a_0 \ln r + b_0, \\ \chi_m &= a_m r^{-m} + b_m r^m \quad \text{for} \quad m \neq 0. \end{aligned}$$

The large solution is unacceptable, so we can choose $a_m = 0$. The boundary condition at the axis is usually given as $\mathbf{d}\ln\chi_m/dr = m/r$ as $r \to 0$.

The Grad-Shafranov equation is a nonlinear equation and generally needs a numerical solution. Since it is a two-dimensional problem with $\phi = const$, one needs to introduce two-dimensional mesh to describe the solution. A poloidal angle coordinate θ_{eq} around the magnetic axis of plasma torus is usually introduced in addition to the radial flux coordinate χ. There are two types of expressions. In the direct method, $\chi(X,Z)$ and $\theta_{eq}(X,Z)$ are given in the (X,Z) grids; In the inverse method, $X(\chi,\theta_{eq})$ and $Z(\chi,\theta_{eq})$ are given in the $(\chi,\ \theta_{eq})$ grids. Further review of the equilibrium computational method can be found in reference [5].

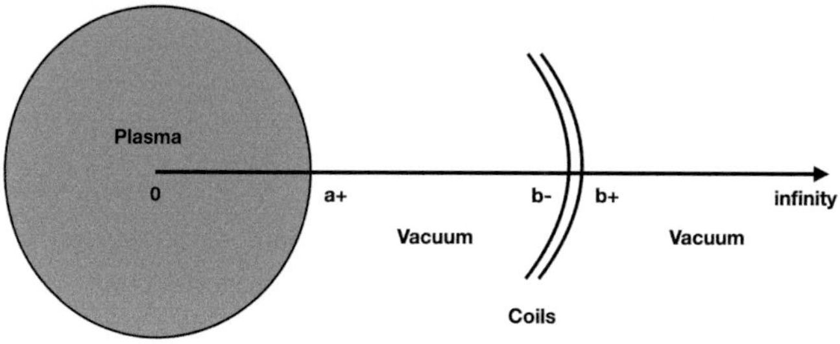

Figure 4.1 Free boundary equilibrium problem.

4.3 FREE BOUNDARY EQUILIBRIUM

In Section 4.2, we have described the Grad-Shafranov equation and its solution as a fixed boundary value problem. In this section, let us discuss the free boundary equilibrium solution. This includes using the fixed boundary solution to determine the vacuum solution and coil structure. It will become apparent later that to keep plasma force balance, it is necessary to have an external coil structure; otherwise, the hoop force from bending the plasma cannot be balanced.

4.3.1 GENERAL MATHEMATICAL FRAMEWORK

Let us first use the simple one-dimensional problem to clarify the mathematical picture. In figure 4.1, one can see that the plasma with minor radius a is surrounded by the vacuum region. Between the inner and outer vacuum regions, there is the coil structure at b. The outer vacuum region extends to infinity. In figure 4.1, the symbol $\pm$ is introduced to represent the inner and outer interfaces. Let us discuss the solutions in each region and then match them together.

First, we consider the solution in the plasma region. In the one-dimensional case, the Grad-Shafranov equation becomes an ordinary differential equation of second order. Therefore, there are two independent homogeneous solutions. The general solution in the plasma region can be written as

$$\chi_{plasma} = a_p\chi_p^{(1)} + b_p\chi_p^{(2)} + \chi_p^{(s)}, \tag{4.37}$$

where $\chi_p^{(1,2)}$ are the homogeneous solutions and $\chi_p^{(s)}$ is the special solution. Applying the boundary conditions at the magnetic axis and at the plasma edge a, the constants a_p and b_p are completely determined. Therefore, χ_{plasma} is completely determined. Here, it is noted that the matchings are actually required for the normal and tangential components of the magnetic field. There is an addable constant for χ. The shape of the plasma-vacuum interface can be freely chosen.

In the inner vacuum region, one also has

$$\chi_I = a_I\chi_I^{(1)} + b_I\chi_I^{(2)}. \tag{4.38}$$

This solution needs to match the solution in the plasma region. Here, we point out that for a differential equation of second order, the matching conditions contain both the value and its derivative. This is different from the boundary condition. In the case without the surface current, both the normal and tangential components have to be continuous across the plasma-vacuum interface at a. Note that because $\nabla \cdot \mathbf{B} = 0$, only one tangential component matching is required. After the matching between equations (4.37) and (4.38), both a_I and b_I are also completely determined by the fixed boundary solution in the plasma region.

In the outer vacuum region, one also has

$$\chi_{II} = a_{II}\chi_{II}^{(1)}. \tag{4.39}$$

Here, the boundary condition at infinity $\chi_{II} \to 0$ (or small) has been taken into account so that only one independent solution is left.

One needs now to match the solutions in equations (4.38) and (4.39) across the coil structure. Since a_I and b_I are completely fixed, there is only one constant a_{II} left. It is determined by the continuity of the normal component of magnetic field across the coil structure at b. In order to match the tangential component of magnetic field, there must be the current in the coils. The procedure determines the coil current from the fixed boundary solution in the plasma region.

This procedure also shows that by adjusting the coil structure, one can control the shape of plasma vacuum interface. This paves the way for the feedback control of the plasma confinement.

In the general two or three-dimensional cases, the situation is similar. One just needs to match the multiple Fourier components. Here, it should be pointed out that as soon as the fixed boundary solution in the plasma region is given, the solution in the inner vacuum region is determined. One can simply treat the two continuity conditions on the plasma-vacuum interface as the two initial conditions to find the vacuum solution by outward shooting. This determines the inner vacuum solution at a given location of the coil structure. The matching with the outer vacuum solution then determines the coil current. This is particularly useful for stellarator coil design.

We have considered constructing the free boundary equilibrium from the inside to the outside with a given fixed boundary equilibrium. Conversely, one may consider the equilibrium problem from outside to inside with the coil current given. This shows the mechanism for feedback control. The boundary condition at infinity and the currents at a given coil structure determine χ and $\partial\chi/\partial n$ at a given plasma-vacuum interface. In this case, the mathematical problem in the plasma region is overdetermined in general since there are two boundary conditions on the plasma edge and one at the magnetic axis. Only a particular toroidal current profile can satisfy the boundary conditions. Note that the Grad-Shafranov equation is nonlinear.

Other factors may also help to establish the equilibrium, such as the shape of the plasma-vacuum interface and the position of the magnetic axis.

4.3.2 GREEN'S FUNCTION METHOD

In the previous subsection, we discussed the free boundary equilibrium in a two-dimensional case. In this subsection, we will treat a general axisymmetric case using the Green function method. It is assumed that the fixed boundary solution in the plasma region has been obtained, which is denoted as U. The Green function method is then used to determine the coil structure at a given surface.

Note that the Biot-Savart law can be used to compute the magnetic field from a given electric current distribution. In the axisymmetry case, the current in a circular coil is considered. It is intended to solve the so-called Green's function $G(\mathbf{R},\mathbf{R}')$, which satisfies the following equation

$$\Delta^* G = RI\delta(R-R',Z-Z'), \tag{4.40}$$

where (R',Z') is the source coordinate and I is a constant. Reference [14] describes the solution of Green's function for the vector potential A_ϕ. Noting that $\chi = -RA_\phi$ equation (4.40) can be solved to obtain the Green function as follows [15, 16, 17]

$$G(\mathbf{R},\mathbf{R}') = \frac{1}{2\pi}\frac{\sqrt{RR'}}{k}\left[(2-k^2)K(m)-2E(m)\right]. \tag{4.41}$$

Here, $K(m)$ and $E(m)$ are the complete elliptic integrals of the first and second kinds, respectively, with the argument $m = k^2$ and

$$k^2 = \frac{4RR'}{(R+R')^2+(Z-Z')^2}.$$

To solve the free boundary equilibrium problem, one also needs the so-called Green's theorem [15, 16, 17]

$$\begin{aligned}
&\nabla\cdot\left[\chi\frac{1}{R^2}\nabla G(\mathbf{R},\mathbf{R}')\right]-\nabla\cdot\left[G(\mathbf{R},\mathbf{R}')\frac{1}{R^2}\nabla\chi\right]\\
&= \frac{1}{R^2}\chi\Delta^* G(\mathbf{R},\mathbf{R}')-\frac{1}{R^2}G(\mathbf{R},\mathbf{R}')\Delta^*\chi,
\end{aligned} \tag{4.42}$$

We will use the Green theorem in equation (4.42) to obtain the solution χ in the inner and outer vacuum regions. Similar to figure 4.1, the plasma-vacuum interface is denoted by Γ_a, and the coil structure in between Γ_b^- and Γ_b^+.

We first consider the inner vacuum region between Γ_a and Γ_b^-. Integrating equation (4.42) over the inner vacuum region yields

$$\begin{aligned}
\chi(\mathbf{R}') = &-\int_{\Gamma_a} d\mathbf{S}\cdot\left[\chi\frac{1}{R^2}\nabla G(\mathbf{R},\mathbf{R}')\right]+\int_{\Gamma_a} d\mathbf{S}\cdot\left[G(\mathbf{R},\mathbf{R}')\frac{1}{R^2}\nabla\chi\right]\\
&+\int_{\Gamma_b^-} d\mathbf{S}\cdot\left[\chi\frac{1}{R^2}\nabla G(\mathbf{R},\mathbf{R}')\right]-\int_{\Gamma_b^-} d\mathbf{S}\cdot\left[G(\mathbf{R},\mathbf{R}')\frac{1}{R^2}\nabla\chi\right].
\end{aligned}$$

Here, it has been assumed that $d\mathbf{S}$ is always pointed radially outward. Using the continuity conditions across Γ_a, one obtains

$$\begin{aligned}\chi(\mathbf{R}') &= -\int_{\Gamma_a} d\mathbf{S}\cdot\left[U\frac{1}{R^2}\nabla G(\mathbf{R},\mathbf{R}')\right]+\int_{\Gamma_a} d\mathbf{S}\cdot\left[G(\mathbf{R},\mathbf{R}')\frac{1}{R^2}\nabla U\right]\\ &\quad +\int_{\Gamma_b^-} d\mathbf{S}\cdot\left[\chi\frac{1}{R^2}\nabla G(\mathbf{R},\mathbf{R}')\right]-\int_{\Gamma_b^-} d\mathbf{S}\cdot\left[G(\mathbf{R},\mathbf{R}')\frac{1}{R^2}\nabla\chi\right] \quad (4.43)\end{aligned}$$

We next consider the outer vacuum region between Γ_a and Γ_b^-. Integrating equation (4.42) over the inner vacuum region yields

$$\chi(\mathbf{R}') = -\int_{\Gamma_b^-} d\mathbf{S}\cdot\left[\chi\frac{1}{R^2}\nabla G(\mathbf{R},\mathbf{R}')\right]+\int_{\Gamma_b^-} d\mathbf{S}\cdot\left[G(\mathbf{R},\mathbf{R}')\frac{1}{R^2}\nabla\chi\right]. \quad (4.44)$$

Here, $\chi = G = 0$ at infinity has been considered. Using equation (4.43) for plasma-vacuum interface, one obtains

$$\begin{aligned}U(\Gamma_a) &= -\int_{\Gamma_a} d\mathbf{S}\cdot\left[U\frac{1}{R^2}\nabla G(\mathbf{R},\mathbf{R}')\right]+\int_{\Gamma_a} d\mathbf{S}\cdot\left[G(\mathbf{R},\mathbf{R}')\frac{1}{R^2}\nabla U\right].\\ &\quad \int_{\Gamma_b^-} d\mathbf{S}\cdot\left[\chi\frac{1}{R^2}\nabla G(\mathbf{R},\mathbf{R}')\right]-\int_{\Gamma_b^-} d\mathbf{S}\cdot\left[G(\mathbf{R},\mathbf{R}')\frac{1}{R^2}\nabla\chi\right]. \quad (4.45)\end{aligned}$$

Using equation (4.44) for the inner vacuum-coli structure interface, Γ_b^-, one obtains

$$\begin{aligned}\chi(\Gamma_b^-) &= -\int_{\Gamma_a} d\mathbf{S}\cdot\left[U\frac{1}{R^2}\nabla G(\mathbf{R},\mathbf{R}')\right]+\int_{\Gamma_a} d\mathbf{S}\cdot\left[G(\mathbf{R},\mathbf{R}')\frac{1}{R^2}\nabla U\right].\\ &\quad \int_{\Gamma_b^-} d\mathbf{S}\cdot\left[\chi\frac{1}{R^2}\nabla G(\mathbf{R},\mathbf{R}')\right]-\int_{\Gamma_b^-} d\mathbf{S}\cdot\left[G(\mathbf{R},\mathbf{R}')\frac{1}{R^2}\nabla\chi\right]. \quad (4.46)\end{aligned}$$

Using equation (4.43) for the inner vacuum-coli structure interface, Γ_b^+, one obtains

$$\chi(\Gamma_b^+) = -\int_{\Gamma_b^-} d\mathbf{S}\cdot\left[\chi\frac{1}{R^2}\nabla G(\mathbf{R},\mathbf{R}')\right]+\int_{\Gamma_b^-} d\mathbf{S}\cdot\left[G(\mathbf{R},\mathbf{R}')\frac{1}{R^2}\nabla\chi\right]. \quad (4.47)$$

From equations (4.46) and (4.47), one obtains, respectively,

$$\frac{\partial\chi(\Gamma_b^-)}{\partial n} = \mathscr{T}^-\chi(\Gamma_b^-), \quad (4.48)$$

$$\frac{\partial\chi(\Gamma_b^+)}{\partial n} = \mathscr{T}^+\chi(\Gamma_b^+). \quad (4.49)$$

Here, we have introduced the symbolic descriptions with the operators $\mathscr{T}^\pm$. In the one-dimensional case, $\mathscr{T}^\pm$ are just constants; In the general case, one can use the Fourier decomposition in the poloidal direction to obtain the explicit representations for equations (4.48) and (4.49).

Note that the normal and tangential components of the magnetic field are determined by χ and $\partial\chi/\partial n$. One can require $\chi(\Gamma_b^-) = \chi(\Gamma_b^+)$ in equations (4.48) and (4.48). However, there will be a jump between $\partial\chi(\Gamma_b^-)/\partial n$ and $\partial\chi(\Gamma_b^+)/\partial n$ since $\mathscr{T}^- \neq \mathscr{T}^+$. By requiring the normal component of magnetic field to be continuous, one can determine the current required on the coil structure by using the jump of the tangential component of magnetic field as required by Ampere's law. The procedure is similar to the resistive wall mode description in the stability analyses [18]. Here, we have discussed the axisymmetric case. A similar procedure can be developed for the nonaxisymmetric case in stellarators.

4.4 QUASISYMMETRIC STELLARATORS

In this section, we discuss the stellarator equilibrium, especially the quasi-helical symmetric stellarators [7, 2]. We first describe the so-called Boozer coordinates from the covariant representation of the magnetic field and then the Lagrangian in this coordinate system. Using the Lagrangian, the quasi-symmetry concept is described. There is another stellarator optimization approach, the omnigeneous magnetic configuration [19]. A vanishing time-averaged magnetic drift of particle trajectories away from a given magnetic surface is sought for. In this way, particles stay confined to a given magnetic surface on average. Since the guiding center of a circulating particle moves primarily along the magnetic field line, the concern about the particle drifting away from a magnetic surface is for the trapped particle confinement.

Unlike the toroidal symmetry case as described in Section 4.2, the existence of the magnetic field in the stellarator equilibrium is not guaranteed. In the following description of quasisymmetric stellarators, the existence of the flux surfaces is assumed, so that the contravariant representation of magnetic field is expressed as follows

$$2\pi\mathbf{B} = \nabla\psi \times \nabla\theta + \iota\nabla\psi \times \nabla\zeta. \tag{4.50}$$

Here, the rotational transform of field line $\iota = 1/q$ and 2π is introduced to keep in line with the definition introduced originally by A. Boozer. The corresponding vector potential is

$$2\pi\mathbf{A} = \psi\nabla\theta - \chi\nabla\zeta + \nabla g. \tag{4.51}$$

Because the equilibrium equation requires the current density to be on the magnetic surfaces as well, one has the following contravariant representation of current density as well

$$2\pi\mathbf{j} = \frac{\partial G_{tot}}{\partial\psi}\nabla\psi \times \nabla\theta + \iota\frac{\partial I_{tot}}{\partial\psi}\nabla\psi \times \nabla\zeta, \tag{4.52}$$

where

$$\frac{\partial G_{tot}}{\partial\psi} = \frac{\partial G(\psi)}{\partial\psi} + \frac{\partial\omega(\psi,\theta,\zeta)}{\partial\zeta},$$

$$\frac{\partial I_{tot}}{\partial\psi} = \frac{\partial I(\psi)}{\partial\psi} + \frac{\partial\omega(\psi,\theta,\zeta)}{\partial\theta},$$

where G and I are the poloidal and toroidal currents in the region external to a constant ψ surface and ω is just a free function. The contravariant representation of $\mathbf{j}$ in equation (4.52) and Ampere's law $\mu_0\mathbf{j} = \nabla \times \mathbf{B}$ leads to the following covariant representation of the magnetic field

$$\mathbf{B} = \frac{\mu_0}{2\pi}\left[G(\psi)\nabla\zeta + I(\psi)\nabla\theta - \omega\nabla\psi + \nabla F\right]. \tag{4.53}$$

Similar to the free function ω in the contravariant representation of $\mathbf{j}$ in equation (4.52), there is a free function ν in the contravariant representation of $\mathbf{B}$ in equation (4.50) as discussed in the axisymmetric case in Section 4.1.2. The representation in equation (4.50) is invariant if one defines $\theta = \theta_n + \iota\nu$ and $\zeta = \zeta_n + \nu$. Inserting these into the covariant representation $\mathbf{B}$ in equation (4.53), one finds that the representation is unchanged if the transformations are made $\omega_n = \omega + (dG/d\psi + \iota dI/d\psi)\nu$ and $F_n = F + (G + \iota I)\nu$. There are two straightforward choices. First is $\omega_n = 0$, which corresponds to the Hamada coordinates [6]

$$\mathbf{B} = \frac{\mu_0}{2\pi}\left[G(\psi)\nabla\zeta_H + I(\psi)\nabla\theta_H + \nabla F\right]. \tag{4.54}$$

This can be confirmed by its Jacobiian by dotting the contravariant and covariant magnetic field representations. The other choice is $F = 0$, which leads to the Boozer coordinates [7]

$$\mathbf{B} = \frac{\mu_0}{2\pi}\left[G(\psi)\nabla\zeta_B + I(\psi)\nabla\theta_B + \beta_*\nabla\psi\right]. \tag{4.55}$$

By dotting the contravariant and covariant magnetic field representations, one obtains the Jacobian of the Boozer coordinates

$$\mathscr{J}_B = \mu_0\frac{G + \iota I}{(2\pi B)^2}. \tag{4.56}$$

As described in reference [2], the Taylor Lagrangian is used to describe the guiding cention motion, which is given as follows [20]

$$L_T(\mathbf{X}, \dot{\mathbf{X}}) = \frac{1}{2}m\left(\dot{\mathbf{X}}\cdot\mathbf{b}\right)^2 + Ze\dot{\mathbf{X}}\cdot\mathbf{A} - (\mu B + Ze\Phi). \tag{4.57}$$

One can use the Boozer coordinates to reduce it. Note that

$$\dot{\mathbf{X}}(t;\psi,\theta,\zeta) = \frac{\partial\mathbf{X}}{\partial t} + \frac{\partial\mathbf{X}}{\partial\psi}\dot{\psi} + \frac{\partial\mathbf{X}}{\partial\theta}\dot{\theta} + \frac{\partial\mathbf{X}}{\partial\zeta}\dot{\zeta}.$$

Therefore

$$\mathbf{b}\cdot\dot{\mathbf{X}} = \frac{\mu_0}{2\pi B}\left(G(\psi)\dot{\zeta} + I(\psi)\dot{\theta}\right),$$

$$\mathbf{A}\cdot\dot{\mathbf{X}} = \frac{1}{2\pi}\left(\psi\dot{\theta} - \chi\dot{\zeta}\right).$$

Here, ψ, which describes the drift from the magnetic surface, is assumed to be small as compared to $\dot{\theta}_B$ and $\dot{\zeta}_B$. Using them, the Taylor Lagrangian in equation (4.57) is transformed to

$$L_T = \frac{m}{2}\left(\mu_0 \frac{G\dot{\zeta}_B + I\dot{\theta}_B}{2\pi B}\right)^2 + Ze\frac{\psi\dot{\theta}_B - \chi\dot{\zeta}_B}{2\pi} - \mu B - Ze\Phi. \tag{4.58}$$

From equation (4.58), one can see that the guiding-center Lagrangian depends only on the magnitude of magnetic field, B, instead of the magnetic field vector, **B**. This leads reference [21] to consider the magnetic field strength B is a function of ψ and $\zeta_{helical} = M\theta_H - N\zeta_H$, i.e., being helically symmetric, and Φ is a function of ψ and B. This is the concept of quasisymmetric stellarators.

In the quasisymmetric stellarator configuration, one can choose a new coordinate system $(\psi, \theta_{helical}, \zeta_H)$. The Taylor Lagrangian in equation (4.58) is independent of ζ_H. The Euler-Lagrange equation becomes

$$\frac{d}{dt}\frac{\partial L_T}{\partial \dot{\zeta}_B} - \frac{\partial L_T}{\partial \zeta_B} = \frac{d}{dt}\frac{\partial L_T}{\partial \dot{\zeta}_B} = 0. \tag{4.59}$$

This indicates that the canonical momentum $p_{\zeta_H} = \partial L_T / \partial \dot{\zeta}_B$ is an invariant, which assures the confinement of the charged particles. This is different from the axisymmetry case, in which the canonical momentum is conserved. Here, the conservation is in the guiding center picture.

Note here that reference [22] shows that β_* assumes the same symmetry in the quasisymmetric configuration, and the assumption of small ψ can be released. Alternatively, one may use the Littlejohn phase space Lagrangian in equation (4.25) to prove the quasisymmetry. Nevertheless, the canonical coordinates as reviewed in subsection 4.1.4 are better to be considered for a Hamiltonian description of guiding center motion. The most challenging part of stellarator optimization lies in the unproven assumption of the existence of nested magnetic surfaces.

4.5 LAGRANGIAN AND HAMILTONIAN DESCRIPTIONS OF MAGNETIC FIELD LINE FLOW

Similar to the guiding center description of charged particle orbit as mentioned in subsection 4.1.4, the magnetic-field-line flow can also be represented as a Hamiltonian system. The main reference to this approach can be found in reference [23]. As compared to the non-Hamiltonian description, for example, in reference [24], the Lagrangian and Hamiltonian descriptions are based on a scalar, the Lagrangian, to represent the flow and therefore are invariant with coordinate transformation. It allows the properties of Lagrangian and Hamiltonian formalism to be fully exploited. Only the equilibrium theory or lowest order theory is described here. In the perturbation theory of next order, the systematic Lie transform formulation can be applied, which can be found in reference [23] with the modification for the singular case given in reference [25].

We first introduce the magnetic-field-line flow mathematically by considering the Lagrangian

$$\delta \int d\lambda L = \delta \int d\lambda \mathbf{A}(\mathbf{x}) \cdot \frac{d\mathbf{x}}{d\lambda}.$$

Variation of this Lagrangian integral yields the so-called Lagrangian equation

$$\begin{aligned} \frac{\partial}{\partial \lambda}\left(\frac{\partial L}{\partial \dot{x}^i}\right) - \frac{\partial L}{\partial x^i} &= \frac{\partial A_i}{\partial \lambda} - \frac{\partial A_j}{\partial x^i}\frac{dx^i}{d\lambda} \\ &= \frac{\partial x^j}{\partial \lambda}\frac{\partial A_i}{\partial x^j} - \frac{\partial A_j}{\partial x^i}\frac{dx^i}{d\lambda} \\ &= (\nabla \times \mathbf{A}) \times \frac{d\mathbf{x}}{d\lambda} = \mathbf{B} \times \frac{d\mathbf{x}}{d\lambda} = 0, \end{aligned} \tag{4.60}$$

Equation (4.60) shows that the magnetic-field-line flow $d\mathbf{x}/d\lambda$ is parallel to $\mathbf{B}$ as it should be.

In fact, for an MHD equilibrium, one not only needs to consider the magnetic field line flow, but also the current-density-line flow. The force balance $\mathbf{j} \times \mathbf{B} = \nabla p$ requires both $\mathbf{j}$ and $\mathbf{B}$ to be on the same surface. The current density line flow is described by the Lagrangian

$$\delta \int d\lambda L_c = \frac{1}{\mu_0}\delta \int d\lambda \mathbf{B}(\mathbf{x}) \cdot \frac{d\mathbf{x}}{d\lambda}.$$

Its Lagrangian equation reads

$$\frac{\partial}{\partial \lambda}\left(\frac{\partial L_c}{\partial \dot{x}^i}\right) - \frac{\partial L_c}{\partial x^i} = \frac{1}{\mu_0}\nabla \times \mathbf{B} \times \frac{d\mathbf{x}}{d\lambda} = \mathbf{j} \times \frac{d\mathbf{x}}{d\lambda} = 0. \tag{4.61}$$

With the concepts of magnetic-field-line and current-density-line flows introduced, we will next describe the magnetic field line pattern in the azimuthally symmetric system. The cylinder coordinates (R, ϕ, Z) with ϕ the azimuthal angle are used.

For the magnetic-field-line flow, one has

$$\gamma = A_R(Z,R)dR + A_Z(Z,R)dZ + A_\phi(Z,R)d\phi. \tag{4.62}$$

Here, it is assumed that the azimuthally symmetric system is considered with ϕ being the ignorable coordinate. According to Noether's theorem, one has that A_ϕ is conserved by the flow, and we may denote it as χ. To be specific, we consider the toroidal systems such as tokamak or stellarators. Using Stokes' theorem, one can prove that χ is the poloidal magnetic flux. In toroidal system one can introduce the coordinates $(\chi, \theta_{eq}, \phi)$ with θ_{eq} the poloidal angle. The Lagrangian in equation (4.62) is then transformed to

$$\gamma = A_\chi(\chi, \theta_{eq})d\chi + A_{\theta_{eq}}(\chi, \theta_{eq})d\theta_{eq} - \chi d\phi. \tag{4.63}$$

Following equation (4.60), the magnetic field is expressed as

$$\mathbf{B} = \nabla\phi \times \nabla\chi + \left(\frac{\partial A_{\theta_{eq}}}{\partial \chi} - \frac{\partial A_\chi}{\partial \theta_{eq}}\right) \nabla\chi \times \nabla\theta_{eq}. \tag{4.64}$$

Similar to equation (4.63), for the current-density flow, one has

$$\gamma_c = B_\chi(\chi, \theta_{eq})d\chi + B_{\theta_{eq}}(\chi, \theta_{eq})d\theta_{eq} - g(\chi)d\phi. \tag{4.65}$$

It is noted here that it is because of considering the current-density flow that one obtains $B_\phi = g(\chi)$, instead of $B_\phi = g(\chi)$.

Note that $B_\phi = g(\chi)$ is required for MHD equilibrium. Without it, the surface is not a true equilibrium magnetic surface. Using equations (4.64) and (4.65) one obtains

$$\frac{\partial A_{\theta_{eq}}}{\partial \chi} - \frac{\partial A_\chi}{\partial \theta_{eq}} = \frac{g(\chi)|\nabla\phi|^2}{\nabla\chi \times \nabla\theta_{eq} \cdot \nabla\phi} = q\frac{\partial \theta_{pest}}{\partial \theta_e q},$$

where θ_{pest} is the poloidal angle in the PEST coordinates in equation (4.11). Using this result, equation (4.64) is reduced to

$$\mathbf{B} = \nabla\phi \times \nabla\chi + q\nabla\chi \times \nabla\theta_{pest}. \tag{4.66}$$

This is the lowest order equilibrium solution. The derivation is an alternative to that in Section 4.1.2.

4.6 X POINT EQUILIBRIUM

In this section, we discuss the MHD equilibrium at the so-called X point. We first discuss the so-called 360-degree issue of X point equilibrium and then describe the asymptotic equilibrium solution at X point tip [26]. The dual-poloidal-region q coordinates will be described afterward [27]. The X point physics is not only for the equilibrium, but also for the so-called tearing mode stability, the formation of magnetic islands, and the stochastic field structure.

To have a basic picture of X point in the tokamak or stellarator equilibrium, let us use a Solovév equilibrium to show the existence of X points [28]. Assuming that

$$-\mu_0 P'_\chi = a \quad \text{and} \quad -ff'_\chi = bX^2,$$

one can obtain the exact Solovév equilibrium solution

$$\chi = \left[(b+c_0)X_0^2 + c_0(X^2 - X_0^2)\right]\frac{Z^2}{2} + \frac{1}{8}(a - c_0)(X^2 - X_0^2)^2, \tag{4.67}$$

where X_0 is the major radius of magnetic axis and a, b, c_0 are constant parameters. This solution can be verified by inserting it directly into the Grad-Shafranov equation (4.34). Using equation (4.67), a typical cross section can be plotted in figure 4.2.

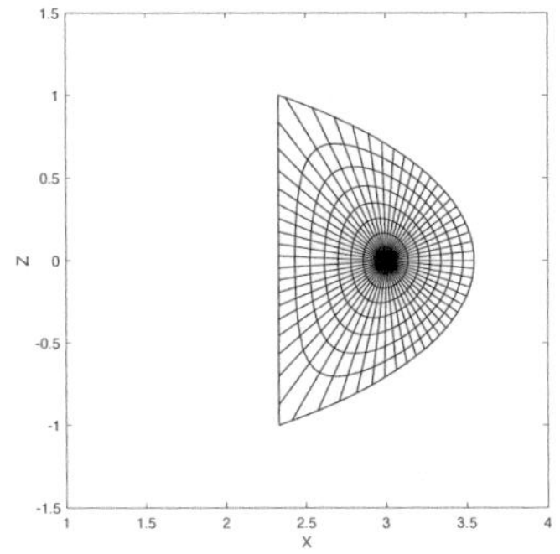

Figure 4.2 Typical cross-section of Solovév equilibrium.

From figure 4.2 one can see that the X points appear on the top and bottom of the cross section. The X points are featured by the vanishing of the poloidal field. From the representation of magnetic field in equation (4.7), one can see that the X point appears when $\nabla\chi = 0$. This can be obtained by the stationary points of χ:

$$\begin{aligned} \frac{\partial \chi}{\partial Z} &= 0 \rightarrow X^2 - X_0^2 = -\frac{b+c_0}{c_0} X_0^2, \\ \frac{\partial \chi}{\partial (X^2 - X_0^2)} &= 0 \rightarrow Z^2 = -\frac{1}{2}\frac{a-c_0}{c_0}(X^2 - X_0^2). \end{aligned}$$

These determine the X-point coordinates (X_s, Z_s) in the Solovév equilibrium in equation (4.67).

4.6.1 ASYMPTOTIC EQUILIBRIUM SOLUTION AT THE X-POINT TIP

In this subsection, we describe the asymptotic equilibrium solution at the X-point tip. We consider the case with the X point lying at the plasma edge, *i.e.*, on the plasma-vacuum interface. The theory can be extended to the case with X point inside the plasma. We first show the 360-degree issue for X-point equilibrium and then use the conformal transform to find the X-point equilibrium solution.

4.6.1.1 360 degree issue for X point equilibrium

In this subsection, we first prove that there is a 360-degree issue for X point equilibrium at the plasma edge. This is shown both in rectangular and polar coordinates.

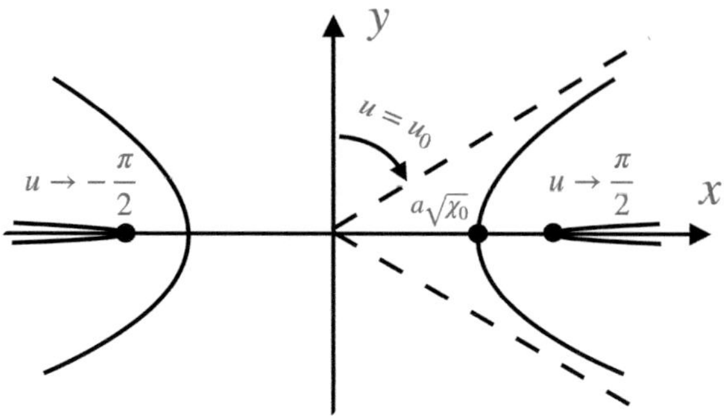

Figure 4.3 The local coordinate system and the hyperbolic surface structure at the X point. Reproduced from reference [26] with the permission of AIP Publishing.

To study the asymptotic solution at the X point, the toroidal current in the Grad-Shafranov equation (4.34) is expanded as follows:

$$-R^2 p' - ff' = s_0 + s_1\chi + \cdots,$$

We also introduce the local coordinate system (x, y) around the X point $(x, y) = (0, 0)$ as shown in figure 4.3. The expansion in the vicinity of the X point leads the Grad-Shafranov equation in equation (4.34) to be reduced to

$$\frac{\partial^2 \chi}{\partial x^2} + \frac{\partial^2 \chi}{\partial y^2} = s_0. \tag{4.68}$$

This equation describes the plasma-filled segment with finite toroidal current. The vacuum region is instead described by the equation

$$\frac{\partial^2 \chi}{\partial x^2} + \frac{\partial^2 \chi}{\partial y^2} = 0. \tag{4.69}$$

In the plasma-filled segment, the solution of equation (4.68) is

$$\chi = \frac{x^2}{a^2} - \frac{y^2}{b^2}, \tag{4.70}$$

where

$$\frac{1}{a^2} = \frac{s_0}{2} + c, \quad \frac{1}{b^2} = c \tag{4.71}$$

with c is a constant, which determines the plasma boundary shape,

$$\chi = \frac{x^2}{a^2} - \frac{y^2}{b^2} = 0. \tag{4.72}$$

In the vacuum, equation (4.69) can also be solved, yielding

$$\chi = \frac{x^2}{d^2} - \frac{y^2}{d^2}. \tag{4.73}$$

where d is a constant.

Let us now analyze the solutions. From the boundary condition in equation (4.72), one can see that the plasma-vacuum boundaries are defined by the two straight lines $x/a \pm y/b = 0$. Because $a \neq b$ in the plasma-filled segment, the included angle of two straight-line boundaries on the plasma side is not a right angle. In the vacuum regions, instead, the solution in equation (4.69) indicates that three vacuum segments must have right angles. Obviously, as soon as $s_0 \neq 0$ in the plasma region, the sum of four segment angles are not 2π. This indicates that the asymptotic solution at the X-point tip does not exist.

Alternatively, one can also analyze the problem in polar coordinates. The vacuum equation in polar coordinates ($x = r\cos\theta,\ y = r\sin\theta$) can be expressed

$$\frac{1}{r}\frac{\partial}{\partial r}\left(r\frac{\partial \chi}{\partial r}\right) + \frac{1}{r^2}\frac{\partial^2 \chi}{\partial \theta^2} = 0. \tag{4.74}$$

The solution of equation (4.74) is

$$\chi = \sum_{\nu} c_\nu r^\nu \cos \nu\theta. \tag{4.75}$$

Here, c_ν and ν are constants. They will be determined by the boundary conditions.

According to equation (4.70), the plasma region solution can be transformed to the polar coordinate as follows

$$\chi = c_p r^2 \left(\cos 2\theta - \cos 2\theta_0\right), \tag{4.76}$$

where c_p is a constant, and θ_0 is the included angle of the plasma-vacuum boundaries. The continuity conditions on the plasma-vacuum interface are that the poloidal flux and its radial derivative are continuous, assuming that there is no surface current. The plasma segment solution in equation (4.76) gives rise to two boundary conditions on the boundaries $\theta = \pm\theta_0$:

$$\chi|_{\theta=\theta_0} = 0, \tag{4.77}$$

$$\left.\frac{\partial \chi}{\partial \theta}\right|_{\theta=\theta_0} = 2c_p r^2 \sin 2\theta. \tag{4.78}$$

To match the vacuum solution in equation (4.75) to the boundary condition in equation (4.78) uniformly for all r, the index ν in equation (4.75) must be 2. Therefore, the vacuum solution in equation (4.75) is reduced to

$$\chi = c_v r^2 \cos 2\theta. \tag{4.79}$$

The solution in equation (4.79) also indicates that three vacuum segments must have right angles. Therefore, we also find that, as soon as $s_0 \neq 0$ in the plasma region, the sum of four segment angles is not 2π, which indicates that the asymptotic solution at the X point tip does not exist.

4.6.1.2 Asymptotic equilibrium solution at X point tip

In the previous subsection, it was shown that the X point problem at the plasma edge with plasma boundaries delineated by a wedge formed by two straight lines has no solution. Note that a wedge formed by two straight lines can be the asymptote of a hyperbola. This leads us to consider a hyperbolic boundary of plasma to approach a wedge formed by two straight lines [26]. Because the vacuum region is described by the Laplace equation (4.69), we will use the conformal transform to map the boundary to a straight line to solve this problem.

The conformal transform is introduced as follows

$$z = A\sin w, \tag{4.80}$$

where $z = x + iy$ and $w = u + iv$. The real and imaginary parts of the transform in equation (4.80) are as follows

$$x = A\sin u\cosh v, \tag{4.81}$$

$$y = A\cos u\sinh v. \tag{4.82}$$

These give (x,y) for given (u,v).

We also need the inverse transform, *i.e.*, obtaining (u,v) from a given (x,y). To obtain u, it is noted that $\cosh^2 v - \sinh^2 v = 1$, one obtains from equations (4.81) and (4.82) that

$$1 = \frac{x^2}{A^2\sin^2 u} - \frac{y^2}{A^2\cos^2 u}, \tag{4.83}$$

i.e.,

$$x^2(1-\sin^2 u) - y^2\sin^2 u = A^2\sin^2 u\left(1-\sin^2 u\right).$$

One can solve it for $\sin^2 u$,

$$\sin^2 u = \frac{(x^2+y^2+A^2) - \sqrt{(x^2+y^2+A^2)^2 - 4A^2x^2}}{2A^2} \equiv E(x,y). \tag{4.84}$$

Here, we have chosen the negative sign before the square root to ensure that $\sin u$ is less than unity. From equation (4.84) one can solve $\sin u$ and subsequently invert it for u. The result is

$$u = sign\{x\}\sin^{-1}\sqrt{E(x,y)}, \tag{4.85}$$

where we have taken the sign of u to be the same as x.

To obtain v, it is noted that $\cos^2 u + \sin^2 u = 1$. One then obtains from equations (4.81) and (4.82) that

$$1 = \frac{x^2}{A^2\cosh^2 v} + \frac{y^2}{A^2\sinh^2 v}.$$

This can be reduced to

$$\sinh^2 v = \frac{(x^2+y^2-A^2)+\sqrt{(x^2+y^2-A^2)^2+4A^2y^2}}{2A^2} \equiv F(x,y). \quad (4.86)$$

Here, we have chosen the positive sign before the square root to ensure $\sinh^2 v$ is positive. One can solve equation (4.86) for $\sinh v$ and subsequently v. The result is

$$v = sign\{y\}\sinh^{-1}\sqrt{F(x,y)}, \quad (4.87)$$

where we have taken the sign of v to be the same as y.

We have shown how the transformation in equation (4.80) simplifies the plasma-vacuum boundary in a hyperbola to a constant line $u = u_0$ in the (u,v) space. Next, we determine the boundary conditions in the transformed space. There are two boundary conditions: χ and $\partial\chi/\partial n$.

The change of first boundary condition about χ is trivial. In the (u,v) space, the first boundary condition becomes

$$\chi|_{u_0} = \chi_0. \quad (4.88)$$

To derive the second boundary condition that the normal derivative of χ on the boundary $u = u_0$ is continuous, we first determine the coordinate gradient transform. From equations (4.81) and (4.82) one obtains

$$\begin{aligned} \nabla x &= A\cos u\cosh v\nabla u + A\sin u\sinh v\nabla v, \\ \nabla y &= -A\sin u\sinh v\nabla u + A\cos u\cosh v\nabla v. \end{aligned}$$

Using these transformations, one can calculate the gradient of χ in the plasma-filled segment on the hyperbolic plasma-vacuum boundary

$$\begin{aligned} \nabla\chi &= \frac{2x}{A^2\sin^2 u}\nabla x - \frac{2y}{A^2\cos^2 u}\nabla y \\ &= \frac{2A\sin u\cosh v}{A^2\sin^2 u}(A\cos u\cosh v\nabla u + A\sin u\sinh v\nabla v) \\ &\quad - \frac{2\cos u\sinh v}{A^2\cos^2 u}(-A\sin u\sinh v\nabla u + A\cos u\cosh v\nabla v) \\ &= 2\left(\cot u\cosh^2 v + \tan u\sinh^2 v\right)\nabla u. \end{aligned}$$

The continuity of normal derivative across the plasma-vacuum interface can be used to obtain the normal derivative of χ at the boundary in the (u,v) space

$$\begin{aligned} \left.\frac{\partial\chi}{\partial u}\right|_{u=u_0} &= 2\left(\cot u_0\cosh^2 v + \tan u_0\sinh^2 v\right) \\ &= 2\left(\cot u_0 + (\tan u_0 + \cot u_0)\sinh^2 v\right) \\ &= 2\left(\cot u_0 + (\tan u_0 + \cot u_0)\frac{1}{2}(\cosh 2v - 1)\right) \\ &= \underbrace{\cot u_0 - \tan u_0}_{part\ I} + \underbrace{(\tan u_0 + \cot u_0)\cosh 2v}_{part\ II}. \end{aligned} \quad (4.89)$$

Here, the underbraces have been introduced to specify the two parts of the boundary conditions for later analyses.

Equations (4.88) and (4.89) form the complete boundary conditions for determining the asymptotic solution in the vacuum in the (u, v) space. With the boundary conditions obtained, we proceed to derive the vacuum solution. Because of the conformal transform, the Laplace equation in the (u, v) space is preserved

$$\frac{\partial^2 \chi}{\partial u^2} + \frac{\partial^2 \chi}{\partial v^2} = 0. \tag{4.90}$$

In view of that the boundary condition in equation (4.89) can be split into two parts, we therefore solve the Laplace equation, equation (4.90), accordingly in two separated parts:

$$\chi = \chi_I + \chi_{II}. \tag{4.91}$$

with χ_I and χ_{II} satisfying the respective boundary conditions.

Noting that $\nabla^2 (u - u_0) = 0$, one obtains the solution of part I satisfying the boundary conditions in equation (4.88) and the first part of equation (4.89) as follows

$$\chi_I = \chi(u_0) + (\cot u_0 - \tan u_0)(u - u_0). \tag{4.92}$$

To obtain the solution of part II, we introduce

$$\chi_{II}(u, v) = U(u)V(v). \tag{4.93}$$

Inserting it into equation (4.90) yields

$$\frac{\partial^2 V}{\partial v^2} - 2^2 V = 0 \quad \text{and} \quad \frac{\partial^2 U}{\partial u^2} + 2^2 U = 0.$$

One can obtain their solutions satisfying the boundary conditions in equation (4.88) and the second part of equation (4.89) as follows

$$\begin{aligned} V &= (\tan u_0 + \cot u_0)\cosh 2v, \\ U &= \frac{1}{2}\sin 2(u - u_0). \end{aligned}$$

They completely determine the second part of solution in equation (4.93).

Combining two parts of solutions in equations (4.92) and (4.93) yields the complete solution in the (u, v) space according to equation (4.91)

$$\begin{aligned} \chi &= \chi(u_0) + (\cot u_0 - \tan u_0)(u - u_0) \\ &\quad + \frac{1}{2}(\tan u_0 + \cot u_0)\cosh 2v \sin 2(u - u_0). \end{aligned} \tag{4.94}$$

The solution in the (x, y) space can be obtained by using the coordinate transform formula in equations (4.85) and (4.87).

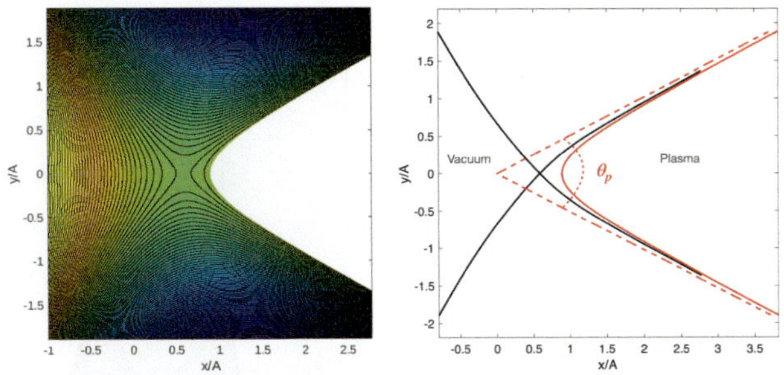

Figure 4.4 X point structure with an acute angle of the plasma-filled segment. Reproduced from reference [26] with the permission of AIP Publishing.

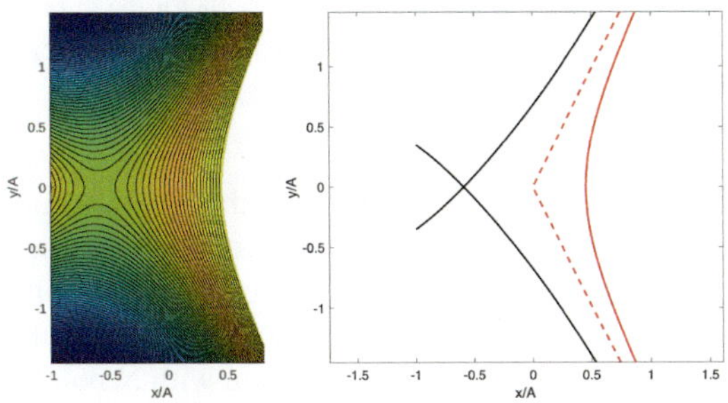

Figure 4.5 X point structure with an obtuse angle of the plasma-filled segment. Reproduced from reference [26] with the permission of AIP Publishing.

Equation (4.94) can be used to determine the X point structure [26]. The typical results are plotted in figures 4.4 and 4.5. The results show that the X point moves beyond the plasma boundary to fall in the vacuum region. An acute angle of the plasma-filled segment results in an obtuse opposite angle of vacuum segment, and vice versa. The case with the equal opposite angles occurs only in the case with the angle of the plasma-filled segment being a right angle. In the case with an acute angle of the plasma-filled segment, the new X point shifts inside the X point formed by the asymptotes of hyperbolic boundary; In the case with an obtuse angle of the plasma-filled segment, the new X point shifts outside the X point formed by the asymptotes of hyperbolic plasma boundary.

4.6.2 DUAL POLOIDAL REGION Q DESCRIPTION

In this subsection, we describe the dual-poloidal-region q coordinates [27]. This is particularly useful for describing the equilibria with X points for stability analyses and the feedback control.

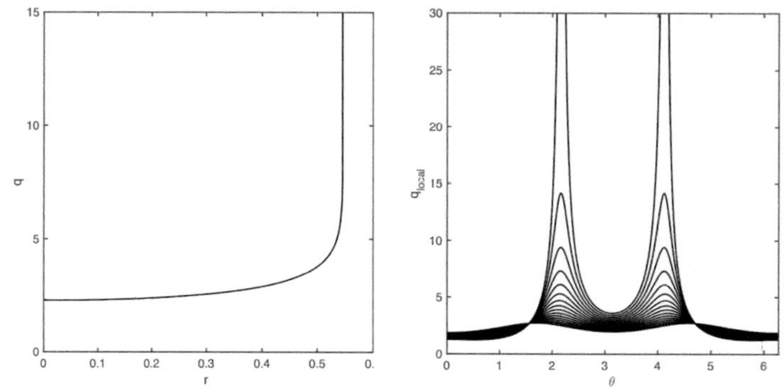

Figure 4.6 The safety factor q (left) and the local safety factor q_{local} on sequential magnetic surfaces with the top curve corresponding to the plasma edge (right). Reproduced from reference [27] with the permission of AIP Publishing.

At the beginning of this section, we have used the Solovév equilibrium in equation (4.67) as an example to discuss the X point features. Because the poloidal field vanishes at the X point, i.e., $\nabla\chi = 0$ there. One can see that this causes the Jacobian to become infinite at the X point. From the definition of the safety factor in equation (4.14), one can see that in the configuration with X points, the safety factor becomes infinite at the plasma edge. In fact, only the local safety factor q_{local} is infinite at the X point, and it is the surface average that makes the safety factor become infinite everywhere on the last closed flux surface, as shown in figure 4.6.

Let us first describe the dual-poloidal-region q coordinates in the PEST coordinates. In equation (4.9), we show that the magnetic field can be represented as follows

$$\mathbf{B} = \nabla\phi \times \nabla\chi + q_{local}(\chi, \theta_{eq})\nabla\chi \times \nabla\theta_{eq}, \tag{4.95}$$

where $q_{local}(\chi, \theta_{eq}) \equiv g\mathscr{J}_{eq}/X^2$. To introduce the dual-poloidal-region q coordinate system, equation (4.95) is rewritten as

$$\mathbf{B} = \nabla\phi \times \nabla\chi + \frac{g\hat{\mathscr{J}}_{eq}}{X^2}h(\chi, \theta_{eq})\nabla\chi \times \nabla\theta_{eq}, \tag{4.96}$$

where

$$\hat{\mathscr{J}}_{eq} = \begin{cases} \mathscr{J}_{eq} = \frac{1}{\nabla\chi \times \nabla\theta_{eq} \cdot \nabla\phi}, & \theta_{eq} \in \Theta_{core}; \\ \mathscr{J}_{eq}^{max} = const. & \theta_{eq} \in \Theta_X. \end{cases}$$

$$h(\chi, \theta_{eq}) = \begin{cases} 1, & \theta_{eq} \in \Theta_{core}; \\ \frac{\mathscr{J}_{eq}}{J_{eq}^{max}}, & \theta_{eq} \in \Theta_X. \end{cases}$$

Here, J_{eq}^{max} is a constant and specifies the maximum allowed value for Jacobian, and Θ_{core} and Θ_X are introduced to denote the poloidal core region and the vicinity of X points as shown in figure 4.7. Typical $h(\chi,\theta_c)$ is plotted in figure 4.8. Following the procedure described in Section 4.1.2 by using $\hat{\mathscr{J}}_{eq}$ to define the magnetic coordinates of PEST type, instead of $\mathscr{J}_{eq}$, one obtains

$$\mathbf{B} = \nabla\phi \times \nabla\chi + q_c(\chi)h(\chi,\theta_c)\nabla\chi \times \nabla\theta_c, \tag{4.97}$$

where

$$\theta_c = \frac{g}{q_c}\int_0^{\theta_{eq}} d\theta_{eq}\frac{\hat{J}_{eq}}{X^2}, \tag{4.98}$$

$$q_c(\chi) = \frac{g}{2\pi}\int_0^{2\pi} d\theta_{eq}\frac{\hat{J}_{eq}}{X^2}. \tag{4.99}$$

Equation (4.97) gives the representation of magnetic field in the PEST dual-poloidal-region q coordinates. Let us discuss its features. First, we note that as shown in figure 4.7, the dual q description affects only the edge layer where $\mathscr{J}_{eq}$ is larger than $\mathscr{J}_{eq}^{\max}$ in the vicinity of x points. In this edge layer, the local safety factor is averaged in the poloidal core region Θ_{core}, and therefore the coordinate system (χ,θ_c,ϕ) becomes the straight-field-line flux coordinates with the finite safety factor $q_c(\chi)$ being a function of χ only as in equation (4.99) in this region. In the vicinity of X point, Θ_X, however, $q_c(\chi)h(\chi,\theta_c)$ remains just q_{local}, where the magnetic fields run almost toroidally.

Following the procedure described in Section 4.1.2, the dual q coordinate description in the PEST coordinates in equation (4.97) can be converted to the general magnetic coordinates of dual-poloidal-region q. Letting

$$\zeta_{gc} = \phi + \nu(\chi,\theta_c) \quad \text{and} \quad \theta_{gc} = \theta_c + \nu(\chi,\theta_c)/(q_c h),$$

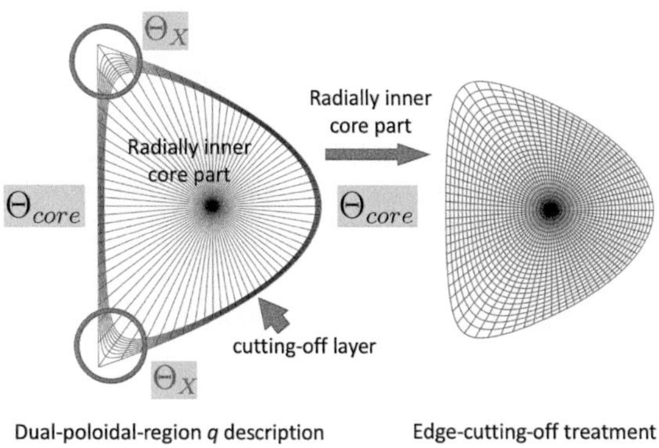

Figure 4.7 Explanation of the dual-poloidal-region q coordinates. Reproduced from reference [27] with the permission of AIP Publishing.

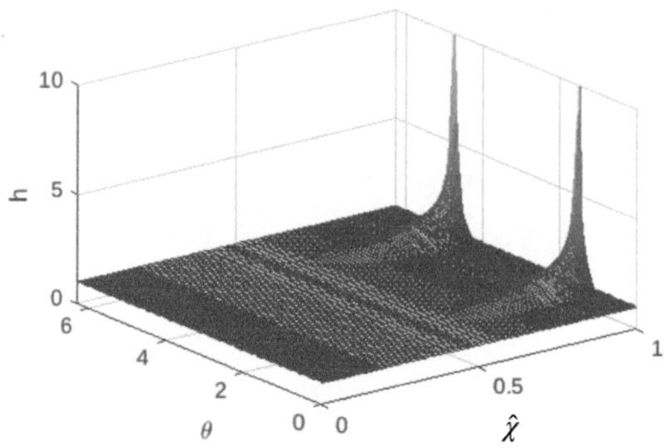

Figure 4.8 Typical value of $h(\chi,\theta_c)$ in the $(\hat{\chi},\theta)$ space. Reproduced from reference [27] with the permission of AIP Publishing.

one obtains

$$\nabla\phi\times\nabla\chi = \nabla\zeta_{gc}\times\nabla\chi - \frac{\partial v}{\partial\theta_c}\nabla\theta_c\times\nabla\chi,$$
$$q_c h(\chi,\theta_c)\nabla\chi\times\nabla\theta_c = q_c h\nabla\chi\times\nabla\theta_{gc}$$
$$-\frac{\partial v}{\partial\theta_c}\nabla\chi\times\nabla\theta_c, -vh\frac{\partial}{\partial\theta_{gc}}\left(\frac{1}{h}\right)\nabla\chi\times\nabla\theta_{gc}.$$

Therefore, the magnetic field in equation (4.97) can be transformed to

$$\mathbf{B} = \nabla\zeta_{cg}\times\nabla\chi + q_c h_g(\chi,\theta_{gc})\nabla\chi\times\nabla\theta_{gc}, \tag{4.100}$$

where

$$h_g(\chi,\theta_{gc}) = \begin{cases} 1, & \theta_{gc}\in\Theta_{core}; \\ \frac{J_{eq}}{\mathscr{J}_{eq}^{max}}\left[1+\frac{v}{q_c}\frac{\partial}{\partial\theta_{gc}}\left(\frac{\mathscr{J}_{eq}^{max}}{\mathscr{J}_{eq}}\right)\right], & \theta_{gc}\in\Theta_X. \end{cases}$$

Equation (4.100) gives the representation of magnetic field in the general dual-poloidal-region q coordinates.

The dual poloidal region q description isolates the singularity of the Jacobian to the vicinity of X points, which otherwise is spread over the whole magnetic surface. This is important for treating X-point physics.

4.7 EQUILIBRIUM OF LARGE-ASPECT-RATIO TOKAMAK

In this section, we describe the equilibrium theory in the large-aspect-ratio tokamak. This is useful for understanding the equilibrium physics picture, such as the Shafranov shift, etc. We will also describe how to construct the magnetic coordinates in the large-aspect-ratio limit as given by J. D. Meiss and R. D. Hazeltine in reference [8]. These analytical equilibrium theories will be used in the stability theories to be described in the later chapters. They are useful for understanding the MHD theory.

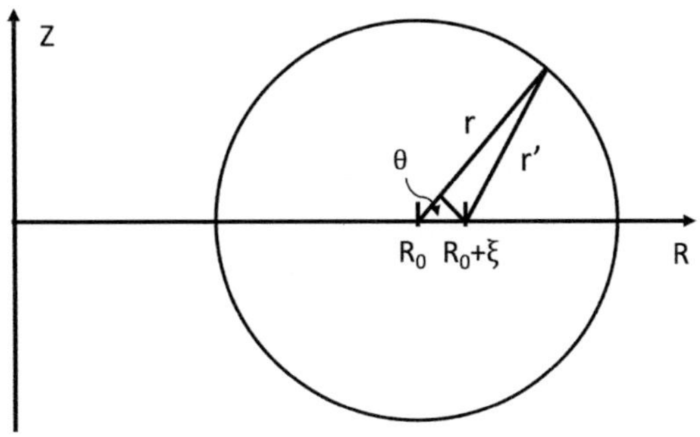

Figure 4.9 Coordinate system for describing large-aspect-ratio equilibrium with r' indicating the shifted magnetic surface and ξ the Shafranov shift.

4.7.1 GRAD-SHAFRANOV EQUATION IN THE ARGE-ASPECT-RATIO LIMIT

In this subsection, we derive the equilibrium solution for the configuration with circular cross-section and large aspect ratio. The solutions have been obtained in various papers, for example, in references [8] and [13]. We will describe the solution from the Grad-Shafranov equation (4.34):

$$\nabla \cdot \left(\frac{\nabla \chi}{R^2} \right) = -\mu_0 p'_\chi - \frac{g g'_\chi}{R^2}. \tag{4.101}$$

We first introduce the local cylindric-like orthogonal coordinates in figure 4.9: (r, θ, ϕ), where r is the minor radius, θ is the poloidal angle, and ϕ is the axisymmetric toroidal angle. In this coordinate system, one has

$$\begin{aligned} R &= R_0 + r\cos\theta \equiv R_0 h_s, \\ Z &= r\sin\theta, \end{aligned}$$

and Jacobian

$$\mathcal{J} = \frac{1}{\nabla r \times \nabla\theta \cdot \nabla\phi} = Rr.$$

We then decompose

$$\begin{aligned} \nabla\chi &= \frac{\partial\chi}{\partial r}\nabla r + \frac{\partial\chi}{\partial\theta}\nabla\theta \\ &= Rr\frac{\partial\chi}{\partial r}\nabla\theta \times \nabla\phi + \frac{R}{r}\frac{\partial\chi}{\partial\theta}\nabla\phi \times \nabla r. \end{aligned} \tag{4.102}$$

Therefore, one has

$$
\begin{aligned}
R^2\nabla\cdot\frac{\nabla\chi}{R^2} &= R^2\mathscr{J}^{-1}\frac{\partial}{\partial r}\left(\frac{1}{R^2}Rr\frac{\partial\chi}{\partial r}\right)+R^2\mathscr{J}^{-1}\frac{\partial}{\partial\theta}\left(\frac{1}{rR^2}R\frac{\partial\chi}{\partial\theta}\right)\\
&= \frac{R}{r}\frac{\partial}{\partial r}\left(\frac{r}{R}\frac{\partial\chi}{\partial r}\right)+\frac{R}{r}\frac{\partial}{\partial\theta}\left(\frac{1}{rR}\frac{\partial\chi}{\partial\theta}\right). \qquad (4.103)
\end{aligned}
$$

Using equation (4.103), we can solve equation (4.101) order by order. In the lowest order, one can obtain that

$$
\begin{aligned}
\frac{d\chi}{dr}R^2\nabla\cdot\frac{\nabla\chi}{R^2} &= \frac{d\chi}{dr}\nabla^2\chi=\frac{d\chi}{dr}\frac{1}{r}\frac{d}{dr}\left(r\frac{d\chi}{dr}\right)\\
&= \frac{1}{r}\left(\frac{d\chi}{dr}\right)^2+\frac{d\chi}{dr}\frac{d^2\chi}{dr^2}. \qquad (4.104)
\end{aligned}
$$

Therefore, the lowest order equilibrium becomes

$$
\frac{1}{2}\frac{dg^2}{dr}+\frac{\partial\chi}{\partial r}\frac{\partial^2\chi}{\partial r^2}+\frac{1}{r}\left(\frac{\partial\chi}{\partial r}\right)^2=-\mu_0R_0^2\frac{dp}{dr}, \qquad (4.105)
$$

i.e.,

$$
\frac{1}{2}\frac{d(B_\phi^2+B_\theta^2)}{dr}+\frac{B_\theta^2}{r}=-\mu_0\frac{dp}{dr}. \qquad (4.106)
$$

To derive the first-order equation, one needs to express d^2p/dr^2 and $d\chi/dr$. From the equilibrium equation of lowest order, equation (4.105), one obtains

$$
\begin{aligned}
\mu_0\frac{d^2p}{dr^2} &= -\frac{1}{2R^2}\frac{d^2(g^2)}{dr^2}-\frac{1}{R^2}\left(\frac{\partial^2\chi}{\partial r^2}\right)^2-\frac{1}{R^2}\frac{\partial\chi}{\partial r}\frac{d^3\chi}{dr^3}+\frac{1}{R^2r^2}\left(\frac{\partial\chi}{\partial r}\right)^2\\
&\quad -2\frac{1}{R^2}\frac{1}{r}\frac{\partial\chi}{\partial r}\frac{\partial^2\chi}{\partial r^2}. \qquad (4.107)
\end{aligned}
$$

Next, let us consider the first order. As reference [13], the magnetic surface is denoted by $r_f=r-\xi\cos\theta$, which can be proved by the vector relation $\mathbf{r}_f=\boldsymbol{\xi}+\mathbf{r}$ as shown in figure 4.9. It is pointed out that R_0 in this notation represents the major radius of the geometric center, instead of the magnetic axis, and therefore is independent of equilibrium parameters, such as plasma beta, etc. This representation is related to the so-called Shafranov shift by $\xi=-\Delta$. Then, the surface functions can be expanded as follows:

$$
\chi(r_f) = \chi(r-\xi\cos\theta)=\chi_0-\xi\cos\theta\frac{\partial\chi_0}{\partial r}. \qquad (4.108)
$$

The same expansion holds for $p(r_f)$ and $g(r_f)$.

We reduce the Grad-Shafranov equation (4.101) term by term. We first reduce the first term. Using equation (4.103), the first term is reduced as follows

$$\begin{aligned} R^2\nabla\cdot\frac{\nabla\chi}{X^2} &= \frac{R_0^2(1+(r/R_0)\cos\theta)^2}{r(1+(r/R_0)\cos\theta)}\left\{\frac{\partial}{\partial r}\left[\frac{rh_s}{R_0^2(1+(r/R_0)\cos\theta)^2}\frac{\partial\chi}{\partial r}\right]\right. \\ &\quad\left.+\frac{\partial}{\partial\theta}\left[\frac{h_s}{rR_0^2(1+(r/R_0)\cos\theta)^2}\frac{\partial\chi}{\partial\theta}\right]\right\} \\ &= \frac{R_0^2(1+(r/R_0)\cos\theta)}{r}\left\{\frac{\partial}{\partial r}\left[\frac{r}{R_0^2(1+(r/R_0)\cos\theta)}\frac{\partial\chi}{\partial r}\right]\right. \\ &\quad\left.+\frac{\partial}{\partial\theta}\left[\frac{1}{rR_0^2(1+(r/R_0)\cos\theta)}\frac{\partial\chi}{\partial\theta}\right]\right\}. \end{aligned} \tag{4.109}$$

The first term on the right in equation (4.109) can be further reduced as

$$\begin{aligned} &\frac{R_0^2(1+(r/R)\cos\theta)}{r}\frac{\partial}{\partial r}\left[\frac{r}{R_0^2(1+(r/R)\cos\theta)}\frac{\partial\chi}{\partial r}\right] \\ = &\frac{1+(r/R)\cos\theta}{r}\frac{\partial}{\partial r}\left\{\frac{r}{1+(r/R)\cos\theta}\left[\frac{\partial\chi_0}{\partial r}-\cos\theta\frac{\partial}{\partial r}\left(\xi\frac{\partial\chi_0}{\partial r}\right)\right]\right\} \\ = &\frac{1}{R}\cos\theta\frac{\partial}{\partial r}\left\{r\frac{\partial\chi_0}{\partial r}\right\} \\ &+\frac{1}{r}\frac{\partial}{\partial r}\left\{r\left[-(r/R)\cos\theta\frac{\partial\chi_0}{\partial r}-\cos\theta\frac{\partial}{\partial r}\left(\xi\frac{\partial\chi_0}{\partial r}\right)\right]\right\} \\ = &\underbrace{-\frac{1}{R}\cos\theta\frac{\partial\chi_0}{\partial r}}_{-\frac{1}{R}B_\theta^2\cdot R^2/\frac{d\chi}{dr}\ G5}-\cos\theta\frac{1}{r}\frac{\partial}{\partial r}\left\{r\frac{\partial}{\partial r}\left(\xi\frac{\partial\chi_0}{\partial r}\right)\right\}, \end{aligned} \tag{4.110}$$

in which the last term becomes

$$\begin{aligned} -\frac{1}{r}\frac{\partial}{\partial r}\left\{r\frac{\partial}{\partial r}\left(\xi\frac{\partial\chi_0}{\partial r}\right)\right\}\cos\theta &= -\frac{1}{r}\frac{\partial}{\partial r}\left\{r\left(\frac{d\xi}{dr}\frac{\partial\chi_0}{\partial r}+\frac{\partial^2\chi}{\partial r^2}\xi\right)\right\}\cos\theta \\ = -\left(\underbrace{\frac{d^2\xi}{dr^2}\frac{\partial\chi_0}{\partial r}}_{-B_\theta^2\frac{d^2\xi}{dr^2}\cdot R^2/\frac{d\chi}{dr}}\right. &+ \underbrace{\frac{d\xi}{dr}\frac{\partial^2\chi_0}{\partial r^2}+\frac{\partial^2\chi}{\partial r^2}\frac{d\xi}{dr}}_{-2\frac{1}{R_0^2}\frac{\partial\chi_0}{\partial r}\frac{\partial^2\chi_0}{\partial r^2}\frac{d\xi}{dr}\cdot R^2/\frac{d\chi}{dr}}\left.+\underbrace{\frac{d^3\chi}{dr^3}\xi}_{x1}\right)\cos\theta \\ -\frac{1}{r}\left(\underbrace{\frac{d\xi}{dr}\frac{\partial\chi_0}{\partial r}}_{-\frac{1}{r}\frac{1}{R_0^2}\left(\frac{\partial\chi_0}{\partial r}\right)^2\frac{d\xi}{dr}\cdot R^2/\frac{d\chi}{dr}}\right. &\left.+\underbrace{\frac{\partial^2\chi}{\partial r^2}\xi}_{x3}\right)\cos\theta. \end{aligned} \tag{4.111}$$

The second term on the right in equation (4.109) can be further reduced as

$$\begin{aligned}
&\frac{R_0^2(1+(r/R)\cos\theta)}{r}\frac{\partial}{\partial\theta}\left[\frac{1}{rR_0^2(1+(r/R)\cos\theta)}\frac{\partial\chi}{\partial\theta}\right]\\
=\;&\frac{R_0^2(1+(r/R)\cos\theta)}{r}\frac{\partial}{\partial\theta}\left[\frac{1}{rR_0^2}\frac{\partial\chi_0}{\partial\theta}-\frac{1}{rR_0^2}\frac{\partial\chi_0}{\partial\theta}(r/R)\cos\theta+\frac{1}{rR_0^2}\xi\sin\theta\frac{\partial\chi_0}{\partial r}\right]\\
=\;&\cos\theta\xi\underbrace{\frac{1}{r^2}\frac{\partial\chi_0}{\partial r}}_{x2},
\end{aligned}\tag{4.112}$$

where it has been noted that $\frac{\partial\chi_0}{\partial\theta}=0$.

The last term in equation (4.101) can be reduced as follows:

$$\begin{aligned}
\frac{1}{2}\frac{dg^2}{d\chi} &= \frac{1}{2}\frac{1}{\frac{d\chi_0}{dr}-\cos\theta\frac{d}{dr}\left(\xi\frac{\partial\chi_0}{\partial r}\right)}\left[\frac{dg_0^2}{dr}-\cos\theta\frac{d}{dr}\left(\xi\frac{\partial g_0^2}{\partial r}\right)\right]\\
&= \frac{1}{2}\frac{1}{\frac{d\chi_0}{dr}}\left\{1+\frac{1}{\frac{d\chi_0}{dr}}\cos\theta\frac{d}{dr}\left(\xi\frac{\partial\chi_0}{\partial r}\right)\right\}\left[\frac{dg_0^2}{dr}-\cos\theta\frac{d}{dr}\left(\xi\frac{\partial g_0^2}{\partial r}\right)\right]\\
&\to^{1st\ order} \frac{1}{2}\frac{1}{\frac{d\chi_0}{dr}}\left\{\frac{\frac{dg_0^2}{dr}}{\frac{d\chi_0}{dr}}\cos\theta\frac{d}{dr}\left(\xi\frac{\partial\chi_0}{\partial r}\right)-\cos\theta\frac{d}{dr}\left(\xi\frac{\partial g_0^2}{\partial r}\right)\right\}\\
&= \underbrace{\frac{1}{2}\frac{1}{\frac{d\chi_0}{dr}}\left\{\frac{\frac{dg_0^2}{dr}}{\frac{d\chi_0}{dr}}\cos\theta\xi\frac{\partial^2\chi_0}{\partial r^2}-\cos\theta\xi\frac{\partial^2 g_0^2}{\partial r^2}\right\}}_{x6}.
\end{aligned}$$

Similarly, the first term on the right-hand side of equation (4.101) can be reduced as follows:

$$\begin{aligned}
\frac{dp}{d\chi} &= \frac{1}{\frac{d\chi_0}{dr}-\cos\theta\frac{d}{dr}\left(\xi\frac{\partial\chi_0}{\partial r}\right)}\left[\frac{dp_0}{dr}-\cos\theta\frac{d}{dr}\left(\xi\frac{\partial p_0}{\partial r}\right)\right]\\
&= \frac{1}{\frac{d\chi_0}{dr}}\left\{1+\frac{1}{\frac{d\chi_0}{dr}}\cos\theta\frac{d}{dr}\left(\xi\frac{\partial\chi_0}{\partial r}\right)\right\}\left[\frac{dp_0}{dr}-\cos\theta\frac{d}{dr}\left(\xi\frac{\partial p_0}{\partial r}\right)\right]\\
&= \frac{1}{\frac{d\chi_0}{dr}}\frac{dp_0}{dr}+\frac{1}{\frac{d\chi_0}{dr}}\left\{\frac{\frac{dp_0}{dr}}{\frac{d\chi_0}{dr}}\cos\theta\frac{d}{dr}\left(\xi\frac{\partial\chi_0}{\partial r}\right)-\cos\theta\frac{d}{dr}\left(\xi\frac{\partial p_0}{\partial r}\right)\right\}
\end{aligned}$$

$$
\begin{aligned}
&= \frac{1}{\frac{d\chi_0}{dr}}\frac{dp_0}{dr} + \frac{1}{\frac{d\chi_0}{dr}}\left\{\frac{\frac{dp_0}{dr}}{\frac{d\chi_0}{dr}}\cos\theta\xi\frac{\partial^2\chi_0}{\partial r^2} - \cos\theta\xi\frac{\partial^2 p_0}{\partial r^2}\right\} \\
&= \frac{1}{\frac{d\chi_0}{dr}}\frac{dp_0}{dr} + \frac{1}{\frac{d\chi_0}{dr}}\left\{\frac{\partial^2\chi_0}{\partial r^2}\left(-\frac{1}{\mu_0}\frac{1}{2R_0^2\frac{d\chi}{dr}}\frac{\partial g^2}{\partial r} - \underbrace{\frac{1}{\mu_0 R^2}\frac{\partial^2\chi}{\partial r^2}}_{x4} - \underbrace{\frac{1}{\mu_0 R^2}\frac{1}{r}\frac{\partial\chi}{\partial r}}_{x5}\right)\cos\theta\xi\right\} \\
&\quad - \frac{1}{\frac{d\chi_0}{dr}\mu_0}\left\{\left(-\frac{1}{2R_0^2}\frac{d^2g^2}{dr^2}\right.\right. \\
&\quad \left.\left. \underbrace{-\frac{1}{R^2}\left(\frac{\partial^2\chi}{\partial r^2}\right)^2}_{x4} - \frac{1}{R^2}\frac{\partial\chi}{\partial r}\frac{d^3\chi}{dr^3} + \frac{1}{R^2r^2}\left(\frac{\partial\chi}{\partial r}\right)^2 - \underbrace{2}_{x5}\frac{1}{R^2}\frac{1}{r}\frac{\partial\chi}{\partial r}\frac{\partial^2\chi}{\partial r^2}\right)\cos\theta\xi\right\} \\
&= \frac{1}{\frac{d\chi_0}{dr}}\frac{dp_0}{dr} - \frac{1}{2\mu_0 R_0^2}\frac{1}{\frac{d\chi_0}{dr}}\left\{\frac{\frac{dg_0^2}{dr}}{\frac{d\chi_0}{dr}}\cos\theta\xi\frac{\partial^2\chi_0}{\partial r^2} - \cos\theta\xi\frac{\partial^2 g_0^2}{\partial r^2}\right\}. \\
&\quad + \frac{1}{\mu_0}\left\{\cos\theta\xi\left(\frac{1}{R^2}\frac{d^3\chi}{dr^3} - \frac{1}{R^2r^2}\left(\frac{\partial\chi}{\partial r}\right) + \frac{1}{R^2}\frac{1}{r}\frac{\partial^2\chi}{\partial r^2}\right)\right\},
\end{aligned}
$$

where equations (4.105) and (4.107) have been used. Therefore, one has

$$
\begin{aligned}
-\mu_0 R^2\frac{dp}{d\chi} &= \underbrace{-2\mu_0 R^2\frac{r}{R}\cos\theta\frac{1}{\frac{d\chi_0}{dr}}\frac{dp_0}{dr}}_{-2\mu_0(r/R)\frac{dp}{dr}\cdot R^2/\frac{d\chi}{dr}} + \underbrace{\frac{1}{2}\frac{1}{\frac{d\chi_0}{dr}}\left(\frac{\frac{dg_0^2}{dr}}{\frac{d\chi_0}{dr}}\cos\theta\xi\frac{\partial^2\chi_0}{\partial r^2} - \cos\theta\xi\frac{\partial^2 g_0^2}{\partial r^2}\right)}_{x6} \\
&\quad -\cos\theta\xi\left(\underbrace{\frac{d^3\chi}{dr^3}}_{x1} - \underbrace{\frac{1}{r^2}\left(\frac{\partial\chi}{\partial r}\right)}_{x2} + \underbrace{\frac{1}{r}\frac{\partial^2\chi}{\partial r^2}}_{x3}\right).
\end{aligned}
$$

Noting the cancellations between the same underbrace indices "xi" and combining equations (4.109)–(4.113) yields

$$
\frac{1}{r}\frac{d}{dr}\left(rB_\theta^2\frac{d\xi}{dr}\right) + \frac{B_\theta^2}{R} - 2\mu_0\frac{r}{R}\frac{dp}{dr} = 0. \tag{4.113}
$$

This is the equation that governs the Shafranov shift ξ. The solution of this equation also depends on the boundary conditions. Since it is a second-order differential equation, two deterministic conditions are required. Since it is the first-order equation, one can regard the radial coordinate r to be the magnetic surface r' by neglecting the second-order contributions. The boundary condition at the magnetic axis is simply $\xi = 0$. One can specify another boundary condition at the plasma edge $r' = a$. Note that equation (4.113) is an inhomogeneous equation. One can formally add a homogeneous solution to make the solution satisfy the boundary conditions. The homogeneous solution corresponds to adding the vacuum solution to the equilibrium equation in the plasma region. Reference [3] discussed the effects of adding a vertical vacuum field on the balance of hoop force in the toroidal confinement.

4.7.2 MAGNETIC COORDINATES IN THE LARGE-ASPECT-RATIO TOKAMAK

To apply the equilibrium solution in the large-aspect-ratio limit in the previous subsection 4.7.1, one needs to construct the magnetic coordinates or the straight-field-line coordinates in this limit. In this subsection, the theory developed by J. D. Meiss and R. D. Hazeltine in reference [8] will be described. Both the magnetic coordinates and metric parameters are derived. They will be used in the later chapters for stability analyses in the large-aspect-ratio limit. The theory in the large-aspect-ratio limit is important for understanding tokamak physics, as well as the mathematical treatment of toroidal geometry.

We limit ourselves to the case with the toroidal axisymmetric angle ϕ chosen as the toroidal angle. In this case, the magnetic coordinates of the PEST type as described in Section 4.1.2 is used. According to equation (4.11), one has

$$\theta_{pest} = \frac{1}{q}\int_0^{\theta_{eq}} d\theta_{eq} \frac{g\mathcal{J}_{eq}}{X^2}, \tag{4.114}$$

where the Jacobian is given by

$$\mathcal{J} = \frac{1}{\nabla\chi \times \nabla\theta_{eq} \cdot \nabla\phi}. \tag{4.115}$$

Here, θ_{eq} is the original poloidal angle before transforming to the straight-field-line coordinates. In the large-aspect-ratio limit, the original coordinates usually are the so-called "Shafranov coordinates" with $\theta_{eq} = \theta_s$ and

$$X = X_0 - \Delta(r_s) + r_s \cos\theta_s, \tag{4.116}$$

$$Z = r_s \sin\theta_s, \tag{4.117}$$

where X_0 is the major radius of the magnetic axis, the magnetic surfaces are assumed to be circles with the center shifted by Δ. In this case, r_s is considered to be the minor radius of the flux surface, i.e., $r_f \approx r_s$.

To obtain the flux coordinates denoted by $(r_f, \theta_f = \theta_{pest}, \phi)$ using equation (4.115), the ordering assumption that $\Delta/r \sim r/R \sim \varepsilon$ is adopted. We will only carry out the calculation till the first order. We first compute the Jacobian in equation (4.115). From equations (4.116) and (4.117) one obtains

$$\frac{\partial X}{\partial r_s} = -\Delta' + \cos\theta_s, \tag{4.118}$$

$$\frac{\partial X}{\partial \theta_s} = -r_s \sin\theta_s, \tag{4.119}$$

$$\frac{\partial Z}{\partial r_s} = \sin\theta_s, \tag{4.120}$$

$$\frac{\partial Z}{\partial \theta_s} = r_s\cos\theta_s. \tag{4.121}$$

Using the partial differential chain rule

$$\frac{\partial(r,\theta)}{\partial(X,Z)}\frac{\partial(X,Z)}{\partial(r,\theta)} = I \rightarrow \frac{\partial(r,\theta)}{\partial(X,Z)} = \left\{\frac{\partial(X,Z)}{\partial(r,\theta)}\right\}^{-1}, \tag{4.122}$$

one obtains the Jacobian using equations (4.118)–(4.121)

$$\begin{aligned}\mathscr{J} &= \frac{1}{\chi'_{r_f}}\frac{1}{\nabla r_f\times\nabla\theta_s\cdot\mathbf{e}_\phi|\nabla\phi|} = R\nabla X\times\nabla Y\cdot\mathbf{e}_\phi \\ &= \frac{1}{\chi'_{r_f}}X\left(\frac{\partial X}{\partial r_s}\frac{\partial Z}{\partial\theta_s} - \frac{\partial X}{\partial\theta_s}\frac{\partial Z}{\partial r_s}\right) \\ &= \frac{1}{\chi'_{r_f}}r_f X\left(1-\Delta'_{r_f}\cos\theta_s\right).\end{aligned} \tag{4.123}$$

Inserting the Jacobian in equation (4.123) into equation (4.114), one obtains the poloidal angle in the PEST magnetic coordinates

$$\begin{aligned}\theta_f &= \int_0^{\theta_s} d\theta\left[1-(r/R+\Delta')\cos\theta\right] \\ &= \theta_s - \left(r_f/R+\Delta'\right)\sin\theta_s.\end{aligned} \tag{4.124}$$

Therefore, in the cylindrical coordinates, the magnetic surface (R,Z,ϕ) is determined as follows:

$$\begin{aligned}R &= R_0-\Delta+r_s\cos\theta_s \\ &= R_0-\Delta+r_f\cos\left[\theta_f+\left(r_f/R+\Delta'\right)\sin\theta_f\right] \\ &= R_0-\Delta+r_f\left\{\cos\theta_f\cos\left[\left(r_f/R+\Delta'\right)\sin\theta_f\right]-\sin\theta_f\sin\left[\left(r_f/R+\Delta'\right)\sin\theta_f\right]\right\} \\ &= R_0-\Delta+r_f\cos\theta_f-r_f\left(r_f/R+\Delta'\right)\sin^2\theta_f \\ &= R_0-\Delta+r_f\cos\theta_f-\frac{1}{2}r_f\left(r_f/R+\Delta'\right)\left(1-\cos2\theta_f\right),\end{aligned} \tag{4.125}$$

$$\begin{aligned}Z &= r_s\sin\theta_s \\ &= r_f\sin\left[\theta_f+\left(r_f/R+\Delta'\right)\sin\theta_f\right] \\ &= r_f\sin\theta_f\cos\left[\left(r_f/R+\Delta'\right)\sin\theta_f\right]+r_f\cos\theta_f\sin\left[\left(r_f/R+\Delta'\right)\sin\theta_f\right] \\ &= r_f\sin\theta'+r_f\left(r_f/R+\Delta'\right)\cos\theta_f\sin\theta_f \\ &= r_f\sin\theta_f+\frac{1}{2}r_f\left(r_f/R+\Delta'\right)\sin2\theta_f.\end{aligned} \tag{4.126}$$

We will use the differential chain rule in equation (4.122) to obtain the metric parameters. In this case, we need to obtain various derivatives from equations (4.125) and (4.126)

$$\begin{aligned}
\frac{\partial R}{\partial r_f} &= -\Delta' + \cos\theta_f - \left(\frac{r_f}{R} + \frac{1}{2}(r_f\Delta')'\right)(1-\cos 2\theta_f), \\
\frac{\partial R}{\partial \theta_f} &= -r_f \sin\theta_f - r_f\left(r_f/R + \Delta'\right)\sin 2\theta_f, \\
\frac{\partial Z}{\partial r_f} &= \sin\theta_f + \left(\frac{r_f}{R} + \frac{1}{2}(r_f\Delta')'\right)\sin 2\theta, \\
\frac{\partial Z}{\partial \theta_f} &= r_f\cos\theta_f + r_f\left(r_f/R + \Delta'\right)\cos 2\theta_f.
\end{aligned}$$

To compute the matrix inversion on the right-hand of equation (4.122), one needs to evaluate the following determinant

$$\begin{aligned}
T &= \begin{vmatrix} \frac{\partial R}{\partial r_f} & \frac{\partial R}{\partial \theta_f} \\ \frac{\partial Z}{\partial r_f} & \frac{\partial Z}{\partial \theta_f} \end{vmatrix} \\
&= \left[-\Delta' + \cos\theta_f - \left(\frac{r_f}{R} + \frac{1}{2}(r_f\Delta')'\right)(1-\cos 2\theta_f)\right] r_f\cos\theta_f \\
&\quad + \left[-\Delta' + \cos\theta_f - \left(\frac{r_f}{R} + \frac{1}{2}(r_f\Delta')'\right)(1-\cos 2\theta_f)\right] r_f\left(r_f/R + \Delta'\right)\cos 2\theta_f \\
&\quad + \left[\sin\theta_f + \left(\frac{r_f}{R} + \frac{1}{2}(r_f\Delta_f)'\right)\sin 2\theta_f\right] r_f\sin\theta_f \\
&\quad + \left[\sin\theta_f + \left(\frac{r_f}{R} + \frac{1}{2}(r_f\Delta')'\right)\sin 2\theta_f\right] r_f\left(r_f/R + \Delta'\right)\sin 2\theta_f \\
&= r_f\left(1 + \frac{r_f}{R}\cos\theta_f\right). \qquad (4.127)
\end{aligned}$$

Here, only the terms up to the first order are kept. Using the chain rule in equation (4.122) and the determinant in equation (4.127) for matrix inversion, one obtains

$$\begin{pmatrix} \frac{\partial r}{\partial R} & \frac{\partial r}{\partial Z} \\ \frac{\partial \theta}{\partial R} & \frac{\partial \theta}{\partial Z} \end{pmatrix} = \frac{1}{T}\begin{pmatrix} \frac{\partial Z}{\partial \theta} & -\frac{\partial R}{\partial \theta} \\ -\frac{\partial Z}{\partial r} & \frac{\partial R}{\partial r} \end{pmatrix} =$$

$$\frac{1}{T}\begin{pmatrix} r\cos\theta + r(r/R+\Delta')\cos 2\theta & r\sin\theta + r(r/R\Delta')\sin 2\theta \\ -\sin\theta - \left(\frac{r}{R} + \frac{1}{2}(r\Delta')'\right)\sin 2\theta & -\Delta' + \cos\theta - \left(\frac{r}{R} + \frac{1}{2}(r\Delta')'\right)(1-\cos 2\theta) \end{pmatrix}. \qquad (4.128)$$

Here and in the following, we drop the subscript "f" for magnetic coordinates. Using equation (4.128), one can calculate the metric parameters as follows

$$\nabla r \cdot \nabla r = \frac{\partial r}{\partial R}\frac{\partial r}{\partial R} + \frac{\partial r}{\partial Z}\frac{\partial r}{\partial Z}$$

$$= \frac{1}{T^2}\left\{ \left[r\cos\theta + r\left(r/R+\Delta'\right)\cos 2\theta\right]^2 + \left[r\sin\theta + r\left(r/R+\Delta'\right)\sin 2\theta\right]^2 \right\}$$

$$= \frac{r^2}{T^2}\left\{ \cos^2\theta + 2\left(r/R+\Delta'\right)\cos 2\theta\cos\theta + \sin^2\theta + 2\left(r/R+\Delta'\right)\sin 2\theta\sin\theta \right\}$$

$$= 1 - 2\frac{r}{R}\cos\theta + 2\left(r/R+\Delta'\right)\cos\theta$$

$$= 1 + 2\Delta'\cos\theta,$$

$$\nabla\theta \cdot \nabla\theta = \frac{\partial\theta}{\partial R}\frac{\partial\theta}{\partial R} + \frac{\partial\theta}{\partial Z}\frac{\partial\theta}{\partial Z}$$

$$= \frac{1}{T^2}\left\{ -\sin\theta - \left(\frac{r}{R} + \frac{1}{2}(r\Delta')'\right)\sin 2\theta \right\}^2$$

$$+ \frac{1}{T^2}\left\{ -\Delta' + \cos\theta - \left(\frac{r}{R} + \frac{1}{2}(r\Delta')'\right)(1-\cos 2\theta) \right\}^2$$

$$= \frac{1}{T^2}\left\{ \sin^2\theta - 2\Delta'\cos\theta + 2\left(\frac{r}{R} + \frac{1}{2}(r\Delta')'\right)\sin 2\theta\sin\theta \right.$$

$$\left. + \cos^2\theta - 2\left(\frac{r}{R} + \frac{1}{2}(r\Delta')'\right)\cos\theta + 2\left(\frac{r}{R} + \frac{1}{2}(r\Delta')'\right)\cos 2\theta\cos\theta \right\}$$

$$= \frac{1}{r^2}\left\{ 1 - 2\frac{r}{R}\cos\theta - 2\Delta'\cos\theta \right\}$$

$$= \frac{1}{r^2}\left\{ 1 - 2\left(\frac{r}{R} + \Delta'\right)\cos\theta \right\},$$

$$\nabla r \cdot \nabla\theta = \frac{\partial r}{\partial R}\frac{\partial\theta}{\partial R} + \frac{\partial r}{\partial Z}\frac{\partial\theta}{\partial Z}$$

$$= \frac{1}{T^2}\left\{ \left[r\cos\theta + r\left(r/R+\Delta'\right)\cos 2\theta\right]\left[-\sin\theta - \left(\frac{r}{R} + \frac{1}{2}(r\Delta')'\right)\sin 2\theta\right] \right.$$

$$\left. + \left[r\sin\theta + r\left(r/R+\Delta'\right)\sin 2\theta\right]\left[\xi' + \cos\theta - \left(\frac{r}{R} + \frac{1}{2}(r\Delta')'\right)(1-\cos 2\theta)\right] \right\}$$

$$= \frac{1}{T^2}\left\{ r\cos\theta\left[\underbrace{-\sin\theta}_{x1} \underbrace{- \left(\frac{r}{R} + \frac{1}{2}(r\Delta')'\right)\sin 2\theta}_{o1}\right] + \underbrace{r\left(r/R+\Delta'\right)\cos 2\theta\left[-\sin\theta\right]}_{o2} \right.$$

$$
\begin{aligned}
&+r\sin\theta\left[\xi'+\underbrace{\cos\theta}_{x1}-\left(\frac{r}{R}+\frac{1}{2}(r\Delta')'\right)+\underbrace{\left(\frac{r}{R}+\frac{1}{2}(r\Delta')'\right)\cos 2\theta}_{o1}\right]\\
&\left.\underbrace{+r\left(r/R+\Delta'\right)\sin 2\theta\left[\cos\theta\right]}_{o2}\right\}\\
=\;&\frac{r}{T^2}\left\{-\left(\frac{r}{R}+\frac{1}{2}(r\Delta')'\right)\sin\theta+\left(r/R+\Delta'\right)\sin\theta\right.\\
&\left.-\sin\theta\left[\Delta'+\left(\frac{r}{R}+\frac{1}{2}(r\Delta')'\right)\right]\right\}\\
=\;&-\frac{1}{r}\sin\theta\left(\frac{r}{R}+\frac{1}{2}(r\Delta')'\right).
\end{aligned}
$$

Using equation (4.128), the Jacobian in the magnetic coordinates can be expressed as

$$
\begin{aligned}
\mathcal{J} &= \frac{1}{|\nabla\chi|\nabla r\times\nabla\theta\cdot\mathbf{e}_\phi|\nabla\phi|}=\frac{R}{|\nabla\chi|}\nabla X\times\nabla Y\cdot\mathbf{e}_\phi\\
&= \frac{r}{\chi'}\left(1+\frac{r}{R}\cos\theta\right)\left[R_0+\xi+r\cos\theta-\frac{1}{2}r\left(r/R-\xi'\right)\left(1-\cos 2\theta\right)\right]\\
&= \frac{Rr}{\chi'}\left(1+\frac{r}{R}\cos\theta\right)\left(1+\frac{r}{R}\cos\theta\right)\\
&= \frac{Rr}{\chi'}\left(1+2\frac{r}{R}\cos\theta\right)
\end{aligned}
$$

These review the theory given in reference [8]. The results will be used for stability analyses in the later chapters.

4.8 CONCLUSIONS AND DISCUSSION

In this chapter, the equilibrium theories for axisymmetric and nonaxisymmetric systems are described. In the axisymmetric system, for example tokamak, the construction of magnetic coordinates or straight-field-line coordinates and the derivation of the Grad-Shafranov equation are described. The canonical magnetic coordinates are also described for Hamiltonian description of guiding center motion of charged particles in a magnetic field. In the non-axisymmetric system, for example, the stellarator, the quasisymmetric stellarator configuration is discussed using the Boozer coordinates and the guiding center Lagrangian.

The procedure to determine the coil structure from a fixed boundary equilibrium solution is given. This is useful for the engineering design of the coil structure, which is particularly useful for stellarator coil design. The coil structure in a stellarator is complicated. The procedure gives room to optimize the location of coils.

The Lagrangian and Hamiltonian descriptions of magnetic-field-line flow are also described in this chapter. It shows that not only the magnetic-field-line flow but also the current-density flow are needed for a complete description of the magnetic flux structure.

Besides, the X-point equilibrium is described. This helps to understand the field-line flow toward the divertors. It is also useful for understanding the magnetic island and stochastic field line structure in the tearing modes. The dual-poloidal-region q coordinates are also discussed, which are particularly important for the design of the so-called resonance magnetic perturbations and stability analyses with non-ideal MHD effects taken into account.

In the end, the equilibrium of a large-aspect-ratio tokamak is described. The magnetic coordinates or the straight-field-line coordinates in the large-aspect-ratio limit are also given. They help understand the Shafranov shift in toroidal equilibrium are useful for further stability analyses in the later chapters.

The review of the equilibrium computational method can be found in reference [5]. The description of the adaptive equilibrium code ATEQ can be found in reference [4].

References

1. Hinton F L, Hazeltine R D 1976 Theory of plasma transport in toroidal confinement systems, *Rev. Mod. Phys. 48*, 239. DOI: https://doi.org/10.1103/RevModPhys.48.239
2. Boozer A H 2005 Physics of magnetically confined plasmas *Rev. Mod. Phys. 76*, 1071. https://doi.org/10.1103/RevModPhys.76.1071
3. Linjin Zheng 2015 IOP Concise physics: Advanced tokamak stability theory (San Rafael: Morgan & Claypool Publishers & London: IOP Publishing). https://doi.org/10.1088/978-1-6270-5423-2
4. Linjin Zheng 2019 Magnetically Confined Fusion Plasma Physics, deal MHD theory (London: IOP Publishing). https://doi.org/10.1088/978-1-64327-138-5
5. Grimm R C, Greene J M, Johnson J L 1976 *Methods of Computational Physics* (New York: Academic).
6. Hamada S 1962 Hydromagnetic equilibria and their proper coordinates *Nucl. Fusion 2*, 23. DOI 10.1088/0029-5515/2/1-2/005
7. Boozer A H 1982 Establishment of magnetic coordinates for a given magnetic field *Phys. Fluids 25*, 520–1. https://doi.org/10.1063/1.863765
8. Meiss J D, Hazeltine R D 1990 Canonical coordinates for guiding center particles, *JPhys. Fluids B 2*, 2563–2567. https://doi.org/10.1063/1.859380
9. Meng L, Breizman B N, Linjin Zheng 2016 Canonical straight field line magnetic coordinates for tokamaks, *J. Comp. Phys. 326*, 334–341. https://doi.org/10.1016/j.jcp.2016.09.004
10. Littlejohn R G 2014 Hamiltonian formulation of guiding center motion, *Phys. Fluids 24*, 1730–1749. https://doi.org/10.1063/1.863594
11. Zheng L-J, Kotschenreuther M 2006 AEGIS: An adaptive ideal-magnetohydrodynamics shooting code for axisymmetric plasma stability *J. Comput. Phys. 211*, 748–66. https://doi.org/10.1016/j.jcp.2005.06.009

12. Grad H and Rubin H 1958 Hydromagnetic equilibria and force-free fields *Proc. 2nd UN Conf. on the Peaceful Uses of Atomic Energy* vol **31** (Geneva: International Atomic Energy Agency) p 190.
13. Shafranov V D 1966 Plasma equilibrium in a magnetic field *Reviews of Plasma Physics* vol **2** (New York: Consultants Bureau) p 103.
14. Jackson J D 1975 *Classical electrodynamics, second edition* (New York: Wiley).
15. Lackner K 1976 Computation of ideal MHD equilibria *Comput. Phys. Comm. 12*, 33–44.
16. Johson J L, Dalhed H E, Greene J M, Grimm R C, Hsieh Y Y, Jardin S C, Manickam J, Okabayashi M, Storer R G, Todd A M M, Voss D E, and Weimer K E 1979 Numerical determination of axisymmetric toroidal magnetohydrodynamic equilibria *J. Comput. Phys 32*, 212–34.
17. Jardin S 2010 *Computational methods in plasma physics* (Boca Raton: CRC Press).
18. Chance M S 1997 Vacuum calculations in azimuthally symmetric geometry *Phys. Plasmas 4*, 2161–80. https://doi.org/10.1063/1.872380
19. Cary J R, Shasharina S G 1997 Omnigenity and quasihelicity in helical plasma confinement systems, *Physics of Plasmas 4*, 3323–33. https://doi.org/10.1063/1.872473
20. Taylor, J. B., 1964 Equilibrium and Stability of Plasma in Arbitrary Mirror Fields *Phys. Fluids 7*, 767. https://doi.org/10.1063/1.1711283
21. Nührenberg J, Zille R, 1988 Quasi-helically symmetric toroidal stellarators *Phys. Lett. A 129*, 113. https://doi.org/10.1016/0375-9601(88)90080-1
22. Imbert-Gérard L, Paul E J, Wright A M 2025 An Introduction to Stellarators (SIAM, Philadelphia). https://lmig-math.github.io/AnIntroductiontoStellarators-Imbert-Gérard-et-al.pdf
23. Cary J R, Littlejohn R G 1983 Noncanonical Hamiltonian mechanics and its application to magnetic field line flow *Annals of Phys. 151*, 1–34. https://doi.org/10.1016/0003-4916(83)90313-5
24. Pies G T, Lortz D 1971 Asymptotic magnetic surface *Plasma Phys. 13*, 799–816. DOI 10.1088/0032-1028/13/9/010
25. Zheng Linjin 2023 Modification of Lie's transform perturbation theory for charged particle motion in a magnetic field *Phys. Plasmas 30*, 042515. https://doi.org/10.1063/5.0142104
26. Zheng Linjin, Kotschenreuther M T, Waelbroeck F L 2023 Asymptotic vacuum solution at tokamak X-point tip featured *Phys. Plasmas 30*, 112508. https://doi.org/10.1063/5.0173656
27. Zheng Linjin, Kotschenreuther M T, Waelbroeck F L, Austin M E 2025 X-point effects on the ideal MHD modes in tokamaks in the description of dual-poloidal-region safety factor *Phys. Plasmas 32*, 012501. https://doi.org/10.1063/5.0227541
28. Solovév L S 1967 The theory of hydrodynamic stability of toroidal plasma configurations *Sov. Phys. JETP 26*, 400.

5 Linear ideal MHD stability theory in cylindrical geometry

In this chapter, the linear ideal MHD theory in cylindrical geometry is described. We first describe the MHD equilibrium in cylindrical geometry. The basic set of linear equations for stability analyses is given next. The equations will be used for MHD spectrum and stability analyses. The ideal MHD analyses in cylindrical geometry will help the analyses in toroidal geometry and the resistive MHD analyses in the following chapters.

5.1 EQUILIBRIUM

In this section, we first describe the equilibrium of the cylinder system. This is needed for the linear analyses of the MHD modes in the cylinder geometry.

In the cylinder limit, we use the coordinate system (r, θ, z), where r is the minor radius, θ denotes azimuthal or poloidal angle, and z is the longitudinal coordinate. The longitudinal coordinate z is related to the toroidal angle by $z = R\phi$ in the toroidal system, where $R = const.$ is the major radius in the cylinder model. It is also defined that $\psi = r^2/a^2$, and the dot is used to represent the derivative with respect to r. In this coordinate system, one has $\dot{\chi} = B_\theta/R$, $g = RB_z$, and $q = rB_z/RB_\theta$.

In equation (4.105) in Section 4.7.1, it has been shown that the Grad-Shafranov equation in the cylinder geometry with circular cross-section is reduced to

$$\frac{1}{2}\frac{dg^2}{dr} + \frac{\partial \chi}{\partial r}\frac{\partial^2 \chi}{\partial r^2} + \frac{1}{r}\left(\frac{\partial \chi}{\partial r}\right)^2 = -\mu_0 R_0^2 \frac{dp}{dr}. \tag{5.1}$$

It can be further reduced to

$$\frac{1}{2}\frac{d(B_\phi^2 + B_\theta^2)}{dr} + \frac{B_\theta^2}{r} = -\mu_0 \frac{dp}{dr}. \tag{5.2}$$

This equilibrium equation will be used for linear analyses in the next sections.

5.2 BASIC SET OF LINEAR MHD EQUATIONS

With the equilibrium described in Section 5.1, we proceed to introduce the basic set of linear MHD equations in the cylinder geometry with circular cross-section. The MHD equations in cylinder geometry are obtained by reducing the toroidal theory to be derived in Section 6.1 [1]. They will be used for the spectrum and stability

DOI: 10.1201/9781003522133-5

analyses in the following sections. More general frameworks will be laid out later in Section 6.1. We just need to reduce them to a cylinder geometry. The energy principles and their Euler-Lagrange equations, both in the plasma and vacuum regions, are also described.

5.2.1 ENERGY PRINCIPLE IN PLASMA REGION

In this subsection, we describe the energy principle and its Euler-Lagrange equation in the plasma region in the cylinder geometry with circular cross-section. We derive them by reducing the toroidal formulation to be described in Chapter 6 to the cylindrical one. We only outline the key results. The details can be found in reference [1].

As to be proved in equation (6.36), the energy principle in axisymmetric toroidal geometry can be expressed as

$$\delta W_p + \delta W_k = \frac{1}{2\mu_0}\int d\psi \,(\Xi_\psi'^\dagger \quad \Xi_\psi^\dagger)\begin{pmatrix}\mathscr{F} & \mathscr{K}\\ \mathscr{K}^\dagger & \mathscr{G}\end{pmatrix}\begin{pmatrix}\Xi_\psi'\\ \Xi_\psi\end{pmatrix}, \tag{5.3}$$

with it's Euler-Lagrangian given in equation (6.37)

$$\frac{d}{d\psi}\left(\mathscr{F}\frac{d\Xi_\psi}{d\psi}+\mathscr{K}\,\Xi_\psi\right)-\mathscr{K}^\dagger\frac{d\Xi_\psi}{d\psi}-\mathscr{G}\Xi_\psi=0. \tag{5.4}$$

In the cylinder limit, the matrices in equations (5.3) and (5.4) are reduced to scalars.

Notie that $\psi = r^2/a^2$, $|\nabla\psi|^2 = 4r^2/a^4$, $|\nabla\theta|^2 = 1/r^2$, $\nabla\psi\cdot\nabla\theta = 0$, $\mathscr{J}_g = 1/\nabla\psi\times\nabla\theta\cdot\nabla\phi = Ra^2/2$. We can reduce the unknown Ξ as follows

$$\xi_\psi = \frac{2r}{a^2}\xi_r, \quad \frac{d\xi_\psi}{d\psi} = \frac{d\xi_r}{dr}+\frac{1}{r}\xi_r,$$

One can also reduce the metric parameters: $|\nabla\psi|^2 = 4r^2/a^4$, $|\nabla\theta|^2 = 1/r^2$, $\nabla\psi\cdot\nabla\theta = 0$, $\mathscr{J}_g = 1/\nabla\psi\times\nabla\theta\cdot\nabla\phi = Ra^2/2$, and

$$\mathscr{G}_{11} = \mathscr{J}_g|\nabla\theta\times\nabla\phi|^2 = \frac{1}{2R}, \tag{5.5}$$

$$\mathscr{G}_{22} = \mathscr{J}_g|\nabla\phi\times\nabla\psi|^2 = \frac{2r^4}{Ra^4}, \tag{5.6}$$

$$\mathscr{G}_{33} = \mathscr{J}_g|\nabla\psi\times\nabla\theta|^2 = \frac{Rr^2}{a^4}, \tag{5.7}$$

and $\mathscr{G}_{12} = \mathscr{G}_{31} = \mathscr{G}_{23} = 0$. It is also defined $k_z = n/R$. Now, we reduce the plasma equilibrium parameters in equation (5.3). Using equations (5.5)–(5.7), one obtains

$$\begin{aligned}
\mathscr{F}_p &= \left(\frac{\chi'}{n}\right)^2 \mathscr{Q}^2\mathscr{G}_{33}\left\{1-\frac{m^2\mathscr{G}_{33}}{\mathscr{A}_p}\right\} = \frac{a^2R}{2}\frac{\left(k_zB_z+\frac{m}{r}B_\theta\right)^2}{k_z^2+\frac{m^2}{r^2}},\\
\mathscr{K}_p &= \mathscr{Q}\frac{\chi'}{n}\left\{im\frac{\mathscr{G}_{33}\mathscr{C}_p}{\mathscr{A}_p}-(q\chi')'\mathscr{G}_{33}+g'\right\}\\
&= -\mathscr{J}_g\frac{\mathscr{J}_g}{Rr}\left(k_zB_z+\frac{m}{r}B_\theta\right)\frac{2(m/r)B_\theta/r}{k_z^2+\frac{m^2}{r^2}},\\
\mathscr{G}_p &= \mathscr{H}_p-\frac{|\mathscr{C}_p|^2}{\mathscr{A}_p}\\
&= \frac{\mathscr{J}_g^3}{R^2r^2}\left[-2J_z\frac{B_\theta}{r}+\left(k_zB_z+\frac{m}{r}B_\theta\right)^2+\frac{4(mB_\theta/r)^2}{k_z^2+\frac{m^2}{r^2}}\left(\frac{1}{r}\right)^2\right].
\end{aligned}$$

where

$$J_z = \frac{1}{r}\frac{d}{dr}(rB_\theta) = \frac{dB_\theta}{dr}+\frac{B_\theta}{r}.$$

Using these results, the plasma energy integral in equation (5.3) is reduced to

$$\begin{aligned}
\delta W_p &= \frac{2\pi^2}{\mu_0}\int \mathscr{J}_g d\psi\frac{1}{\mathscr{J}_g}\left[\mathscr{F}_p\left(\frac{d\xi_\psi}{d\psi}\right)^2+2\mathscr{K}_p\frac{d\xi_\psi}{d\psi}\xi_\psi+\mathscr{G}_p\xi_\psi^2\right]\\
&= \frac{2\pi^2X}{\mu_0}\int_0^a rdr\left\{\frac{\left(k_zB_z+\frac{m}{r}B_\theta\right)^2}{k_z^2+\frac{m^2}{r^2}}\left(\frac{d\xi_r}{dr}\right)^2+\frac{k_zB_z+\frac{m}{r}B_\theta}{k_z^2+\frac{m^2}{r^2}}\left(k_zB_z-\frac{m}{r}B_\theta\right)2\frac{d\xi_r}{dr}\frac{\xi_r}{r}\right.\\
&\quad\left.+\left[-2J_z\frac{B_\theta}{r}+\left(k_zB_z+\frac{m}{r}B_\theta\right)^2+\frac{(k_zB_z-(m/r)B_\theta)^2}{k_z^2+\frac{m^2}{r^2}}\left(\frac{1}{r}\right)^2\right]\xi_r^2\right\}\\
&= \frac{2\pi^2X}{\mu_0}\left[\int_0^a dr\left(f\left|\frac{d\xi_r}{dr}\right|^2+g|\xi_r|^2\right)+W_a\right], \qquad (5.8)
\end{aligned}$$

where

$$f = \frac{r\left(k_zB_z+\frac{m}{r}B_\theta\right)^2}{k_z^2+\frac{m^2}{r^2}}, \qquad (5.9)$$

$$\begin{aligned}
g &= -\frac{d}{dr}\frac{(k_zB_z)^2-(mB_\theta/r)^2}{k_z^2+\frac{m^2}{r^2}}-\frac{2B_\theta}{r}\frac{d(rB_\theta)}{dr}+r\left(k_zB_z+\frac{m}{r}B_\theta\right)^2 \qquad (5.10)\\
&\quad+\frac{1}{r}\frac{(k_zB_z-(m/r)B_\theta)^2}{k_z^2+\frac{m^2}{r^2}},
\end{aligned}$$

$$W_a = \left.\frac{(k_zB_z)^2-(mB_\theta/r)^2}{k_z^2+\frac{m^2}{r^2}}|\xi_r|^2\right|_a. \qquad (5.11)$$

The equilibrium parameter g can be further reduced using the cylinder equilibrium equation (5.2), yielding

$$
\begin{aligned}
g &= \frac{2k_z^2}{k_z^2+(m/r)^2}\mu_0\frac{dp}{dr}+r\left(k_zB_z+\frac{m}{r}B_\theta\right)^2\frac{k^2+(m/r)^2-(1/r)^2}{k^2+(m/r)^2} \\
&+\frac{(2k_z^2/r)(k_z^2B_z^2-(m/r)^2B_\theta^2)}{(k_z^2+(m/r)^2)^2}.
\end{aligned} \tag{5.12}
$$

Similarly, the inertia energy in equation (5.3) can be reduced as follows

$$
\begin{aligned}
\delta W_k &= \frac{\gamma_N^2}{2}\frac{B_0^2}{X_0^2}\int \mathscr{J}_g d\phi d\theta d\psi\left\{\xi_s^*\left(\frac{\rho_N}{B^2}|\nabla\psi|^2\right)\xi_s+\xi_\psi^*\left(\chi'^2q^2\frac{\rho_N}{B^2}|\nabla\theta|^2\right)\xi_\psi\right\} \\
&= -\frac{2\pi^2X}{\mu_0}\omega_N^2\int dr\frac{r}{X^2}\left\{\left|\frac{\frac{d\xi_r}{dr}\left(k_zB_\theta-\frac{m}{r}B_z\right)-\frac{\xi_r}{r}\left(k_zB_\theta+\frac{m}{r}B_z\right)}{k_z^2+\frac{m^2}{r^2}}\right|^2+B_z^2|\xi_r|^2\right\} \\
&= -\frac{2\pi^2X}{\mu_0}\omega_N^2\left[\int_0^a dr\left(f_k\left|\frac{d\xi_r}{dr}\right|^2+g_k|\xi_r|^2\right)+W_{ak}\right],
\end{aligned} \tag{5.13}
$$

where

$$
f_k = \frac{r}{X^2}\frac{\left(k_zB_\theta-\frac{m}{r}B_z\right)^2}{\left(k_z^2+\frac{m^2}{r^2}\right)^2}, \tag{5.14}
$$

$$
g_k = \frac{1}{X^2r}\frac{\left(k_zB_\theta+\frac{m}{r}B_z\right)^2}{\left(k_z^2+\frac{m^2}{r^2}\right)^2}+\frac{rB_z^2}{X^2}+\frac{1}{X^2}\frac{d}{dr}\frac{(k_zB_\theta)^2-\left(\frac{m}{r}B_z\right)^2}{\left(k_z^2+\frac{m^2}{r^2}\right)^2}, \tag{5.15}
$$

$$
W_{ak} = -\frac{1}{2X^2}\frac{(k_zB_\theta)^2-\left(\frac{m}{r}B_z\right)^2}{\left(k_z^{+}\frac{m^2}{r^2}\right)^2}|\xi|^2\Bigg|_a. \tag{5.16}
$$

One can find the Euler-Lagrange equation by varying $\delta W_k+\delta W_p$ with respect to ξ_r

$$
\frac{d}{dr}\left(f-\omega_N^2f_k\right)\frac{d\xi_r}{dr}-(g-\omega_N^2g_k)\xi_r=0. \tag{5.17}
$$

These recover the cylinder results, for example, in reference [2]. These results will be used in the next two sections for spectrum and stability analyses.

5.2.2 ENERGY PRINCIPLE IN VACUUM REGION

In this subsection, we describe the energy principle and its Euler-Lagrange equation in the vacuum region in the cylinder geometry with circular cross-section. We derive them by reducing the toroidal formulation to be described later in Chapter 6 to the cylindrical one. We only outline the key results. The details can be found in reference [1]

In equation (6.41) in Section 6.1.4.2, we have proved that the energy principle in the vacuum region in axisymmetric toroidal geometry can be expressed as

$$\delta W_v = \frac{1}{2\mu_0}\int d\psi \begin{pmatrix} \frac{\partial \boldsymbol{\varphi}^\dagger}{\partial \psi} & \boldsymbol{\varphi}^\dagger \end{pmatrix} \begin{pmatrix} \mathscr{F}_v & \mathscr{K}_v \\ \mathscr{K}_v^\dagger & \mathscr{G}_v \end{pmatrix} \begin{pmatrix} \frac{\partial \boldsymbol{\varphi}}{\partial \psi} \\ \boldsymbol{\varphi} \end{pmatrix}, \tag{5.18}$$

where $\mathscr{F}_v = \langle \mathscr{J} |\nabla\psi|^2\rangle$, $\mathscr{K}_v = i\langle \mathscr{J}\nabla\psi\cdot\nabla\theta\rangle\mathscr{M}$, and $\mathscr{G}_v = \mathscr{M}\langle \mathscr{J}|\nabla\theta|^2\rangle\mathscr{M} + n^2\langle \mathscr{J}/X^2\rangle$. Similar to the plasma region treatment, the minimization of the energy integral in equation (5.18) with respect to $\boldsymbol{\varphi}^\dagger$ leads to the following Euler-Lagrange equations in the Fourier space as given in equation (6.42)

$$\frac{d}{d\psi}\left(\mathscr{F}_v \frac{d\varphi}{d\psi} + \mathscr{K}_v\varphi\right) - \mathscr{K}_v^\dagger \frac{d\varphi}{d\psi} - \mathscr{G}_v\varphi = 0. \tag{5.19}$$

Similar to the reductions in the plasma region in Section 5.2.1, one can find the vacuum equilibrium parameters in equation (5.18) can be reduced to

$$\begin{aligned} \mathscr{F}_v &= \langle \mathscr{J}|\nabla\psi|^2\rangle = 2R\frac{r^2}{a^2}, \\ \mathscr{K}_v &= i\langle \mathscr{J}\nabla\psi\cdot\nabla\theta\rangle\mathscr{M} = 0, \\ \mathscr{G}_v &= \mathscr{M}\langle \mathscr{J}|\nabla\theta|^2\rangle\mathscr{M} + n^2\langle \mathscr{J}/X^2\rangle = m^2\frac{Ra^2}{2r^2} + n^2\frac{a^2}{2R}. \end{aligned} \tag{5.20}$$

Therefore, one has

$$\begin{aligned} \delta W_v &= \frac{4\pi^2}{a^2\mu_0}\int dr\, r\left[2R\frac{r^2}{a^2}\frac{1}{4r^2/a^4}\left|\frac{d\varphi}{dr}\right|^2 + \left(m^2\frac{Ra^2}{2r^2} + n^2\frac{a^2}{2R}\right)|\varphi|^2\right] \\ &= \frac{2\pi^2 R}{\mu_0}\int dr\left(r\left|\frac{d\varphi}{dr}\right|^2 + \frac{m^2}{r}|\varphi|^2\right). \end{aligned} \tag{5.21}$$

One can obtain the Euler-Lagrange equation as follows

$$\frac{1}{r}\frac{d}{dr}\left(r\frac{d\varphi}{dr}\right) - \frac{m^2}{r^2}\varphi = 0. \tag{5.22}$$

These recover the cylinder results, for example, in reference [2]. These results will be used in the next two chapters for spectrum and stability analyses.

5.3 MHD SPECTRUM IN CYLINDER GEOMETRY

In this section, we analyze the MHD spectrum in the cylinder geometry. More extensive discussion of this subject can be found in references [3], [4], and [1]. This helps understand how the MHD modes are distributed in the ω space in general. We first outline the mathematical theory for a typical singular differential equation. The discretized unstable modes and continuum spectrum will be discussed next, which also explains the so-called accumulating points. In the end, the initial value problem is described, which explains the phase mixing phenomenon. Although the analyses are based on the cylinder model, the results reflect the fundamental features of the shear Alfvén spectrum in the toroidal geometry.

5.3.1 SINGULAR DIFFERENTIAL EQUATION

To understand the MHD equation feature, it is worthwhile to analyze the singular feature of equation (5.17) in this subsection. Equation (5.17) belongs to a more general type of singular differential equations given as follows

$$\frac{d^2\xi}{dz^2}+\frac{1}{z}p(z)\frac{d\xi}{dz}-\frac{1}{z^2}h(x,\omega)\xi=0. \tag{5.23}$$

which has been studied extensively in many mathematical method books, for example, reference [5]. The connection of equation (5.23) to equation (5.17) can be seen as follows. First, it is noted that this equation can be analytically extended to the entire complex plane.$z=x+iy$. Next, it is noted that in the case with ω being negligible in the first term of equation (5.17), equation (5.23) is related to equation (5.17) by letting $p(z)=2z$; In the case with $\omega\neq 0$, noting that $x^2-\omega^2=(x+\omega)(x-\omega)\approx 2\omega z$ in the vicinity of singular point: $z=0$, equation (5.23) can also be related to equation (5.17) by letting $z=x-\omega$.

In the ideal MHD case, functions $p(z)$ and $h(z)$ are usually analytical and therefore can be expanded in the Taylor series as follows:

$$p(z) = p_0+p_1z+\cdots,\quad h(z)=h_0+h_1z+\cdots.$$

Therefore, in the vicinity of a singular point $z=0$, the solution of equation (5.23) can be derived by the Frobenius method of expansion:

$$\xi = z^\nu\sum_{n=0}^{\infty}a_nz^n. \tag{5.24}$$

Inserting it into equation (5.23) yields

$$\sum_{n=0}^{\infty}(\nu+n)(\nu+n-1)a_nz^{\nu+n-2}+(p_0+p_1z+\cdots)\sum_{n=0}^{\infty}(\nu+n)a_nz^{\nu+n-2}$$
$$-(h_0+h_1z+\cdots)\sum_{n=0}^{\infty}a_nz^{\nu+n-2}=0.$$

One can obtain the recurrence relations by equating the coefficients of the power series:

$$z^{\nu-2} : [\nu^2+(p_0-1)\nu-h_0]a_0=0, \tag{5.25}$$
$$z^{\nu-1} : [(\nu+1)\nu+p_0(\nu+1)-h_0]a_1=(-\nu p_1+h_1)a_0. \tag{5.26}$$
$$\cdots$$

To ensure the existence of a nontrivial solution, the coefficient of a_0 in equation (5.25) must be set to vanish. This yields two index solutions

$$\nu_{1,2} = \frac{1-p_0 \pm \sqrt{(p_0-1)^2+4h_0}}{2}. \tag{5.27}$$

In general, the two indices $\nu_{1,2}$ are different. In this case, two sets of the coefficients a_n can be obtained from equations (5.25) and (5.26) respectively with $\nu_{1,2}$ determined. Therefore, using these coefficients, two independent solutions can be constructed by equation (5.24).

However, there is a degenerate case. When $p_0 = 1$ and $h_0 = 0$, equation (5.27) gives only a single degenerate solution $\nu_1 = \nu_2 = 0$. In this case, the expansion in equation (5.24) only leads to a single solution

$$\xi_1(z) = u(z). \tag{5.28}$$

Note that $u(z)$ is regular in this case. One needs to find another independent solution. This is realized by assuming that

$$\xi_2(z) = u(z)\ln z + v(z) = u(z)\ln z + \sum_{n=0}^{\infty} b_n z^n. \tag{5.29}$$

Plugging this expansion into equation (5.23) and noting that $p_0 = 1$ and $g_0 = 0$, one obtains

$$\begin{aligned}
&\sum_{n=1}^{\infty} n(n-1)b_n z^{n-2} + (1+p_1 z+\cdots)\sum_{n=1}^{\infty} nb_n z^{n-2} - (h_1 z+\cdots)\sum_{n=0}^{\infty} b_n z^{n-2} \\
= \; & -2\sum_{n=1}^{\infty} na_n z^{n-2} - (p_1 z+\cdots)\sum_{n=0}^{\infty} na_n z^{n-2}.
\end{aligned} \tag{5.30}$$

Here, it has been considered that $u(z)$ is the solution to equation (5.23) and consequently the terms containing $\ln z$ vanish. One can determine the coefficients b_n by equating the coefficients of z^n, so that $v(z)$ in equation (5.29) is determined. It is noted that $v(z)$ is also regular. We, therefore, obtain two independent solutions: ξ_1 in equation (5.28) and ξ_2 in equation (5.29). The solutions ξ_1 and ξ_2 are referred to as the small and large solutions, respectively. The general mathematical theory outlined here will be used in the spectrum analyses in the following sections.

5.3.2 UNSTABLE MODES AND ACCUMULATING POINTS

In this subsection, we describe the solutions for unstable modes with the positive growth rate $\gamma = i\omega$ [6]. It explains the so-called accumulating points as shown in figure 5.1, which shows that the eigenvalues are accumulated as ω^2 approaches zero. This corresponds to the case with $h_0 < 0$ in equation (5.23).

We directly work on the singular layer equation

$$\frac{d}{dx}\left[(x^2+\gamma^2)\frac{d\xi}{dx}\right] + g(x)\xi = 0. \tag{5.31}$$

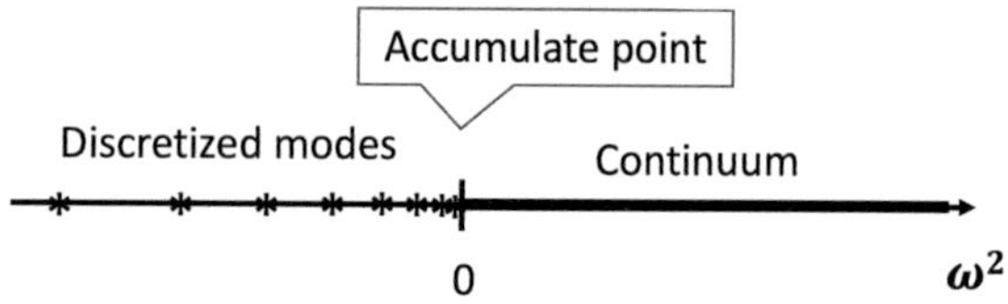

Figure 5.1 MHD mode spectrum in the cylinder geometry.

This corresponds to equation (5.17) with the $\omega_N^2 g_k$ term neglected. In the low frequency or growth rate case, the inertia effect is only important in the field line bending term, the first term in equation (5.31). We use the matching technique to solve this equation. The growth rate γ is assumed to be small. Three regions are considered: two outer regions: $x_0 \leq x \leq 0^+$ and $0^+ \leq x \leq x_a$; and the singular layer: $0^- \leq x \leq 0^+$, where x_0 and x_a represent, respectively, the magnetic axis and plasma boundary, and $x = 0$ is the singular layer.

In the outer regions, the growth rate can be neglected, and equation (5.31) is reduced to

$$\frac{d}{dx}\left(x^2 \frac{d\xi}{dx}\right) + g\xi = 0. \tag{5.32}$$

This is the Euler equation. Its solution can be assumed to be x^ν. Inserting this form of solution into equation (5.32) yields the so-called index equation

$$\nu(\nu + 1) + g = 0.$$

The solution of this index equation is simply

$$\nu = \frac{-1 \pm \sqrt{1-4g}}{2}.$$

The independent solutions in the outer region can then be written as

$$\begin{aligned}\xi_\pm &= \exp\{\nu \ln x\} = x^{-1/2} \exp\left\{\pm \frac{i}{2}(4g-1)^{1/2} \ln x\right\} \\ &= x^{-1/2} \cos\left[\frac{1}{2}(4g-1)^{1/2} \ln x\right] \pm i x^{-1/2} \sin\left[\frac{1}{2}(4g-1)^{1/2} \ln x\right].\end{aligned}$$

They can be used to form the general solution for the outer region

$$\xi = c x^{-1/2} \exp\left\{i\left[\frac{u}{2} \ln x + \lambda\right]\right\} + cc. \tag{5.33}$$

Here, c and λ are constants to be determined by the boundary and matching conditions, $u = (4g-1)^{1/2}$, and "cc" stands for the complex conjugate.

In the singular layer, one needs to include the inertia effect to consider the complete singular layer equation (5.31). Using the slow-varying variable $\tau = x/\omega$, equation (5.31) becomes

$$\frac{d}{d\tau}\left[(\tau^2+1)\frac{d\xi}{d\tau}\right] + g\xi = 0. \tag{5.34}$$

There are two independent solutions in the inner regions. To discuss the eigenmode feature, one does not need to work out the solution. One can expect that in the outer limit $\tau \to \infty$ the inner solution has the following form, as compared to the outer solutions in equation (5.33),

$$\xi = \alpha\tau^{-1/2}\exp\left\{i\left[\frac{u}{2}\ln\tau + \eta\right]\right\} + cc.$$

with α and η are constants.

Now, we match the inner solution with the outer solutions. Since it is a second-order differential equation, the matching requires the continuations of the perturbation and its derivative. Here, it is noted that the asymptotic matching is different from the exact solution matching. In the asymptotic matching, one needs to have a common region in which both solutions overlap, instead of a point matching. This is assured as equation (5.34) converges to equation (5.31) in the limit $\tau \to \infty$. On the left-hand side $x = 0^-$, one obtains

$$c_- = \alpha|\omega|^{1/2}, \tag{5.35}$$

$$\lambda_- = \eta - \frac{u}{2}\ln|\omega| - j_-\pi. \tag{5.36}$$

Here, j_- is an integer. Similarly, on the right hand side $x = 0^+$, one has

$$c_+ = \alpha|\omega|^{1/2}, \tag{5.37}$$

$$\lambda_+ = \eta - \frac{u}{2}\ln|\omega| - j_+\pi. \tag{5.38}$$

Here, j_+ is an integer. Note that there is a boundary condition in each outer region: $r = 0$ and $r = a$. This is a set of six homogeneous conditions for determining six constants: $\lambda_\pm$, $\eta_\pm$, and $c_\pm$. Their solubility condition gives rise to the so-called dispersion relation. Note that equations (5.36) and (5.38) can be transformed to

$$|\omega| = \exp\left\{\frac{2}{u}(\eta - \lambda_- - j_-\pi)\right\},$$

$$|\omega| = \exp\left\{\frac{2}{u}(\eta - \lambda_+ - j_+\pi)\right\}.$$

Equating them yields

$$\lambda_+ - \lambda_- = -(j_+ - j_-)\pi. \tag{5.39}$$

Multiplying them, one obtains

$$\omega^2 = \exp\left\{\frac{2}{u}(2\eta - \lambda_+ - \lambda_- - j\pi)\right\}. \tag{5.40}$$

where $j = j_+ + j_-$. Therefore, equations (5.35) and (5.37), (5.39) and (5.40) constitute a new set of equations. Their solubility condition determines the eigenfrequencies.

It is interesting to point out that the dispersion relation in equation (5.40) allows multiple roots with the possible value of the integer $j \to \infty$. As j becomes larger and larger, the difference between the nearby eigenfrequencies becomes smaller and smaller. The unstable eigenfrequencies accumulate at $\omega^2 \to 0$ as shown in figure 5.1.

5.3.3 CONTINUUM SPECTRUM

In this subsection, we describe the MHD continuum theory in the real space description [3, 4]. This explains the continuum part in figure 5.1. It shows how the Alfvén singularity in the MHD equation leads to a continuum frequency spectrum.

We start with the singular layer equation for Alfvén modes, for instance, equation (5.17),

$$\frac{d}{dx}\left(\omega^2 - \omega_p^2(x)\right)\frac{d\xi}{dx} - k^2\left(\omega^2 - \omega_p^2(x)\right)\xi = 0. \tag{5.41}$$

Here, k is a constant. This is similar to the electrostatic case in reference [3] except $\omega_p(x)$ here corresponds to the field line bending effects in the presence of magnetic shear. Equation (5.41) is singular when $\omega_p(0) = \omega$. Without losing generality, we assume that the singular point is located at $x = 0$ here. It is further assumed that $\omega_p(x)$ is monotonic in the domain: $(0^-, 0^+)$. In this case equation (5.41) can be written as follows

$$(F\xi')' - G\xi = 0, \quad \text{with} \quad F = F_1 x + \cdots \quad \text{and} \quad G = G_1 x + \cdots. \tag{5.42}$$

Comparing to equation (5.23), equation (5.42) corresponds to the case with $p = 1$ and $g_0 = 0$. Thus, its two independent solutions are as follows:

$$\xi_1 = u(x,\omega), \tag{5.43}$$

$$\xi_2 = u(x,\omega)\ln|x| + v(x,\omega). \tag{5.44}$$

The independent solutions in equations (5.43) and (5.44) can be used to construct the general solution

$$\xi = [A_1 u + C_1(u\ln|x| + v)]H(-x) + [A_2 u + C_2(u\ln|x| + v)]H(x), \tag{5.45}$$

where $H(x)$ is the Heaviside step function

$$H(x) = \begin{cases} 0, & x < 0; \\ 1/2, & x = 0; \\ 1, & x > 1, \end{cases}$$

where $A_{1,2}$ and $C_{1,2}$ are constants to be determined by the boundary conditions at both ends and the connection conditions at $x = 0$.

In the non-degenerate case, the two boundary conditions at each end and two connection conditions at $x = 0$ constitute a set of linear homogeneous equations for determining the four constants: $A_{1,2}$ and $C_{1,2}$. The solubility condition for the set of homogeneous equations constitutes the dispersion relation, which determines the eigenmodes and eigenvalues ω. This is the usual fully deterministic case for discretized eigenvalues. However, it is noted that, because equation (5.42) is singular, both $f_\pm = uH(\pm x)$ are the global solutions, leading the problem to become underdetermined, i.e., there is one more freedom in determining the four constants. It can be proved using the Heaviside and delta function properties: $H'(x) = \delta(x)$ and $x\delta(x) = 0$. Noting that

$$\begin{aligned} f'_\pm &= u'H(\pm x) \pm u\delta(x), \\ Ff'_\pm &= Fu'H(\pm x) \pm F\delta(x) = Fu'H(\pm x), \\ (Ff'_\pm)' &= (Fu')'H(\pm x) \pm Fu'\delta(x) = (Fu')'H(\pm x), \end{aligned}$$

one has

$$(Ff'_\pm)' - Gf_\pm = \left[(Fu')' - Gu\right] H(\pm x) = 0. \tag{5.46}$$

Here, it has been noted that u is the solution of equation (5.42). Equation (5.46) shows that both $uH(\pm x)$ are the global solutions valid both in the $x \leq 0$ and $x \geq 0$ regions. One therefore does not need to match $A_1 u$ and $A_2 u$. The only requirement is to match C_1 and C_2, leading to $C_1 = C_2 = C$. The general solution in equation (5.45), after the continuity condition at $x = 0$ is imposed, becomes

$$\xi = [A_1 u + C(u \ln|x| + v)]\, H(-x) + [A_2 u + C(u \ln|x| + v)]\, H(x). \tag{5.47}$$

With the general solution in equation (5.47) obtained, one can discuss the determination of the solution. Equation (5.47) contains three constants $A_{1,2}$ and C. However, only two boundary conditions are required to be satisfied at each end. Because there is an extra constant, a solution can always be constructed for any given frequency in the domain $\omega_- < \omega < \omega_+$. It is different from the regular eigenvalue case, where two constants are required to satisfy two boundary conditions, leading to a dispersion relation that determines the discretized eigenfrequency. Therefore, it is referred to as the continuum spectrum.

One may also perform the complex analysis, which can be found, for example, in reference [1].

5.3.4 INITIAL VALUE PROBLEM: PHASE MIXING

In this subsection, we will review the theory for an initial value problem [4, 7, 8]. The initial values, which are left out in the Fourier transformation, are picked up in the Laplace transform. Consequently, the phase-mixing phenomenon is revealed. Since the detail has been described in reference [1], we only briefly outline the procedure.

Corresponding to the MHD framework in the cylinder model described in subsection 5.2.1, the following initial value problem is considered

$$\frac{d}{dx}\tilde{p}(x,t)\frac{d\tilde{\xi}(x,t)}{dx} - \tilde{q}(x,t)\tilde{\xi}(x,t) = 0. \tag{5.48}$$

Here, to be specific, it is assumed that $\tilde{p}(x,t) = x^2 + \partial^2/\partial t^2$ and the boundary conditions are assumed to be homogeneous

$$B_{1,2}(\xi(x_{1,2})) = \alpha_{1,2}\xi(x_{1,2}) + \beta_{1,2}\xi'(x_{1,2}) = 0, \tag{5.49}$$

where x_1 and x_2. denote the left and right boundaries, respectively.

Let us introduce the Laplace transform:

$$\xi(x,\omega) \;=\; \int_0^\infty \tilde{\xi}(x,t)e^{-st}dt = \int_0^\infty \tilde{\xi}(x,t)e^{i\omega t}dt, \tag{5.50}$$

with the inversion

$$\tilde{\xi}(x,t) \;=\; \frac{1}{2\pi i}\int_{C_B} \xi(x,s)e^{st}ds = -\frac{1}{2\pi}\int_{C_B} \xi(x,\omega)e^{-i\omega t}d\omega. \tag{5.51}$$

Here, $s = -i\omega = -i(\omega_r + i\gamma)$ and "C_B" represents the Bromwich contour in the complex s space as shown in figure 5.2. Using the Laplace transformation in equation (5.50), equation (5.48) becomes

$$\mathscr{L}_\omega\{\xi(z,\omega)\} \equiv \frac{d}{dz}p(z,\omega)\frac{d\xi(z,\omega)}{dz} - q(z,\omega)\,\xi(z,\omega) = \Xi(z), \tag{5.52}$$

where $p(z,\omega) = z^2 - \omega^2$ and

$$\Xi(x) \;\equiv\; \frac{d}{dx}\left(s\frac{d\xi(x,0)}{dx} + \frac{d\xi_t'(x,0)}{dx}\right) + s\xi(x,0) + \xi_t'(x,0). \tag{5.53}$$

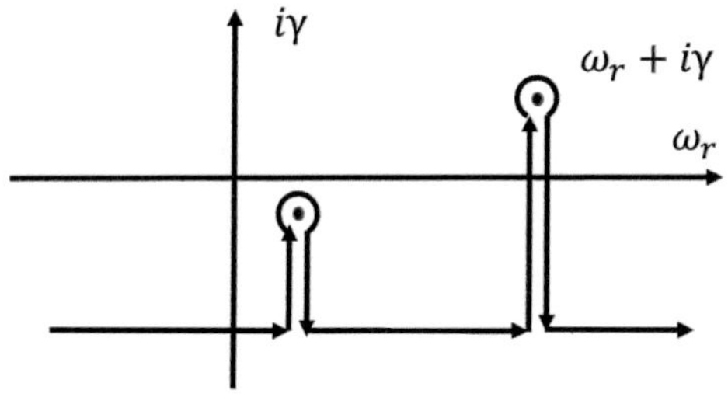

Figure 5.2 Browmwich contour.

One can solve equation (5.52) using the Green's function method with the corresponding Green's function $g(z,\tau)$ satisfying the following equation [9]

$$-\mathscr{L}_\omega\{g\} = (pg'_z)'_z - qg = -\delta(z-\tau). \tag{5.54}$$

and the boundary conditions in equation (5.49). With the Green's function, the solution of equation (5.52) can be expressed as

$$\xi(z) = \int \Xi(\tau) g(z,\tau) d\tau. \tag{5.55}$$

Introduce the so-called Wronskian

$$\Delta(w_1, w_2) \equiv p(z)(w'_1(z)w_2(z) - w'_2(z)w_1(z)).$$

where $w_{1,2}$ are the homogeneous solutions to equation (5.52), which satisfy respectively the left and right boundary conditions in equation (5.49)

$$B_{1,2}(w_{1,2}) = \alpha_{1,2}w_{1,2} + \beta_{1,2}w'_{1,2} = 0. \tag{5.56}$$

Noting that

$$\begin{aligned} 0 &= \int_a^b \left\{ w_2\left[(pw'_1)' - qw_1\right] - w_1\left[(pw'_2)' - qw_2\right]\right\} dz \\ &= -\left[w_2pw'_1 - w_1pw'_2\right]_a^b = \Delta(a) - \Delta(b) \end{aligned}$$

for two arbitrary and distinctive points a and b, one can conclude that $\Delta'_z = 0$. With the Wronskian, one can prove directly that the Green's function is given as follows [9].

$$g(z,\tau) = \frac{w_1(z)w_2(\tau)H(\tau - z) + w_2(z)w_1(\tau)H(z-\tau)}{\Delta(w_2, w_1)}. \tag{5.57}$$

From equation (5.55), the solution for equation (5.52) can be obtained with the Green's function in equation (5.57)

$$\xi(z,\omega) = \frac{w_1(z)\int_z^{x_{b2}} \Xi(\tau)w_2(\tau)d\tau + w_2(z)\int_{x_{b1}}^z \Xi(\tau)w_1(\tau)d\tau}{\Delta(w_2,w_1)}. \tag{5.58}$$

It can be proved directly that ξ in equation (5.58) satisfies the boundary conditions in equation (5.56), noting that $w_{1,2}$ are constructed to satisfy the respective boundary conditions in equation (5.56). We can now make the inverse Laplace transform in equation (5.51) to obtain the initial value solution

$$\tilde{\xi}(z,t) = \int_{C_B} d\omega e^{-i\omega t} \frac{w_1(z)\int_z^{z_{b2}} \Xi(\tau)w_2(\tau)d\tau + w_2(z)\int_{x_{b1}}^z \Xi(\tau)w_1(\tau)d\tau}{\Delta(w_2,w_1)}. \tag{5.59}$$

Note that the phase mixing is a long-term behavior. For this, we just need to study the pole contributions along the Bromwich contour: C_B in figure 5.2. Similar to the

Landau damping case, the contributions from other parts of the contour vanish as $t \to \infty$. The long-term contributions derive from multiple parts. We examine them one by one.

The first is the pole contribution from $\Delta(\omega) = 0$. Note that $\Delta(\omega) = 0$ is actually the dispersion relation governing the eigenfrequency ω_{eg} for quasi modes. Since the ideal-MHD eigenvalues ω^2 have to be real, complex eigenvalues do not satisfy the causality condition as discussed in reference [1]. The expansion at the pole $\Delta(\omega_{eg}) = 0$ leads to $\Delta(\omega) = \Delta'_\omega(\omega - \omega_{eg})$. Using Cauchy's residue theorem, one obtains the asymptotic behavior for this type of poles

$$\begin{aligned}\tilde{\xi}(z,t) &= e^{-i\omega_{eg}t} \\ &\times \frac{w_1(z,\omega_{eg})\int_z^{x_{b2}} \Xi(\tau) w_2(\tau,\omega_{eg}) d\tau + w_2(z,\omega_{eg})\int_{x_{b1}}^z \Xi(\tau) w_1(\tau,\omega_{eg}) d\tau}{\Delta'_\omega}.\end{aligned}$$

Nevertheless, as discussed earlier, this solution can only be regarded as the quasi-modes.

To study the remaining contributions, we carry out the integration by part to equation (5.59)

$$\begin{aligned}\tilde{\xi}(z,t) = \frac{1}{it}\int d\omega e^{-i\omega t}\frac{\partial}{\partial\omega} \\ \times \left\{ \frac{w_1(z,\omega)\int_z^{x_{b2}} \Xi(\tau) w_2(\tau,\omega) d\tau + w_2(z,\omega)\int_{x_{b1}}^z \Xi(\tau) w_1(\tau,\omega) d\tau}{\Delta(w_2,w_1)} \right\}.\end{aligned} \tag{5.60}$$

There are three parts of contributions from $\partial/\partial\omega$ in the integrand. The contribution from $\Delta(w_2, w_1)$ in the denominator relating to the quasi-modes has been discussed earlier. Only two contributions: $\partial w_{1,2}(z,\omega)/\partial\omega$ outside the inner integration and $\partial w_{1,2}(\tau,\omega)/\partial\omega$ inside the integrations, remain. It is proved in subsection 5.3.1 that the homogeneous solution at the singular point behaves as $w_{1,2}(x,\omega) \sim (a_{1,2})\ln(z-\omega)$, so that

$$\frac{\partial}{\partial\omega}w_{1,2}(z,\omega) \to \frac{a_{1,2}}{z-\omega}, \tag{5.61}$$

where $a_{1,2}$ are constants. Note that the contribution from $\partial w_{1,2}(\tau,\omega)/\partial\omega$ inside the τ integration is of order $\mathcal{O}(\ln(z-\omega))$, which is much smaller than the $\mathcal{O}(1/(z-\omega))$ contribution from $\partial w_{1,2}(z,\omega)/\partial\omega$ outside the τ integration. Therefore, the contribution from $\partial w_{1,2}(z,\omega)/\partial\omega$ outside the inner τ integration needs to be obtained.

Using Cauchy's residue theorem and equation (5.61), one can reduce equation (5.60) to

$$\begin{aligned}\tilde{\xi}(x,t) &= \frac{1}{it}\int d\omega e^{-i\omega t}\frac{a_1\int_x^{x_{b2}} \Xi(\tau) w_2(\tau,\omega) d\tau + a_2\int_{x_{b1}}^x \Xi(\tau) w_1(\tau,\omega) d\tau}{(x-\omega)\Delta(w_2,w_1)|} \\ &\to \frac{1}{it}e^{-ixt}\frac{a_1\int_x^{x_{b2}} \Xi(\tau) w_2(\tau,x) d\tau + a_2\int_{x_{b1}}^x \Xi(\tau) w_1(\tau,x) d\tau}{\Delta(w_2,w_1)|_{\omega=x}}.\end{aligned} \tag{5.62}$$

This shows that the initial perturbations are damped in $1/t$ and their phase is xt. As time evolves, the solution becomes oscillating with higher and higher frequency in the configuration space, with its amplitude reducing in time. This is referred to as the continuum phase mixing.

5.4 IDEAL MHD STABILITY THEORY

With the MHD mode spectrum addressed in the last section, we start to study the MHD mode stability in the cylinder geometry in this section. It helps to understand the ideal MHD stability in toroidal geometry, as described later in Chapter 6 and the resistive MHD theory in Chapter 7. We first describe the localized MHD modes and derive Suydam's criterion. Then, both the internal and external modes are discussed. One may also find these theories, for example, in reference [10].

5.4.1 SUYDAM'S CRITERION

In this subsection, we derive the Suydam criterion [11], which corresponds to the Mercier criterion in the toroidal geometry. The Suydam criterion is related to the so-called localized interchange modes.

We first analyze the orderings of the integrand in the energy principle in equation (5.8). It can be proved that the orderings of the first and second terms are given as follows

$$f\left|\frac{d\xi_r}{dr}\right|^2 : g\left|\xi_r\right|^2 \sim B_\theta^2 r\left(\frac{d}{dr}\right)^2 : B_\theta^2\frac{1}{r}.$$

For localized modes, one has $d/dr \gg 1/r$. Therefore, the first term in the energy principle in equation (5.8) is much larger than the second term. Note that the first term is stabilizing. To minimize the energy, we then consider the mode to be localized at the so-called resonance surface at $r = r_s$ where

$$k_z B_z + \frac{m}{r}B_\theta = 0.$$

To study the localized modes, the localized coordinate $x = (r - r_s)/r_s$ is introduced, where $x \ll 1$. The definitions of the safety factor $q = rB_z/RB_\theta$ and magnetic shear $s = (r_s/q)(dq/dr)$ are also introduced. Therefore, one has

$$\begin{aligned}
f\left|\frac{d\xi_r}{dr}\right|^2 &= \frac{r}{n^2/R^2+m^2/r^2}\left(k_z B_z + \frac{m}{r}B_\theta\right)^2\left|\frac{d\xi_r}{dr}\right|^2 \\
&= \frac{r}{n^2/R^2+m^2/r^2}\left(\frac{m}{r}B_\theta\right)^2\left(-\frac{n}{m}\frac{rB_z}{RB_\theta}+1\right)^2\left|\frac{d\xi_r}{dr}\right|^2 \\
&= \frac{1}{r_s(n^2/R^2+m^2/r_s^2)}\left(\frac{m}{r_s}B_\theta\right)^2\left[-\frac{n}{m}q_s(1+sx)+1\right]^2\left|\frac{d\xi_r}{dx}\right|^2 \\
&= \frac{1}{r_s(n^2/R^2+m^2/r_s^2)}\left(\frac{m}{r_s}B_\theta\right)^2 s^2x^2\left|\frac{d\xi_r}{dx}\right|^2,
\end{aligned}$$

where $k_z = -n/R$ and $q_s = m/n$ have been used. One can prove that the second and the terms in the definition of g in equation (5.12) are negligible assuming that $d\xi_r/dx \gg m\xi_r$. Therefore,

$$
\begin{aligned}
g|\xi_r|^2 &= \frac{2k_z^2}{k_z^2+(m/r)^2}\mu_0\frac{dp}{dr}|\xi_r|^2 \\
&= \frac{1}{r_s(n^2/R^2+m^2/r_s^2)}\left(\frac{m}{r_s}B_\theta\right)^2\left(\frac{n^2}{m^2}\frac{r_s^2B_z^2}{R^2B_\theta^2}\right)2\mu_0\frac{r_s}{B_z^2}\frac{dp}{dr}|\xi_r|^2 \\
&= \frac{1}{r_s(n^2/R^2+m^2/r_s^2)}\left(\frac{m}{r_s}B_\theta\right)^2(-\alpha)|\xi_r|^2,
\end{aligned}
$$

where $\alpha = -(2\mu_0 r_s/B_z^2)(dp/dr)$. The plasma energy integral in equation (5.8) is reduced to

$$
\delta W_p = U\int_{\Delta x} dx\left(s^2x^2\left|\frac{d\xi_r}{dr}\right|^2 - \alpha|\xi_r|^2\right), \tag{5.63}
$$

where Δ_x denotes the vicinity of resonance surface and

$$
U = \frac{2\pi^2X}{\mu_0}\frac{1}{n^2/R^2+m^2/r_s^2}\left(\frac{m}{r_s}B_\theta\right)^2.
$$

There are two methods to derive the Suydam criterion from the energy expression in equation (5.63). We first describe the method using the Swartz inequality:

$$
\int\left|x\frac{d\xi_r}{dx}\right|^2 dx\int|\xi_r|^2dx \geq \left|\int x\frac{d\xi_r}{dx}\xi_r^* dx\right|^2 = \frac{1}{4}\int|\xi_r|^2dx.
$$

With this inequality, the energy principle is reduced to

$$
\delta W_p \geq U\left(\frac{s^2}{4}-\alpha\right)\int_{\Delta x}|\xi_r|^2dx,
$$

This yields the so-called Suydam criterion for stability.

$$
\frac{1}{4}-\frac{\alpha}{s^2}\geq 0. \tag{5.64}
$$

This shows that the stability is determined by the balance of the destabilizing drive from the pressure gradient and stabilizing effect from the magnetic shear.

The stability criterion in equation (5.64) can be alternatively derived by the Newcomb theorem [12]. Variation of the energy principle in equation (5.63) leads to the Euler-Lagrangian equation

$$
\frac{d}{dx}\left(x^2\frac{d\xi}{dx}\right)+\frac{\alpha}{s^2}\xi = 0. \tag{5.65}
$$

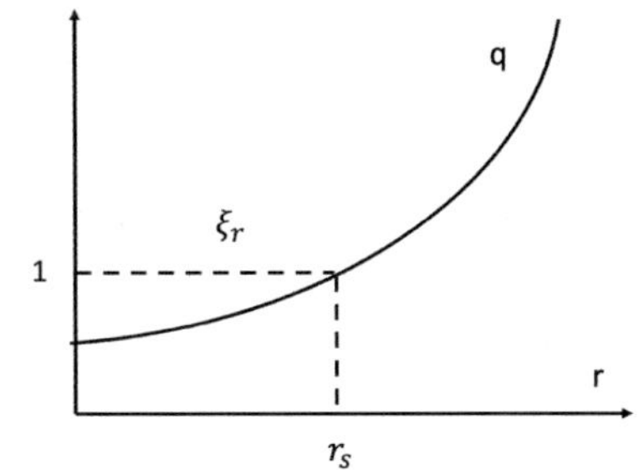

Figure 5.3 Safety factor profile and $n = 1$ internal kink mode.

This is the Euler differential equation with the solution

$$\xi = \xi_0 x^{-\frac{1}{2} \pm \frac{1}{2}\sqrt{1-4D_s}}, \tag{5.66}$$

where ξ_0 is a constant and

$$D_s = \alpha / s^2. \tag{5.67}$$

The Newcomb theorem 5 states that the system is unstable when the solution of the Euler-Lagrangian equation crosses twice the zero [12]. Apparently, to avoid the solution equation (5.66) from becoming oscillatory, which leads the solution to cross the zero multiply, one has a stability condition in equation (5.64).

5.4.2 INTERNAL KINK MODES

In this subsection, we describe the $n = 1$ internal kink mode theory in the cylinder model developed in this field [13, 14]. This is related to the tokamak plasma disruption and sawtooth oscillations [15]. Here, we only review the linear theory. The nonlinear theory [16] will be described in Section 9.4.

The cylinder equilibrium with circular cross-section is considered. As shown in figure 5.3, the safety factor q is assumed to increase monotonically with $q < 1$ in the core region and a rational surface r_s for $n = 1$ mode in the plasma region, where the safety factor $q = 1$.

The energy principle in the cylinder model has been given in equations (5.8) and (5.13)

$$\delta W_p = \frac{2\pi^2 B_z^2}{\mu_0 X} \int_0^r dr \left\{ \left[\left| r\frac{d\xi_r}{dr} \right|^2 + (m^2-1)\xi^2 \right] \left(\frac{n}{m} - \frac{1}{q} \right)^2 r + g|\xi_r|^2 \right\}, \tag{5.68}$$

$$\delta W_k = -\frac{2\pi^2 B_z^2}{\mu_0 X} \frac{\omega_N^2}{m^2} \int_0^a dr \left(r^3 \left| \frac{d\xi_r}{dr} \right|^2 + g_k|\xi_r|^2 \right) \tag{5.69}$$

with g and g_k being in equations (5.12) and (5.15). We consider the $n = m = 1$ modes. To construct the eigenvalue problem, three regions are considered: the inner region

$r < r_s$, the singular layer at r_s, and the outer region $r > r_s$. The solutions in each region are then matched together to form the eigenvalue problem for the global solution.

We first consider the inner region $r < r_s$. The inertial energy is negligible in this region, and one can then minimize the potential energy in equation (5.68) alone. The potential energy can be minimized by the solution of the Euler-Lagrange equation

$$\frac{d}{dr}\left[r^3\left(\frac{n}{m}-\frac{1}{q}\right)^2\right]\frac{d\xi}{dr}+g\xi=0. \tag{5.70}$$

Carrying out the integration by parts and using the solution of equation (5.70), the potential energy in equation (5.68) is reduced to

$$\begin{aligned}\delta W_{p,min} &= \frac{2\pi^2 B_z^2}{\mu_0 X}\int_0^r dr\xi\left\{-\frac{d}{dr}\left[r^3\left(\frac{n}{m}-\frac{1}{q}\right)^2\right]\frac{d\xi}{dr}+g\xi\right\} \\ &\quad +\frac{2\pi^2 B_z^2}{\mu_0 X}\xi r^3\left(\frac{n}{m}-\frac{1}{q}\right)^2\frac{d\xi_r}{dr}\bigg|_{r=r_s} \\ &= \frac{2\pi^2 B_z^2}{\mu_0 X}\xi r^3\left(\frac{n}{m}-\frac{1}{q}\right)^2\frac{d\xi_r}{dr}\bigg|_{r=r_s}.\end{aligned} \tag{5.71}$$

The lowest order solution of the Euler-Lagrange equation (5.70) is

$$\xi_r^{(0)} = const., \quad \text{for} \quad r < r_s. \tag{5.72}$$

This lowest-order solution is shown in figure 5.3. Nevertheless, note here that the potential energy is not simply $\frac{2\pi^2 B_z^2}{\mu_0 X}|\xi_r^{(0)}|^2\int_0^r drg$. There is a remaining contribution from the first term on the right-hand side of equation (5.68) due to the next order solution $\xi_r^{(1)}$. Equation (5.71) yields

$$\frac{d\xi_r^{(1)}}{dr}\bigg|_{r=r_s} = \frac{\mu_0 X}{2\pi^2 B_z^2}\frac{\delta W_{p,min}^{(2)}}{\xi_r^{(0)} r_s^3 s^2 x^2}. \tag{5.73}$$

Here, $\delta W_{p,min}^{(2)}$ represents the potential energy after minimization with the solution of the Euler-Lagrange equation (5.70) with the boundary conditions imposed, $(n/m-1/q)^2 = s^2x^2$, $q = 1+sx$, $s = rq'/q$, and $x = (r-r_s)/r_s$.

We next consider the singular layer. In the singular layer, one needs to include the kinetic energy in equation (5.69). In the singular layer, only the highest-order derivative contribution needs to be kept. Consequently, the Euler-Lagrange equation in the singular layer becomes

$$\frac{d}{dr}\left\{r^3\left[\left(1-\frac{1}{q}\right)^2-\omega_N^2\right]\right\}\frac{d\xi}{dr}=0.$$

Using the local coordinate x, the magnetic shear s, and the normalized growth rate γ_N, this equation is reduced to

$$\frac{d}{dx}(s^2x^2+\gamma_N^2)\frac{d\xi}{dx}=0.$$

This equation can be solved readily

$$\xi = c_0\left[\arctan\left(\frac{sx}{\gamma_N}\right)+c_1\right], \tag{5.74}$$

where $a_{0,1}$ are constants.

Now, we can match the layer solution in equation (5.74) to the solutions in the inner ($r<r_s$) and outer ($r>r_s$) regions. The outer-region solution is simply $\xi\to 0$. This leads to $c_1=-\pi/2$ in equation (5.74) after the matching. Taking the inner limit of the singular layer solution in equation (5.74) yields

$$\xi|_{x\to-\infty} = -c_0\pi, \quad \left.\frac{d\xi}{dx}\right|_{x\to-\infty}=-c_0\frac{\gamma_N}{sx^2}.$$

Matching them to the inner solution in the limit approaching the singular layer in equations (5.72) and (5.73), one obtains $c_0=-\xi_0/\pi$ and the normalized growth rate

$$\gamma_N = -\frac{\mu_0 X}{2\pi B_z^2}\frac{\delta W_{p,min}^{(2)}}{\left|\xi_r^{(0)}\right|^2 r_s^3}. \tag{5.75}$$

Note here that one needs to require $\delta W_{p,min}^{(2)}<0$ for causality. Equation (5.75) gives the growth rate for the $n=1$ internal kink modes in the cylinder model.

5.4.3 EXTERNAL KINK MODES

Now, we consider the external kink modes in this subsection [13, 15]. It is needed to consider the coupling to the vacuum region in this case. We consider the cases without a wall and with a perfectly conducting wall. The resistive wall mode theory can be found in reference [10], which will also be reviewed in the toroidal configuration in Section 6.4.

Let us first reduce the plasma energy in equation (5.8). Considering the case with large aspect ratio $a/R\ll 1$ and low beta $\alpha\ll 1$, the equilibrium functions f and g in equations (5.9) and (5.12) are reduced to

$$f = \frac{r^3B_z^2}{R^2}\left(\frac{n}{m}-\frac{1}{q}\right)^2,$$

$$g = \frac{rB_z^2}{R^2}\left(\frac{n}{m}-\frac{1}{q}\right)^2\left(m^2-1\right).$$

This leads to

$$
\begin{aligned}
&\int_0^a dr\left(f\left|\frac{d\xi_r}{dr}\right|^2+g\left|\xi_r\right|^2\right)\\
=\ &\frac{B_z^2}{R^2}\int_0^a\left(\left|r\frac{d\xi_r}{dr}\right|^2+\left(m^2-1\right)\left|\xi_r\right|^2\right)\left(\frac{n}{m}-\frac{1}{q}\right)^2 rdr.
\end{aligned}
$$

One can also reduce the surface contribution in equation (5.11) and obtains

$$
\begin{aligned}
W_a &= \frac{B_z^2}{R^2}\left[\left(\frac{n}{m}\right)^2-\left(\frac{1}{q}\right)^2\right]\left|\xi_r\right|^2 a^2\\
&= \frac{B_z^2}{R^2}\left[\frac{2}{q}\left(\frac{n}{m}-\frac{1}{q}\right)^2+\left(\frac{n}{m}-\frac{1}{q}\right)^2\right]a^2\left|\xi_r\right|^2.
\end{aligned}
$$

Finally, the plasma energy in equation (5.8). is reduced to

$$
\begin{aligned}
\delta W_p &= \frac{2\pi^2 B_z^2}{\mu_0 R}\left\{\int_0^a\left(\left|r\frac{d\xi_r}{dr}\right|^2+\left(m^2-1\right)\left|\xi_r\right|^2\right)\left(\frac{n}{m}-\frac{1}{q}\right)^2 rdr\right.\\
&\left.+\left[\frac{2}{q}\left(\frac{n}{m}-\frac{1}{q}\right)^2+\left(\frac{n}{m}-\frac{1}{q}\right)^2\right]a^2\left|\xi_r\right|^2\right\}.
\end{aligned}
\tag{5.76}
$$

We next reduce the vacuum energy in equation (5.21). It can be reduced to

$$
\begin{aligned}
\delta W_v &= -\frac{2\pi^2 R}{\mu_0}\int dr\left(r\left|\frac{d\varphi}{dr}\right|^2+\frac{m^2}{r}\left|\varphi\right|^2\right).\\
&= -\frac{2\pi^2 R}{\mu_0}\int dr\varphi^*\left(\frac{d}{dr}\left(r\frac{d\varphi}{dr}\right)-\frac{m^2}{r}\varphi\right)+\frac{2\pi^2 R}{\mu_0}r\,\varphi^*\frac{d\varphi}{dr}\bigg|_a^b.
\end{aligned}
\tag{5.77}
$$

Its Euler-Lagrangian equation is given in equation (5.22) as follows

$$
\frac{1}{r}\frac{d}{dr}\left(r\frac{d\varphi}{dr}\right)-\frac{m^2}{r^2}\varphi=0.
$$

The vacuum energy can be reduced by the solution of the Euler-Lagrangian equation. It is solvable in this case. Suppose there is a perfectly conducting wall at $r=b$, where $B_r=-\partial\varphi/\partial r=0$. The solution of the Euler-Lagrangian equation is as follows

$$
\varphi = d\left(\frac{r^m}{b^{m-1}}+\frac{b^{m+1}}{r^m}\right), \tag{5.78}
$$

$$
\frac{d\varphi}{dr} = md\left(\frac{r^{m-1}}{b^{m-1}}-\frac{b^{m+1}}{r^{m+1}}\right), \tag{5.79}
$$

where d is constant. The vacuum solution needs to match the solution in the plasma region at their interface. The radial magnetic field in the plasma edge $r = a$ is as follows

$$b_r = \nabla r \cdot \nabla \times \boldsymbol{\xi} \times \mathbf{B} = \mathbf{B} \cdot \nabla \xi_r$$
$$= -\frac{mB_z}{R}\left(\frac{n}{m} - \frac{1}{q}\right)\xi_r.$$

Matching it to the vacuum solution in equation (5.79) yields

$$d = \frac{B_z}{R}\frac{1}{\frac{a^{m-1}}{b^{m-1}} - \frac{b^{m+1}}{a^{m+1}}}\left(\frac{n}{m} - \frac{1}{q}\right)\xi_r. \tag{5.80}$$

The vacuum solution in equation (5.78) thus becomes

$$\varphi = \frac{B_z}{R}\frac{\frac{r^m}{b^{m-1}} + \frac{b^{m+1}}{r^m}}{\frac{a^{m-1}}{b^{m-1}} - \frac{b^{m+1}}{a^{m+1}}}\left(\frac{n}{m} - \frac{1}{q}\right)\xi_r. \tag{5.81}$$

Using this solution of the Euler-Lagrange equation, the vacuum energy in equation (5.77) becomes

$$\delta W_v = \frac{2\pi^2 B_z^2}{\mu_0 R} m\lambda \left(\frac{n}{m} - \frac{1}{q}\right)^2 a^2 |\xi_r|^2, \tag{5.82}$$

where

$$\lambda = \frac{1 + (b/a)^{2m}}{1 - (b/a)^{2m}}.$$

Finally, combining equations (5.76) and (5.82), the total potential energy for external kink modes becomes

$$\delta W_{tot} = \frac{2\pi^2 B_z^2}{\mu_0 R}\left\{\int_0^a \left(\left|r\frac{d\xi_r}{dr}\right|^2 + (m^2-1)|\xi_r|^2\right)\left(\frac{n}{m} - \frac{1}{q}\right)^2 rdr\right.$$
$$\left. + \left[\frac{2}{q}\left(\frac{n}{m} - \frac{1}{q}\right)^2 + (1+m\lambda)\left(\frac{n}{m} - \frac{1}{q}\right)^2\right] a^2 |\xi_r|_a^2\right\}. \tag{5.83}$$

Although the energy expression in equation (5.83) is cited widely, there is one more matching condition on the plasma-vacuum interface that needs to be included. Equation (5.80) is obtained by matching b_r. One also needs to match the tangential component of magnetic field. We just made a rough estimate here. The poloidal projection of perturbed magnetic field inside the plasma is

$$b_\theta = \nabla\theta \cdot \nabla \times \boldsymbol{\xi} \times \mathbf{B} \to \mathbf{B} \cdot \nabla(\boldsymbol{\xi} \cdot \nabla\theta).$$

Noting that $\nabla \cdot \mathbf{b} = 0$, one has $\xi_\theta \sim (\partial \xi_r / \partial r)/m$. Matching equation (5.84) to b_θ in the vacuum region yields

$$\frac{1}{m}\frac{\partial \xi_r}{\partial r}\bigg|_a \sim m \frac{B_z}{R} \frac{\frac{a^m}{b^{m-1}} + \frac{b^{m+1}}{a^m}}{\frac{a^{m-1}}{b^{m-1}} - \frac{b^{m+1}}{a^{m+1}}} \xi_r \bigg|_a . \tag{5.84}$$

where equation (5.81) has been used, and it has been assumed that there is no surface current at the plasma edge.

This can be used to determine the stability condition for external kink modes in the low beta case. Noting that $b \geq a$, one has $\lambda < 0$. The vacuum energy contribution, the term proportional to λ in equation (5.83), is destabilizing in general. With equation (5.84), one can study the fixed boundary case, *i.e.*, $b \to a$. In this case, λ tends to infinity. But, equation (5.84) shows that $\xi_r|_a \sim \mathcal{O}(\lambda^{-1})$. Therefore, the last term, the surface contribution, can be dropped, as it should be.

To determine the growth rate, one needs to reduce the kinetic energy in equation (5.13) as well. Note that in the large aspect ratio limit, f_k and g_k in equations (5.14) and (5.15) are reduced to

$$f_k = g_k = \frac{rB_z^2}{R^2}$$

and W_{ak} in equation (5.16) is negligible. Therefore, one obtains

$$\delta W_k = \omega_N^2 \frac{2\pi^2 B_z^2}{\mu_0 R} \int dr r \rho \, |\xi_r|^2 . \tag{5.85}$$

The potential energy in equation (5.83) and the kinetic energy in equation (5.85) determine completely the external modes in the cylinder configuration with low beta.

5.5 CONCLUSIONS AND DISCUSSION

In this chapter, the ideal MHD theory in the cylinder geometry with a circular cross-section is reviewed. Both the mode spectrum and the stability theory are presented.

In the spectrum studies, the growth rates of unstable MHD eigenmodes are investigated. The eigenvalue accumulating point is shown. Subsequently, the continuum spectrum and phase mixing theory are presented. Although the analyses are made in the cylinder geometry, the results are applicable for the toroidal geometry since the singular layer equations in both geometries resemble each other.

In the stability studies, Suydam's criterion, the stability conditions for $n = 1$ internal kink modes and external kink modes are presented. Suydam's criterion corresponds to Mercier's condition in toroidal geometry. The kink modes in toroidal geometry involve the coupling of multiple Fourier components, while only one Fourier component needs to be considered in the cylinder geometry.

The ideal MHD analyses in cylindrical geometry will help the analyses in the toroidal geometry in Chapter 6 and the resistive MHD analyses in Chapter 7.

References

1. Zheng Linjin 2019 Magnetically Confined Fusion Plasma Physics, ideal MHD theory (London: IOP Publishing). https://doi.org/10.1088/978-1-64327-138-5
2. Miyamoto K 1989 Plasma Physics for Nuclear Fusion (New York: The MIT Press).
3. Barston E M 1964 Electrostatic oscillations in inhomogeneous cold plasmas *Annals of Physics* **29** 282–303.
4. Goedbloed J P H, Poedts S 2004 *Principles of Magnetohydrodynamics: With Applications to Laboratory and Astrophysical Plasmas* (Cambridge: Cambridge University Press).
5. Bender, C M, Orszag S A 1978 *Advanced Mathematical Methods for Scientists and Engineers* (Tokyo: McGraw-Hill International Book Company).
6. Greene J M 1976 Introduction to resistive instabilities. Report by Centre de Recherches en Physique des Plasmas, Ecole Polytechnique de Lausanne, LRP 114/76.
7. Sedlacek Z 1971 Electrostatic oscillations in cold inhomogeneous plasma I. Differential equation approach *J. Plasma Physics 5 part 2*, 239–63.
8. Sedlacek Z 1971 Electrostatic oscillations in cold inhomogeneous plasma Part 2. Integral equation approach *J. Plasma Physics 6 part 1*, 187–99.
9. Sokolnikoff I S 1962 Principles and techniques of applied mathematics (New York: John Wiley & Sons, Inc.).
10. Freidberg J P 1987 Ideal Magnetohydrodynamics (New York: Plenum Press).
11. Suydam B R, Stability of a Linear Pinch, in Proceedings of the Second International Conference on the Peaceful Uses of Atomic Energy (United Nations, Geneva, 1958), Vol. 31, p. 157. https://digitallibrary.un.org/record/3808553?ln=en&v=pdf
12. Newcomb W A 1981 Equilibrium and stability of collisionless systems in the paraxial limit *J. Plasma Phys.* **26** 529–84. https://doi.org/10.1017/S0022377800010904
13. Shafranov V D 1970 *Zh. Tekh. Fiz. 40*, 421; 1970 Hydromagnetic Stability of a Current-Carrying Pinch in a Strong Longitudinal Magnetic Field *Sov. Phys. Tech. Phys. 15*, 175.
14. Ara G, Basu B, Coppi B, Laval G, Rosenbluth M N, Waddell B V 1978 Magnetic reconnection and $m = 1$ oscillations in current carrying plasmas *Ann. Phys. 112*, 443–76. DOI: 10.1016/S0003-4916(78)80007-4
15. Wesson J A 1978 Hydromagnetic stability of tokamaks *Nucl. Fusion 18*, 87–132. DOI 10.1088/0029-5515/18/1/010
16. Kadomtsev B B 1975 Disruptive instability in tokamaks *Sov. J. Plasma Phys. 1*, 389.

6 Linear ideal MHD stability theory in toroidal geometry

In this chapter, the cylinder theory in Chapter 5 is extended to the toroidal one. First, we describe the Fourier decomposition of the MHD equations and energy principle. Using the MHD equations in the Fourier space, the MHD spectrum in the toroidal geometry is analyzed, which introduces the continuum and gap concepts in toroidal geometry. Subsequently, the internal modes and the external modes are described in the global description. Both the perfectly conducting and resistive wall cases are investigated, together with the error-field amplification phenomenon. Afterward, the localized mode theories are presented. They include the interchange modes, ballooning modes, and toroidal Alfvén eigenmodes. In the end, the equilibrium with a large aspect ratio and circular cross-section is used to show the theoretical formalism and the application.

As reviewed in Section 3.1, the set of linearized ideal MHD equations is as follows:

$$\rho_m \gamma^2 \boldsymbol{\xi} = \delta \mathbf{J} \times \mathbf{B} + \mathbf{J} \times \delta \mathbf{B} - \nabla \delta P, \tag{6.1}$$

$$\delta \mathbf{B} = \nabla \times \boldsymbol{\xi} \times \mathbf{B}, \tag{6.2}$$

$$\mu_0 \delta \mathbf{J} = \nabla \times \delta \mathbf{B}, \tag{6.3}$$

$$\delta P = -\boldsymbol{\xi} \cdot \nabla P - \Gamma P \nabla \cdot \boldsymbol{\xi}. \tag{6.4}$$

They can be used to form a single ideal MHD equation

$$-\rho_m \omega^2 \boldsymbol{\xi} = \mathbf{F}(\boldsymbol{\xi}), \tag{6.5}$$

where

$$\begin{aligned} \mathbf{F}(\boldsymbol{\xi}) &= \frac{1}{\mu_0}(\nabla \times \mathbf{Q}) \times \mathbf{B} + \frac{1}{\mu_0}(\nabla \times \mathbf{B}) \times \mathbf{Q} + \nabla(\boldsymbol{\xi} \cdot \nabla p) + \Gamma \nabla (p \nabla \cdot \boldsymbol{\xi}) \\ &= -\nabla \left(\frac{\mathbf{B} \cdot \mathbf{Q}}{\mu_0} + p_1 \right) + \frac{1}{\mu_0}(\mathbf{B} \cdot \nabla \mathbf{Q} + \mathbf{Q} \cdot \nabla \mathbf{B}). \end{aligned} \tag{6.6}$$

As reviewed in Section 3.3, the MHD energy principle can be constructed from the linear MHD equations

$$\delta W_k = \delta W(\boldsymbol{\xi}, \boldsymbol{\xi}), \tag{6.7}$$

where

$$\delta W_k = \frac{1}{2} \int \rho_m \left(\frac{\partial \boldsymbol{\xi}}{\partial t} \right)^2 d\mathbf{r}, = -\frac{\omega^2}{2} \int \rho_m |\boldsymbol{\xi}|^2 d\mathbf{r}, \tag{6.8}$$

$$\delta W = \delta W_p + \delta W_s + \delta W_v. \tag{6.9}$$

DOI: 10.1201/9781003522133-6

Here, the plasma energy is given by

$$\delta W_p = \frac{1}{2}\int d\mathbf{r}\left\{\frac{\mathbf{Q}^2}{\mu_0} + \mathbf{j}\cdot(\boldsymbol{\xi}\times\mathbf{Q}) + (\boldsymbol{\xi}\cdot\nabla p)\nabla\cdot\boldsymbol{\xi} + \Gamma p(\nabla\cdot\boldsymbol{\xi})^2\right\} \tag{6.10}$$

and the surface and vacuum energies are respectively given as follows:

$$\delta W_s = -\frac{1}{2}\oint ds\xi_n^2\frac{\partial}{\partial n}\left(p_0 + \frac{B_{0i}^2}{2\mu_0} - \frac{B_{0i}^2}{2\mu_0}\right), \tag{6.11}$$

$$\delta W_v = \frac{1}{2\mu_0}\int_V(\nabla\times\mathbf{A})\cdot(\nabla\times\mathbf{A})d\mathbf{r} = \frac{1}{2\mu_0}\int_V B_1^2 d\mathbf{r}. \tag{6.12}$$

The MHD equation (6.5) has three components. As described in Section 3.5.1, one can relate them to three basic MHD waves: the compressional Alfvén, shear Alfvén, and acoustic waves with their governing equations given, respectively, as follows:

$$\begin{aligned}&\mathbf{e}_\perp\cdot\nabla\left(\frac{\mathbf{B}\cdot\mathbf{Q}}{\mu_0} + \delta P\right) - \rho\omega^2\mathbf{e}_\perp\cdot\boldsymbol{\xi}\\ = &\frac{1}{\mu_0}\mathbf{e}_\perp\cdot(\mathbf{B}\cdot\nabla\mathbf{Q} + \mathbf{Q}\cdot\nabla\mathbf{B}),\end{aligned} \tag{6.13}$$

$$\begin{aligned}&-\mathbf{B}\cdot\nabla\frac{\mathbf{B}\cdot\delta\mathbf{J}}{B^2} + \nabla\cdot\frac{\mathbf{B}}{B^2}\times\rho_m\omega^2\boldsymbol{\xi}\\ = &\delta\mathbf{B}\cdot\nabla\sigma - \mathbf{J}\cdot\nabla\frac{\mathbf{B}\cdot\delta\mathbf{B}}{B^2} + \nabla\times\frac{\mathbf{B}}{B^2}\cdot\nabla\delta P,\end{aligned} \tag{6.14}$$

and

$$\Gamma P\mathbf{B}\cdot\nabla\left(\frac{1}{B^2}\mathbf{B}\cdot\nabla\nabla\cdot\boldsymbol{\xi}\right) + \rho_m\omega^2\nabla\cdot\boldsymbol{\xi} = \rho_m\omega^2\nabla\cdot\boldsymbol{\xi}_\perp, \tag{6.15}$$

where $\mathbf{e}_\perp$ denotes the unit vector perpendicular to the magnetic field.

There is a useful formula from equation (3.118) to simplify the compressional Alfvén mode treatment

$$\frac{1}{B^2}\left(\mathbf{Q} - \mathbf{B}\frac{\mu_0\boldsymbol{\xi}\cdot\nabla p}{B^2}\right)\cdot\mathbf{B} = -\nabla\cdot\boldsymbol{\xi} + \mathbf{B}\cdot\nabla\left(\frac{\boldsymbol{\xi}\cdot\mathbf{B}}{B^2}\right) - 2\boldsymbol{\xi}\cdot\boldsymbol{\kappa}. \tag{6.16}$$

One can also show that

$$\boldsymbol{\kappa} = \mathbf{b}\cdot\nabla\mathbf{b} = \frac{[\mathbf{B}\times\nabla(2\mu_0 P + B^2)]\times\mathbf{B}}{2B^4} \tag{6.17}$$

and

$$\frac{2\mathbf{B}\times\nabla P\cdot\boldsymbol{\kappa}}{B^4} = \frac{\mathbf{B}\times\nabla P\cdot\nabla(2\mu_0 P + B^2)}{B^4} = \mathbf{B}\cdot\nabla\sigma, \tag{6.18}$$

where $\sigma = \mathbf{B}\cdot\mathbf{J}/B^2$.

In the case with the compressional Alfvén mode suppressed ($\nabla_\perp \gg 1/L$), from equation (6.13) $\mathbf{B}\cdot\mathbf{Q}+\mu_0\delta P=0$. Using equation (6.16), one has

$$\nabla\cdot\boldsymbol{\xi}_\perp+2\boldsymbol{\xi}\cdot\boldsymbol{\kappa}+\frac{\mu_0\Gamma p}{B^2}\nabla\cdot\boldsymbol{\xi}=0. \tag{6.19}$$

The equation of the parallel motion is then reduced to

$$\Gamma P\mathbf{B}\cdot\nabla\left(\frac{1}{B^2}\mathbf{B}\cdot\nabla\nabla\cdot\boldsymbol{\xi}\right)+\rho_m\omega^2\left(1+\frac{\mu_0\Gamma P}{B^2}\right)\nabla\cdot\boldsymbol{\xi}=-2\rho_m\omega^2\boldsymbol{\kappa}\cdot\boldsymbol{\xi}. \tag{6.20}$$

These equations will be used for the MHD mode analyses in the toroidal geometry later in this chapter. In the global description, which mainly involves the numerical treatment, the direct projections of MHD equation (6.5) are used. In the analytical analyses, the three-wave decompositions in equations (6.13)–(6.15) are usually used.

6.1 IDEAL MHD THEORY IN THE FOURIER SPACE

In this section, the linear MHD equation (6.5) and energy principle in equation (6.7) are decomposed in the Fourier space for tokomaks. The vacuum and wall equations are also described. The results are matched together to form the eigenvalue problems for various wall conditions. This allows for numerical eigenmode studies, the spectrum analyses, and asymptotic analytical analyses.

6.1.1 VECTOR PROJECTIONS

In this subsection, we will describe the Fourier transformation of the linear MHD equation. The linear MHD equation (6.5) reads

$$\begin{aligned}\mu_0\rho\gamma^2\boldsymbol{\xi} &= (\nabla\times\mathbf{Q})\times\mathbf{B}+(\nabla\times\mathbf{B})\times\mathbf{Q}\\ &\quad-\mu_0\nabla(-\boldsymbol{\xi}\cdot\nabla P-\Gamma P\nabla\cdot\boldsymbol{\xi}),\end{aligned} \tag{6.21}$$

where $\mathscr{Q}=\nabla\times(\boldsymbol{\xi}\times\mathbf{B})$.

Noting that equation (6.21) is a vector equation, we first decompose into three directions. The contra-variant decomposition of the field line displacement as introduced in DCON [1] is as follows:

$$\boldsymbol{\xi} = \xi_\psi\mathscr{J}\nabla\theta\times\nabla\zeta+\xi_\theta\mathscr{J}\nabla\zeta\times\nabla\psi+\xi_\zeta\mathscr{J}\nabla\psi\times\nabla\theta$$

Next, we decompose $\mathbf{Q}$. Noting that

$$\boldsymbol{\xi}\times\mathbf{B} = \xi_s\nabla\psi-\chi' q\xi_\psi\nabla\theta+\chi'\xi_\psi\nabla\zeta=\xi_s\nabla\psi+\xi_\psi\chi'(\nabla\zeta-q\nabla\theta),$$

one has

$$\mathbf{Q} = Q_\psi\nabla\theta\times\nabla\zeta+Q_\theta\nabla\zeta\times\nabla\psi+Q_\zeta\nabla\psi\times\nabla\theta$$

where

$$
\begin{aligned}
Q_\psi &= \chi'\left(\frac{\partial}{\partial\theta}+q\frac{\partial}{\partial\zeta}\right)\xi_\psi = i\chi'(m-nq)\xi_\psi,\\
Q_\theta &= -(\chi'\xi_\psi)'+\frac{\partial\xi_s}{\partial\zeta} = -(\chi'\xi_\psi)'-in\xi_s,\\
Q_\zeta &= -(q\chi'\xi_\psi)'-\frac{\partial\xi_s}{\partial\theta} = -(q\chi'\xi_\psi)'-im\xi_s
\end{aligned}
$$

with

$$
\xi_s = \mathscr{J}(\boldsymbol{\xi}\times\mathbf{B}\cdot\nabla\theta\times\nabla\zeta) = \chi'(q\xi_\theta-\xi_\zeta).
$$

Since [**Q** depends only on two perpendicular components of $\boldsymbol{\xi}$, ξ_ψ and ξ_s are introduced to represent them.

To derive $\delta\mathbf{j}$, one needs the covariant representation of magnetic field

$$
\mathbf{Q} = Q^\psi\nabla\psi+Q^\theta\nabla\theta+Q^\zeta\nabla\zeta.
$$

Its relationship with the contra-variant representation is determined as follows:

$$
\begin{aligned}
Q^\psi &= Q_\psi G_{11}+Q_\theta G_{12}+Q_\zeta G_{31},\\
Q^\theta &= Q_\psi G_{12}+Q_\theta G_{22}+Q_\zeta G_{23},\\
Q^\zeta &= Q_\psi G_{31}+Q_\theta G_{23}+Q_\zeta G_{33},
\end{aligned}
$$

where

$$
\begin{aligned}
G_{11} &= \mathscr{J}(\nabla\theta\times\nabla\zeta)\cdot(\nabla\theta\times\nabla\zeta),\\
G_{22} &= \mathscr{J}(\nabla\zeta\times\nabla\psi)\cdot(\nabla\zeta\times\nabla\psi),\\
G_{33} &= \mathscr{J}(\nabla\psi\times\nabla\theta)\cdot(\nabla\psi\times\nabla\theta),\\
G_{12} &= G_{21} = \mathscr{J}(\nabla\theta\times\nabla\zeta)\cdot(\nabla\zeta\times\nabla\psi),\\
G_{31} &= G_{13} = \mathscr{J}(\nabla\psi\times\nabla\theta)\cdot(\nabla\theta\times\nabla\zeta),\\
G_{23} &= G_{32} = \mathscr{J}(\nabla\zeta\times\nabla\psi)\cdot(\nabla\psi\times\nabla\theta).
\end{aligned}
$$

Subsequently, we can express the perturbed current density as follows:

$$
\begin{aligned}
\mu_0\delta\mathbf{j} = & \left(\frac{\partial}{\partial\theta}Q^\zeta-\frac{\partial}{\partial\zeta}Q^\theta\right)\nabla\theta\times\nabla\zeta+\left(\frac{\partial}{\partial\zeta}Q^\psi-\frac{\partial}{\partial\psi}Q^\zeta\right)\nabla\zeta\times\nabla\psi\\
&+\left(\frac{\partial}{\partial\psi}Q^\theta-\frac{\partial}{\partial\theta}Q^\psi\right)\nabla\psi\times\nabla\theta.
\end{aligned}
$$

There are three projections of the linear MHD equation (6.21). They will be considered one by one.

6.1.2 FOURIER DECOMPOSITIONS

Let us first outline the general Fourier analysis procedure. Because the system is axisymmetric, the toroidal mode number becomes a good quantum number, i.e., no toroidal coupling needs to be considered. Therefore, only poloidal Fourier decomposition needs to be performed. Noting that the linear MHD equations are three-dimensional, they can be formally expressed as

$$\sum_{i=1}^{3} L^{ij}\xi_j = 0, \quad i = 1,2,3. \tag{6.22}$$

Introducing the Fourier decomposition for each component

$$\xi_j = \sum_{m=-\infty}^{\infty} \xi_j^m \frac{1}{\sqrt{2\pi}} \exp\{i(m\theta - n\zeta)\} \tag{6.23}$$

with

$$\xi_j^m = \int_{-\pi}^{\pi} d\theta \xi_j \frac{1}{\sqrt{2\pi}} \exp\{-i(m\theta - n\zeta)\}, \qquad \xi_j^{-m,-n} = \xi_j^{m,n*}.$$

Multiplying equation (6.22) with $\exp\{i(m\theta - n\zeta)\}$ and integrating it over θ, one obtains

$$\sum_{i=1}^{3} \mathscr{L}^{ij}\Xi_j = 0, \quad i = 1,2,3, \tag{6.24}$$

where the Fourier space vector $\Xi_j = \{\xi_j^1, \cdots \xi_j^M\}^T$ with M being the maximum Fourier components considered and the superscript T representing the matrix transpose, and the matrix element of $\mathscr{L}^{ij}$ is defined as follows:

$$\mathscr{L}^{ij}_{mm'} = \frac{1}{2\pi} \int_{-\pi}^{\pi} d\theta L^{ij}(\theta) e^{i(m'-m)\theta}. \tag{6.25}$$

Equations (6.24) are the Fourier transformations of equations (6.22).

In the energy principle description, one has

$$\delta W = \frac{1}{4\pi^2} \int d\chi d\theta d\zeta \sum_{i,j=1}^{3} \xi_i^* L^{ij} \xi_j. \tag{6.26}$$

The transformation in equation (6.23) converts the energy integral in equation (6.26) to

$$\delta W = \int d\chi \sum_{i,j=1}^{3} \Xi_i^\dagger \mathscr{L}^{ij} \Xi_j.$$

These basic decompositions will be used to transform the MHD equations and the energy principle into the Fourier space. We use the calligraphic letters or double square brackets $[\![\cdot]\!]$ to represent the Fourier space matrix, use Ξ for the Fourier space vector for ξ, and the double angles $\langle\langle\cdot\rangle\rangle$ to represent the Fourier space vector.

6.1.3 MHD EQUATIONS IN THE FOURIER SPACE

In this subsection, we use the vector projections and the Fourier decompositions in sections 6.1.1 and 6.1.2 to decompose the MHD equations. The details have been described in the author's earlier book [2]. Here, we just outline the results.

We first consider the parallel projection. Instead of using $\xi_{\parallel}$, we use $\nabla \cdot \boldsymbol{\xi}$ as the independent variable. The $\mathbf{e}_b$ projection of the MHD equation has been obtained in equation (6.15). Using equation (6.16) to express $\nabla \cdot \boldsymbol{\xi}_{\perp}$, the equation of parallel motion, equation (6.15), is transformed to

$$\mathscr{L}_s \langle\langle \nabla \cdot \boldsymbol{\xi} \rangle\rangle \quad = \quad \mathscr{R}_s \Xi_s + \mathscr{R}_s \Xi'_\psi + \mathscr{R}_s \Xi_\psi, \tag{6.27}$$

where

$$\begin{aligned}
\mathscr{L}_s &= -\Gamma P \chi'^2 \left[\left[\frac{1}{\mathscr{J}}\right]\right] \mathscr{Q} \left[\left[\frac{1}{B^2 \mathscr{J}}\right]\right] \mathscr{Q} + \rho_m \omega^2, \\
\mathscr{R}_s &= i\rho_m \omega^2 \chi' \left[\left[\frac{1}{\mathscr{J} B^2}\right]\right] \left[n\left(\mathscr{G}_{22} + q\mathscr{G}_{23}\right) + \left(\mathscr{G}_{23} + q\mathscr{G}_{33}\right) \mathscr{M}\right] - 2\mathscr{K}_s, \\
\mathscr{R}_1 &= \rho_m \omega^2 \chi'^2 \left[\left[\frac{1}{\mathscr{J} B^2}\right]\right] \left[\left(\mathscr{G}_{22} + q\mathscr{G}_{23}\right) + q\left(\mathscr{G}_{23} + q\mathscr{G}_{33}\right)\right], \\
\mathscr{R}_0 &= -\rho_m \omega^2 \chi' \left[\left[\frac{1}{\mathscr{J} B^2}\right]\right] \left[i\chi' \left(\mathscr{G}_{12} + q\mathscr{G}_{31}\right) \mathscr{Q} - \chi'' \left(\mathscr{G}_{22} + q\mathscr{G}_{23}\right)\right. \\
&\quad \left. -(q\chi')' \left(\mathscr{G}_{23} + q\mathscr{G}_{33}\right)\right] + P' \left[\left[\frac{1}{B^2}\right]\right] - 2\mathscr{K}_\psi.
\end{aligned}$$

Here, $\mathscr{K}_s = [[\mathbf{B} \times \mathbf{s} \cdot \boldsymbol{\kappa}/B^2]]$, $\mathscr{K}_\psi = [[\mathbf{B} \times \nabla\psi \cdot \boldsymbol{\kappa}/B^2]]$, $\mathscr{M}_{mm'} = m\mathscr{I}_{mm'}$, $\mathscr{Q}_{mm'} = (m - nq)\mathscr{I}_{mm'}$, $\mathscr{I}$ is the unit matrix, and $\mathbf{s} = \chi'(\nabla\zeta - q\nabla\theta)$. The equation of parallel motion, equation (6.27), can be solved by matrix inversion

$$\langle\langle \nabla \cdot \boldsymbol{\xi} \rangle\rangle \quad = \quad \mathscr{L}_s^{-1} \left(\mathscr{R}_s \Xi_s + \mathscr{R}_s \Xi'_\psi + \mathscr{R}_s \Xi_\psi\right). \tag{6.28}$$

Using equation (6.28) to reduce the acoustic wave coupling, the radial projection $\mathscr{J}^2 \nabla\theta \times \nabla\zeta \cdot \frac{\mathbf{B} \times [\cdots \times \mathbf{B}]}{B^2}$ and the binormal projection $\frac{1}{q\chi'} \mathscr{J}^2 \nabla\zeta \times \nabla\psi \cdot \frac{\mathbf{B} \times [\cdots \times \mathbf{B}]}{B^2}$ of the MHD equation (6.1) are transformed in the Fourier space, respectively, to [3]

$$\left(\mathscr{B}^L \Xi_s + \mathscr{D} \Xi'_\psi + \mathscr{E} \Xi_\psi\right)' - \left(\mathscr{C}^L \Xi_s + \mathscr{E}^L \Xi'_\psi + \mathscr{H} \Xi_\psi\right) = 0, \tag{6.29}$$

$$\mathscr{A} \Xi_s + \mathscr{B} \Xi'_\psi + \mathscr{C} \Xi_\psi = 0. \tag{6.30}$$

Here, the equilibrium matrices in equations (6.29) and (6.30) contain three contributions: the incompressible plasma responses, plasma compressibility effects, and inertia contributions, e.g., $\mathscr{A} = \mathscr{A}_p + \mathscr{A}_c + \gamma^2 \mathscr{A}_i$, where

$$
\begin{aligned}
\mathscr{A}_p &= n(n\mathscr{G}_{22} + \mathscr{G}_{23}\mathscr{M}) + \mathscr{M}(n\mathscr{G}_{23} + \mathscr{G}_{33}\mathscr{M}),\\
\mathscr{B}_p &= -i\chi'\left[n(\mathscr{G}_{22} + q\mathscr{G}_{23}) + \mathscr{M}(\mathscr{G}_{23} + q\mathscr{G}_{33})\right],\\
\mathscr{C}_p &= -i\left[\chi''(n\mathscr{G}_{22} + \mathscr{M}\mathscr{G}_{23}) + (q\chi')'(n\mathscr{G}_{23} + \mathscr{M}\mathscr{G}_{33})\right]\\
&\quad -\chi'(n\mathscr{G}_{12} + \mathscr{M}\mathscr{G}_{31})\mathscr{Q} + i(g'\mathscr{Q} - \mu_0 nP'\mathscr{J}/\chi'),\\
\mathscr{D}_p &= \chi'^2\left[(\mathscr{G}_{22} + q\mathscr{G}_{23}) + q(\mathscr{G}_{23} + q\mathscr{G}_{33})\right],\\
\mathscr{E}_p &= \chi'\left[\chi''(\mathscr{G}_{22} + q\mathscr{G}_{23}) + (q\chi')'(\mathscr{G}_{23} + q\mathscr{G}_{33})\right] - i\chi'^2(\mathscr{G}_{12} + q\mathscr{G}_{31})\mathscr{Q} + \mu_0 p'\mathscr{J},\\
\mathscr{H}_p &= \chi''\left[\chi''\mathscr{G}_{22} + (q\chi')'\mathscr{G}_{23}\right] + (q\chi')'\left[\chi''\mathscr{G}_{23} + (q\chi')'\mathscr{G}_{33}\right]\\
&\quad + i\chi'\left[\chi''(\mathscr{M}\mathscr{G}_{12} - \mathscr{G}_{12}\mathscr{M}) + (q\chi')'(\mathscr{M}\mathscr{G}_{31} - \mathscr{G}_{31}\mathscr{M})\right]\\
&\quad + \chi'^2\mathscr{Q}\mathscr{G}_{11}\mathscr{Q} + \mu_0 p'\chi''\mathscr{J}/\chi' + \mu_0 p'\mathscr{J}' - g'q'\chi'\mathscr{I},\\
\mathscr{A}_c &= i\Gamma P\left\{\frac{\mathscr{J}}{q\chi'}\mathscr{M} - \frac{\chi'}{q}\left[\left[\frac{1}{B^2}\right]\right](\mathscr{G}_{22} + q\mathscr{G}_{23})\mathscr{Q}\right\}\mathscr{L}_s^{-1}\mathscr{R}_s,\\
\mathscr{B}_c &= i\Gamma P\left\{\frac{\mathscr{J}}{q\chi'}\mathscr{M} - \frac{\chi'}{q}\left[\left[\frac{1}{B^2}\right]\right](\mathscr{G}_{22} + q\mathscr{G}_{23})\mathscr{Q}\right\}\mathscr{L}_s^{-1}\mathscr{R}_1,\\
\mathscr{C}_c &= i\Gamma P\left\{\frac{\mathscr{J}}{q\chi'}\mathscr{M} - \frac{\chi'}{q}\left[\left[\frac{1}{B^2}\right]\right](\mathscr{G}_{22} + q\mathscr{G}_{23})\mathscr{Q}\right\}\mathscr{L}_s^{-1}\mathscr{R}_0,\\
\mathscr{B}_c^L &= -\Gamma P[\![\mathscr{J}]\!]\mathscr{L}_s^{-1}\mathscr{R}_s,\\
\mathscr{D}_c &= -\Gamma P[\![\mathscr{J}]\!]\mathscr{L}_s^{-1}\mathscr{R}_1,\\
\mathscr{E}_c &= -\Gamma P[\![\mathscr{J}]\!]\mathscr{L}_s^{-1}\mathscr{R}_0,\\
\mathscr{C}_c^L &= \Gamma P\left[i\chi'^2\left[\left[\frac{1}{B^2}\right]\right](\mathscr{G}_{12} + q\mathscr{G}_{31})\mathscr{Q} + \mathscr{J}'\right]\mathscr{L}_s^{-1}\mathscr{R}_s,\\
\mathscr{E}_c^L &= \Gamma P\left[i\chi'^2\left[\left[\frac{1}{B^2}\right]\right](\mathscr{G}_{12} + q\mathscr{G}_{31})\mathscr{Q} + \mathscr{J}'\right]\mathscr{L}_s^{-1}\mathscr{R}_1,\\
\mathscr{H}_c &= \Gamma P\left[i\chi'^2\left[\left[\frac{1}{B^2}\right]\right](\mathscr{G}_{12} + q\mathscr{G}_{31})\mathscr{Q} + \mathscr{J}'\right]\mathscr{L}_s^{-1}\mathscr{R}_0,\\
\mathscr{A}_i &= \frac{B_0^2}{R_0^2}\left[\left[\mathscr{J}\frac{\rho_m}{B^2}|\nabla\psi|^2\right]\right],\\
\mathscr{C}_i &= \frac{B_0^2}{R_0^2}\left[\left[\chi'\mathscr{J}\frac{\rho_m}{B^2}(\nabla\psi\cdot\nabla\zeta - q\nabla\psi\cdot\nabla\theta)\right]\right],\\
\mathscr{H}_i &= \frac{B_0^2}{R_0^2}\left[\left[\chi'^2\mathscr{J}\frac{\rho_m}{B^2}(|\nabla\zeta|^2 + q^2|\nabla\theta|^2 - 2\nabla\theta\cdot\nabla\zeta)\right]\right],
\end{aligned}
$$

and $\mathscr{B}_i = \mathscr{D}_i = \mathscr{E}_i = 0$, γ denotes the normalized growth rate. Here, the incompressible plasma matrices and inertia ones are self-adjoint, for example $\mathscr{B}_p^L = \mathscr{B}_p^\dagger$, etc.

One can formally solve equation (6.30),

$$\Xi_s = -\mathscr{A}^{-1}\mathscr{B}\Xi'_\psi - \mathscr{A}^{-1}\mathscr{C}\Xi_\psi.$$

Inserting this solution into equation (6.29), we get a single matrix equation governing the ideal MHD modes as in the DCON formalism [1]:

$$\frac{d}{d\psi}\left(\mathscr{F}\Xi'+\mathscr{K}\Xi\right)-\left(\mathscr{K}^L\Xi'+\mathscr{G}\Xi\right) \;=\; 0, \tag{6.31}$$

where the subscript ψ for $\boldsymbol{\xi}$ has been dropped and

$$\begin{aligned}
\mathscr{F} &= \mathscr{D}-\mathscr{B}^L\mathscr{A}^{-1}\mathscr{B},\\
\mathscr{K} &= \mathscr{E}-\mathscr{B}^L\mathscr{A}^{-1}\mathscr{C},\\
\mathscr{K}^L &= \mathscr{E}^L-\mathscr{C}^L\mathscr{A}^{-1}\mathscr{B},\\
\mathscr{G} &= \mathscr{H}-\mathscr{C}^L\mathscr{A}^{-1}\mathscr{C}.
\end{aligned}$$

As in the DCON code [1], the MHD part can be simplified as follows:

$$\begin{aligned}
\mathscr{F} &= \frac{\chi'^2}{n^2}\left\{\mathscr{Q}\mathscr{G}_{33}\mathscr{Q}+\gamma^2\mathscr{A}_i+\left(\mathscr{A}_c+i\frac{n}{\chi'}\mathscr{B}_c^L\right)+\frac{n^2}{\chi'^2}\left(\mathscr{D}_c-i\frac{\chi'}{n}\mathscr{B}_c\right)\right.\\
&\quad-\left[\gamma^2\mathscr{A}_i+\left(\mathscr{A}_c+i\frac{n}{\chi'}\mathscr{B}_c^L\right)+\mathscr{Q}(n\mathscr{G}_{23}+\mathscr{G}_{33}\mathscr{M})\right]\\
&\quad\left.\times\mathscr{A}^{-1}\left[\gamma^2\mathscr{A}_i+\left(\mathscr{A}_c-i\frac{n}{\chi'}\mathscr{B}_c\right)+(n\mathscr{G}_{23}+\mathscr{M}\mathscr{G}_{33})\mathscr{Q}\right]\right\},\\
\mathscr{K} &= \frac{\chi'}{n}\left\{i\left[\gamma^2\mathscr{A}_i+\left(\mathscr{A}_c+i\frac{n}{\chi'}\mathscr{B}_c^L\right)+\mathscr{Q}(n\mathscr{G}_{23}+\mathscr{G}_{33}\mathscr{M})\right]\mathscr{A}^{-1}\mathscr{C}\right.\\
&\quad\left.-\mathscr{Q}\left[\chi''\mathscr{G}_{23}+(q\chi')'\mathscr{G}_{33}-i\chi'\mathscr{G}_{31}\mathscr{Q}-g'\mathscr{I}\right]-i\gamma^2\mathscr{C}_i\right\}\\
&\quad+\left(\mathscr{E}_c-i\chi'\frac{1}{n}\mathscr{C}_c\right),\\
\mathscr{K}^L &= \frac{\chi'}{n}\left\{-i\mathscr{C}^L\mathscr{A}^{-1}\left[\gamma^2\mathscr{A}_i+\left(\mathscr{A}_c-i\frac{n}{\chi'}\mathscr{B}_c\right)+(n\mathscr{G}_{23}+\mathscr{M}\mathscr{G}_{33})\mathscr{Q}\right]\right.\\
&\quad\left.-\left[\chi''\mathscr{G}_{23}+(q\chi')'\mathscr{G}_{33}+i\chi'\mathscr{Q}\mathscr{G}_{31}-g'\mathscr{I}\right]\mathscr{Q}+i\gamma^2\mathscr{C}_i\right\}\\
&\quad+\left(\mathscr{E}_c^L+i\chi'\frac{1}{n}\mathscr{C}_c^L\right).
\end{aligned}$$

Here, the plasma compressibility matrices can be further reduced as follows:

$$\mathscr{A}_c - i\frac{n}{\chi'}\mathscr{B}_c = i\Gamma P\frac{1}{q}\left[\frac{1}{\chi'}[\![\mathscr{J}]\!]\mathscr{M} - \chi'\left[\!\!\left[\frac{1}{B^2}(G_{22}+qG_{23})\right]\!\!\right]\mathscr{Q}\right]\mathscr{L}_s^{-1}\mathscr{R}_{s1},$$

$$\mathscr{A}_c + i\frac{n}{\chi'}\mathscr{B}_c^L = i\Gamma P\frac{1}{q}\left[\frac{1}{\chi'}[\![\mathscr{J}]\!]\mathscr{Q} - \chi'\left[\!\!\left[\frac{1}{B^2}(G_{22}+qG_{23})\right]\!\!\right]\mathscr{Q}\right]\mathscr{L}_s^{-1}\mathscr{R}_s,$$

$$\left(\mathscr{A}_c + i\frac{n}{\chi'}\mathscr{B}_c^L\right) + \frac{n^2}{\chi'^2}\left(\mathscr{D}_c - i\frac{\chi'}{n}\mathscr{B}_c\right)$$

$$= i\Gamma P\frac{1}{q}\left[\frac{1}{\chi'}\langle\mathscr{J}\rangle\mathscr{Q} - \chi'\left[\!\!\left[\frac{1}{B^2}(G_{22}+qG_{23})\right]\!\!\right]\mathscr{Q}\right]\mathscr{L}_s^{-1}\mathscr{R}_{s1},$$

$$\mathscr{E}_c - i\chi'\frac{1}{n}\mathscr{C}_c = \Gamma P\frac{1}{nq}\left[\langle\mathscr{J}\rangle\mathscr{Q} - \chi'^2\left[\!\!\left[\frac{1}{B^2}(G_{22}+qG_{23})\right]\!\!\right]\mathscr{Q}\right]\mathscr{L}_s^{-1}\mathscr{R}_0,$$

$$\mathscr{E}_c^L + i\chi'\frac{1}{n}\mathscr{C}_c^L = i\Gamma P\frac{\chi'}{n}\left[i\chi'^2\left[\!\!\left[\frac{1}{B^2}(G_{12}+qG_{31})\right]\!\!\right]\mathscr{Q} + \langle\mathscr{J}'\rangle\right]\mathscr{L}_s^{-1}\mathscr{R}_{s1},$$

where $\mathscr{R}_{s1} = \mathscr{R}_s - i(n/\chi')\mathscr{R}_1$.

6.1.4 ENERGY PRINCIPLE IN THE FOURIER SPACE

In this subsection, we use the vector projections and the Fourier decompositions in sections 6.1.1 and 6.1.2 to decompose the MHD energy principle. Four regions need to be considered: the plasma region, the vacuum region between the plasma-vacuum interface and the wall, the wall, and the vacuum region external to the wall. The details have been described in author's earlier book [2]. Here, we just outline the results. We start with the energy principle described in equations (6.7)–(6.12).

6.1.4.1 Plasma energy

The plasma energy expression is given in equation (6.10), which is recited here

$$2\mu_0\delta W = \int d\mathbf{r}\left\{\mathbf{Q}^2 + \mu_0\mathbf{J}\cdot\boldsymbol{\xi}\times\mathbf{Q} + \mu_0(\boldsymbol{\xi}\cdot\nabla P)(\nabla\cdot\boldsymbol{\xi})\right\}. \qquad (6.32)$$

Here, the plasma compressibility contribution has been omitted for simplicity, as it is stabilizing and does not affect the marginal stability condition.

Introducing the Fourier decomposition in equation (6.23) and performing similar analyses in subsection 6.1.3, one can find the plasma energy in equation (6.32) is reduced to

$$\delta W = \frac{1}{2\mu_0}\int d\psi\left(\Xi_s^\dagger \quad \Xi_\psi'^\dagger \quad \Xi_\psi^\dagger\right)\begin{pmatrix}\mathscr{A}_p & \mathscr{B}_p & \mathscr{C}_p\\ \mathscr{B}_p^\dagger & \mathscr{D}_p & \mathscr{E}_p\\ \mathscr{C}_p^\dagger & \mathscr{E}_p^\dagger & \mathscr{H}_p\end{pmatrix}\begin{pmatrix}\Xi_s\\ \Xi_\psi'\\ \Xi_\psi\end{pmatrix}, \qquad (6.33)$$

where the equilibrium matrices are defined in subsection 6.1.3.

Similarly, the perpendicular kinetic energy in equation (6.8) can be reduced to

$$\delta W_k = \gamma_N^2 \frac{1}{2\mu_0} \int d\psi \left(\Xi_s^\dagger \quad \Xi_\psi'^\dagger \quad \Xi_\psi^\dagger \right) \begin{pmatrix} \mathscr{A}_k & 0 & \mathscr{C}_k \\ 0 & 0 & 0 \\ \mathscr{C}_k^\dagger & 0 & \mathscr{H}_k \end{pmatrix} \begin{pmatrix} \Xi_s \\ \Xi_\psi' \\ \Xi_\psi \end{pmatrix}, \quad (6.34)$$

where the equilibrium matrices are defined in subsection 6.1.3.

We neglect the surface energy δW_s in equation (3.41), so that the total energy in the plasma region becomes $\delta W_p + \delta W_k$. Minimization of the total energy $\delta W_p + \delta W_k$ with respect to $\Xi_s^\dagger$ leads to

$$\mathscr{A}\Xi_s + \mathscr{B}\Xi_\psi' + \mathscr{C}\Xi_\psi = 0, \quad (6.35)$$

where $\mathscr{A} = \mathscr{A}_p + \mathscr{A}_k, \cdots$. This recovers the MHD equation (6.30). One can also vary the total energy $\delta W_p + \delta W_k$ with respect to Ξ to recover the MHD equation (6.29). We can insert equation (6.35) back into equations (6.33) and (6.34) to obtain

$$2\mu_0(\delta W_p + \delta W_k) = \int d\psi \left(\Xi_\psi'^\dagger \quad \Xi_\psi^\dagger \right) \begin{pmatrix} \mathscr{F} & \mathscr{K} \\ \mathscr{K}^\dagger & \mathscr{G} \end{pmatrix} \begin{pmatrix} \Xi_\psi' \\ \Xi_\psi \end{pmatrix}, \quad (6.36)$$

where the equilibrium matrices are defined in subsection 6.1.3.

The total plasma energy in equation (6.36) can be further minimized with respect to $\Xi_\psi^\dagger$. It gives rise to the Euler-Lagrange equations in the Fourier space

$$\frac{d}{d\psi}\left(\mathscr{F}\frac{d\Xi_\psi}{d\psi} + \mathscr{K}\Xi_\psi \right) - \mathscr{K}^\dagger \frac{d\Xi_\psi}{d\psi} - \mathscr{G}\Xi_\psi = 0. \quad (6.37)$$

which corresponds to equation (6.31) in subsection 6.1.4.1. Using the equation (6.37), the energy in equation (6.36) is reduced to

$$\delta W_p + \delta W_k = \Xi^\dagger \frac{d}{d\psi}\left(\mathscr{F}\frac{d\Xi_\psi}{d\psi} + \mathscr{K}\Xi_\psi \right)\Bigg|_{\psi_1}^{\psi_2}. \quad (6.38)$$

6.1.4.2 Vacuum energy

For complete determination of system energy, the vacuum energy in equation (6.12) needs to be taken into account. In the vacuum one has $\nabla \times \delta\mathbf{B} = 0$. Instead of using the vector potential $\delta\mathbf{A}$, which assures $\nabla \cdot \delta\mathbf{B} = 0$, one can represent the magnetic field by the scalar potential φ with $\delta\mathbf{B} = -\nabla\varphi$, which assures $\nabla \times \delta\mathbf{B} = 0$, while requiring

$$\nabla^2\varphi = 0. \quad (6.39)$$

Here, it is noted that we have limited ourselves to the $n \neq 0$ case. This is because the representation of vacuum magnetic field with scalar potential in equation (6.39), although being simple, cannot include the consideration of $n = 0$ modes. To study the $n = 0$ modes, one more scalar is needed to represent the vacuum magnetic field. This can be found in reference [3]. The vacuum magnetic energy in this representation becomes

$$\delta W_v = \frac{1}{2\mu_0}\int d\mathbf{r}\,|\nabla\varphi|^2. \tag{6.40}$$

Setting up the radial and poloidal grids (S,θ) at a given toroidal and ϕ and introducing the Fourier decomposition, the expression in equation (6.40) is reduced to energy integral to

$$\delta W_v = \frac{1}{2\mu_0}\int d\psi \begin{pmatrix} \frac{\partial\boldsymbol{\varphi}^\dagger}{\partial\psi} & \boldsymbol{\varphi}^\dagger \end{pmatrix} \begin{pmatrix} \mathscr{F}_v & \mathscr{K}_v \\ \mathscr{K}_v^\dagger & \mathscr{G}_v \end{pmatrix} \begin{pmatrix} \frac{\partial\boldsymbol{\varphi}}{\partial\psi} \\ \boldsymbol{\varphi} \end{pmatrix}, \tag{6.41}$$

where $\mathscr{F}_v = \langle \mathscr{J}|\nabla\psi|^2\rangle$, $\mathscr{K}_v = i\langle \mathscr{J}\nabla\psi\cdot\nabla\theta\rangle\mathscr{M}$, and $\mathscr{G}_v = \mathscr{M}\langle\mathscr{J}|\nabla\theta|^2\rangle\mathscr{M} + n^2\langle\mathscr{J}/X^2\rangle$. Similar to the plasma region treatment, the minimization of the energy integral in equation (6.41) with respect to $\boldsymbol{\varphi}^\dagger$ leads to the following Euler-Lagrange equations in the Fourier space:

$$\frac{d}{d\psi}\left(\mathscr{F}_v\frac{d\varphi}{d\psi} + \mathscr{K}_v\varphi\right) - \mathscr{K}_v^\dagger\frac{d\varphi}{d\psi} - \mathscr{G}_v\varphi = 0. \tag{6.42}$$

This can be equivalently expressed as a set of first-order differential equations

$$\mathbf{u}_v' = \mathscr{L}_v\mathbf{u}_v, \tag{6.43}$$

where

$$\mathbf{u}_v \equiv \begin{pmatrix} \mathbf{u}_{v1} \\ \mathbf{v}_{v2} \end{pmatrix} = \begin{pmatrix} \boldsymbol{\varphi} \\ \mathscr{F}_v\boldsymbol{\varphi}' + \mathscr{K}_v\boldsymbol{\varphi} \end{pmatrix},$$

$$\mathscr{L}_v = \begin{pmatrix} -\mathscr{F}_v^{-1}\mathscr{K}_v & \mathscr{F}_v^{-1} \\ \mathscr{G}_v - \mathscr{K}_v^\dagger\mathscr{F}_v^{-1}\mathscr{K}_v & \mathscr{K}_v^\dagger\mathscr{F}_v^{-1} \end{pmatrix}.$$

Inserting back the solutions of the set of Euler-Lagrange equations in equation (6.43), the energy integral in equation (6.41) is reduced to

$$2\mu_0\delta W_v = -\left.\mathbf{u}_{v1}^\dagger\mathbf{v}\right|_{\psi=\psi_a}, \tag{6.44}$$

Similar to the plasma region in equation (6.38), the vacuum energy is reduced to a contribution at the plasma-vacuum interface.

Alternatively, equation (6.39) can be solved using the free space Green's function $G(\mathbf{r}-\mathbf{r}')$ [4], satisfying

$$\nabla^2 G(\mathbf{r}-\mathbf{r}') = -4\pi\delta(\mathbf{r}-\mathbf{r}'). \tag{6.45}$$

Using the identity $\nabla\cdot(\alpha\nabla\beta) = \alpha\nabla^2\beta + \nabla\alpha\cdot\nabla\beta$, it can be proved

$$4\pi\varphi(\mathbf{r}) = \int_V (G\nabla'^2\varphi - \varphi\nabla'^2 G)d\mathbf{r}' = \int \nabla\cdot(G\nabla'\varphi - \varphi\nabla' G)d\mathbf{r}'.$$

This can be used to derive the so-called Fredholm equation:

$$
\begin{aligned}
&4\pi\varphi(\mathbf{r}) + \int_S \varphi(\mathbf{r}')\nabla' G(\mathbf{r}-\mathbf{r}')\cdot d\mathbf{S}' \\
&= \int_S G(\mathbf{r}-\mathbf{r}')\nabla'\varphi(\mathbf{r}')\cdot d\mathbf{S}', \qquad (6.46)
\end{aligned}
$$

where $d\mathbf{S}$ represents the surface element.

Note that when applying equation (6.46) on the boundary, there is a singularity in the surface integration. The singular contribution can be extracted from the second term in equation (6.46). Denoting S_s as the half sphere around the singular point $\mathbf{r}'$ of $\nabla G(\mathbf{r}-\mathbf{r}')$ and letting $\rho = |\mathbf{r}-\mathbf{r}'|$, one obtains

$$
\int_{S_s} \varphi(\mathbf{r}')\nabla' G(\mathbf{r}-\mathbf{r}')\cdot d\mathbf{S}' = \varphi(\mathbf{r})\left(-\frac{1}{\rho^2}\right)2\pi\rho^2 = -2\pi\varphi(\mathbf{r}).
$$

Consequently, equation (6.46) is reduced to

$$
2\pi\varphi(\mathbf{r}) + P\int_S \varphi(\mathbf{r}')\nabla' G(\mathbf{r}-\mathbf{r}')\cdot d\mathbf{S}' = \int_S G(\mathbf{r}-\mathbf{r}')\nabla'\varphi(\mathbf{r}')\cdot d\mathbf{S}'. \qquad (6.47)
$$

Here, P has been introduced to indicate the principal integration.

In the coordinate system $(\mathscr{S}, \theta, \phi)$, one has

$$
d\mathbf{S} = -\frac{\nabla\mathscr{S}}{\nabla\mathscr{S}\cdot\nabla\theta\times\nabla\phi}d\theta d\phi = -\mathscr{J}\nabla\mathscr{S}d\theta d\phi.
$$

Introducing the Fourier decomposition for the toroidal angle ϕ, equation (6.47) is reduced to

$$
\varphi(\theta) - P\int_{-\pi}^{\pi} \varphi(\theta')\mathscr{K}_n(\theta,\theta')d\theta' = -\int_{-\pi}^{\pi} \mathscr{G}_n(\theta,\theta')\mathscr{B}(\theta')d\theta', \qquad (6.48)
$$

where

$$
\begin{aligned}
\mathscr{G}_n &= \frac{1}{2\pi}\int_{-\pi}^{\pi} d\phi' e^{in(\phi-\phi')}\left[G(\mathbf{r}-\mathbf{r}')\right], \\
\mathscr{K}_n &= \frac{1}{2\pi}\int_{-\pi}^{\pi} d\phi'\left[\mathscr{J}\nabla' G\cdot\nabla\mathscr{S}' e^{in(\phi-\phi')}\right], \\
\mathscr{B} &= \frac{1}{2\pi}\int_{-\pi}^{\pi} d\phi'\left[\mathscr{J}\nabla'\varphi(\mathbf{r}')\cdot\nabla\mathscr{S}' e^{in(\phi-\phi')}\right].
\end{aligned}
$$

Now, we introduce the Fourier transformation for poloidal angle to solve equation (6.48). This yields

$$
\delta_{mm'}\varphi_{m'}(\mathscr{S}_o) - \mathscr{K}_{mm'}\varphi_{m'}(\mathscr{S}_s) + \mathscr{G}_{mm'}\mathscr{B}_{m'}(\mathscr{S}_s) = 0. \qquad (6.49)
$$

Here, $\mathscr{S}_o$ and $\mathscr{S}_s$ denote, respectively, the observation and source surfaces, the summation with respect to m' is implied, and

$$
\begin{aligned}
\mathscr{K}_{mm'} &= \frac{1}{2\pi}P\int_{-\pi}^{\pi} d\theta e^{-im\theta}\int_{-\pi}^{\pi} e^{im'\theta'}\mathscr{K}_n(\theta,\theta')d\theta', \\
\mathscr{G}_{mm'} &= \frac{1}{2\pi}\int_{-\pi}^{\pi} d\theta e^{-im\theta}\int_{-\pi}^{\pi} e^{im'\theta'}\mathscr{G}_n(\theta,\theta')d\theta'.
\end{aligned}
$$

Now, we use equation (6.49) to construct the vacuum solutions in various scenarios. the plasma-vacuum interface, the inner side and outer side of the wall are denoted, respectively, as $\mathscr{S}_a$, $\mathscr{S}_{b-}$, and $\mathscr{S}_{b+}$.

First, we consider the no-wall case. To match the plasma solution, only the vacuum solution on the plasma-vacuum interface is needed. Note that the other boundary is infinity in this case, where $\varphi_{m'}$ and $\mathscr{B}_{m'}$ vanish or to be small. Applying equation (6.49) on the interface, one obtains

$$\delta_{mm'}\varphi_{m'}(\mathscr{S}_a) - \mathscr{K}^{aa}_{mm'}\varphi_{m'}(\mathscr{S}_a) + \mathscr{G}^{aa}_{mm'}\mathscr{B}_{m'}(\mathscr{S}_a) = 0. \tag{6.50}$$

Here, the subscripts have been introduced for $\mathscr{K}$ and $\mathscr{G}$ to denote, respectively, the observation and source locations, with a, b_-, and b_+ being, respectively, the plasma-vacuum interface, inner and outer sides of the wall.

In the presence of the perfectly conducting wall, the boundary at infinite in the no-wall case is replaced by that on the conducting wall. If the wall is resistive, one has to consider three interfaces together with the connection equations on the wall, which will be described in the next subsection. Here, we just give the three equations for describing the respective interface. On the plasma-vacuum interface, one has

$$\begin{aligned}&\delta_{mm'}\varphi_{m'}(\mathscr{S}_a) - \mathscr{K}^{aa}_{mm'}\varphi_{m'}(\mathscr{S}_a) + \mathscr{G}^{aa}_{mm'}\mathscr{B}_{m'}(\mathscr{S}_a)\\ = \;& -\mathscr{K}^{ab_-}_{mm'}\varphi_{m'}(\mathscr{S}_{b-}) + \mathscr{G}^{ab_-}_{mm'}\mathscr{B}_{m'}(\mathscr{S}_{b-}).\end{aligned} \tag{6.51}$$

On the inner side of the wall, one has

$$\begin{aligned}&\delta_{mm'}\varphi_{m'}(\mathscr{S}_{b-}) - \mathscr{K}^{b_-b_-}_{mm'}\varphi_{m'}(\mathscr{S}_{b-}) + \mathscr{G}^{b_-b_-}_{mm'}\mathscr{B}_{m'}(\mathscr{S}_{b-})\\ = \;& -\mathscr{K}^{b_-a}_{mm'}\varphi_{m'}(\mathscr{S}_a) + \mathscr{G}^{b_-a}_{mm'}\mathscr{B}_{m'}(\mathscr{S}_a).\end{aligned} \tag{6.52}$$

On the outer side of the wall, one has

$$\delta_{mm'}\varphi_{m'}(\mathscr{S}_{b+}) - \mathscr{K}^{b_+b_+}_{mm'}\varphi_{m'}(\mathscr{S}_{b+}) + \mathscr{G}^{b_+b_+}_{mm'}\mathscr{B}_{m'}(\mathscr{S}_{b+}) = 0. \tag{6.53}$$

These three equations are sufficient to determine the eigenvalue problem, which will be further discussed in the later chapters.

6.1.5 THE EIGENVALUE PROBLEM

In this subsection, we describe the solutions of the Euler-Lagrangian equations in the plasma and vacuum regions and then match them across the plasma-vacuum and vacuum-wall interfaces to construct the eigenvalue problems. Various wall conditions will be described, including the no-wall case, the perfectly conducting wall, and the resistive wall cases.

We first consider the boundary conditions at the left end ψ_l in the interior of plasma core. Depending on the mode extension, there are three possible types of boundaries: When the modes are extended to the magnetic axis, we use the cylinder limit to describe the boundary condition at the magnetic axis, *i.e.*, $\xi_{\psi,m} \propto r^m$; if the modes end at the left at a singular surface, the small solution of the singular layer equation is used; if the modes end at the left at a regular surface, one simply uses the condition $\Xi_\psi = 0$.

Next, we consider the vacuum side. The solution of plasma region needs to match the vacuum solutions. First, we describe the solution of the Euler-Lagrangian equations (6.36) in the plasma region. The equations are transformed to a set of first-order differential equations by introducing the following $2M$ unknowns $\mathbf{u} = \begin{pmatrix} \Xi_\psi \\ \mathbf{u}_2 \end{pmatrix}$, where $\mathbf{u}_2 = \mathscr{F}\Xi'_\psi + \mathscr{K}\Xi_\psi$, Consequently, equation (6.36) is reduced to the set of $2M$ first-order differential equations

$$\mathbf{u}' = \mathscr{L}\mathbf{u}, \tag{6.54}$$

where $2M \times 2M$ matrix

$$\mathscr{L} = \begin{pmatrix} -\mathscr{F}^{-1}\mathscr{K} & \mathscr{F}^{-1} \\ \mathscr{G} - \mathscr{K}^{\mathrm{L}}\mathscr{F}^{-1}\mathscr{K} & \mathscr{K}^{\mathrm{L}}\mathscr{F}^{-1} \end{pmatrix}.$$

Furthermore, it is noted that the set of eigenmode equations in equation (6.36) can be solved by the independent solution method, together with the multiple region matching technique with the independent solutions determined numerically as described in reference [3]. Imposing M boundary conditions at the left end ψ_l, only M independent solutions remain:

$$\begin{pmatrix} \Xi_p \\ \mathscr{W}_2 \end{pmatrix} \equiv \begin{pmatrix} \boldsymbol{\xi}^1, & \cdots, & \boldsymbol{\xi}^M \\ \mathbf{u}_2^1, & \cdots, & \mathbf{u}_2^M \end{pmatrix},$$

where the superscripts have been introduced to label the independent solutions. Subsequently, the general solution can be constructed as a combination of the M independent solutions,

$$\begin{pmatrix} \boldsymbol{\xi} \\ \mathbf{u}_2 \end{pmatrix} = i \begin{pmatrix} \Xi_p \\ \mathscr{W}_p \end{pmatrix} \mathbf{c}_p. \tag{6.55}$$

Here, $\mathbf{c}_p$ is a constant vector with M elements, which are adjusted to satisfy the boundary conditions at the plasma-vacuum interface.

To the boundary condition, one needs to evaluate the parallel perturbed magnetic field.

$$\begin{aligned}
\langle\langle \mathscr{J}_g \mathbf{B}\cdot\mathbf{Q} \rangle\rangle &= \left[\left[\mathscr{J}_g \chi' g_{12} \chi' \left(\frac{\partial}{\partial\theta} + q\frac{\partial}{\partial\phi} \right) \xi_\psi + \mathscr{J}_g \chi' g_{22} \left[-(\chi'\xi_\psi)' + \frac{\partial \xi_s}{\partial\phi} \right] \right.\right. \\
&\quad \left.\left. + \mathscr{J}_g \chi' q g_{33} \left[-(q\chi'\xi_\psi)' - \frac{\partial\xi_s}{\partial\theta} \right] \right]\right] \\
&= i\chi'^2 \mathscr{G}_{12}\mathscr{Q} + \chi'\mathscr{G}_{22}\left[-(\chi'\Xi_\psi)' - in\Xi_s \right] \\
&\quad + \chi' q\mathscr{G}_{33}\left[-(q\chi'\Xi_\psi)' - i\mathscr{M}\Xi_S \right] \\
&= i\chi'^2 \mathscr{G}_{12}\mathscr{Q} - \chi'^2\left[\mathscr{G}_{22} + q\mathscr{G}_{33} \right]\Xi'_\psi - \left[\chi'\mathscr{G}_{22}\chi'' + \chi' q\mathscr{G}_{33}(q\chi')' \right]\Xi_\psi \\
&\quad - i\chi'\left[n\mathscr{G}_{22} + q\mathscr{G}_{33}\mathscr{M} \right]\left[-\mathscr{A}^{-1}(\mathscr{B}\Xi'_\psi + \mathscr{C}\Xi_\psi) \right] \\
&= -\mathscr{D}\Xi'_\psi - (\mathscr{E} - \mu_0 P' \mathscr{J}_g)\Xi_\psi - \mathscr{B}^\dagger\left[-\mathscr{A}^{-1}(\mathscr{B}\Xi'_\psi + \mathscr{C}\Xi_\psi) \right].
\end{aligned}$$

Therefore

$$\langle\langle \mathbf{B}\cdot\mathbf{Q}\rangle\rangle = -\mathscr{J}_g^{-1}\left[\mathscr{F}\Xi'_\psi + (\mathscr{K} - \mu_0 P'\mathscr{J}_g)\Xi_\psi\right].$$

As shown in equation (6.38) integration by parts of the plasma energy in equation (6.36) confirms this. Consequently, one obtains that Ξ_ψ and $\mathbf{u}_2$ in the plasma region are related to the magnetic field and pressure as follows:

$$\begin{aligned}\langle\langle \mathscr{J}\nabla\psi\cdot\delta\mathbf{B}\rangle\rangle &= i\mathscr{Q}\Xi_\psi,\\ -\langle\langle \mathscr{J}(\mathbf{B}\cdot\delta\mathbf{B} - \boldsymbol{\xi}\cdot\nabla P)\rangle\rangle &= \mathbf{u}_2.\end{aligned}$$

Using equation (6.55) and defining $\mathbf{c}_p = \Xi_p^{-1}\mathbf{c}_p^{new}$ and $\mathscr{W}_p^{new} = \mathscr{W}_p\Xi_p^{-1}$ without loss of generality, one obtains

$$\langle\langle \mathscr{J}\nabla\psi\cdot\delta\mathbf{B}\rangle\rangle = -\mathscr{Q}\mathbf{c}_p, \tag{6.56}$$

$$-\langle\langle \mathscr{J}(\mathbf{B}\cdot\delta\mathbf{B} - \boldsymbol{\xi}\cdot\nabla P)\rangle\rangle = i\mathscr{W}_p\mathbf{c}_p. \tag{6.57}$$

We consider two types of modes: internal and external modes. As the name indicates, the internal modes are the modes that are present inside the plasma. The existence of the internal modes can therefore be determined by whether there is a nontrivial solution that satisfies the boundary condition $\Xi_\psi = 0$ at the right end ψ_r inside the plasma (or Ξ_ψ is small if the right end is a singular surface $\det|\mathscr{F}| = 0$). Numerically, one can simply scan the determinant $\det|\Xi_p|$ from the left end to the right to see whether $\det|\Xi_p|$ vanishes at a certain ψ_r inside the plasma region with an eigenfrequency. If yes, there is an internal mode. The corresponding internal eigen mode is then determined by solving the homogeneous linear equations $\Xi_p\mathbf{c}_p = 0$ (or being small if it is a singular surface) at ψ_r, with one of M components $\mathbf{c}_p$ being the normalization constant. With $\mathbf{c}_p$ determined, the eigenfunction is given by $\Xi_\psi = \Xi_p\mathbf{c}_p$ for $\psi_l \le \psi \le \psi_r$. The procedure is similar to the DCON code [1].

For the external modes, one needs to have the vacuum equations described in subsection 6.1.4.2. In the presence of a wall, there are two vacuum regions inside and outside of the wall. For brevity's sake, we describe the solutions in the inner and outer vacuum regions simultaneously. We also use an independent solution method to describe vacuum solutions. The $2M$ independent solutions of equation (6.43) can be used to construct the following independent solution matrices:

$$\begin{aligned}\begin{pmatrix}\mathscr{U}_1\\ \mathscr{V}_1\end{pmatrix} &\equiv \begin{pmatrix}\mathbf{u}^1, & \cdots, & \mathbf{u}^M\\ \mathbf{v}^1, & \cdots, & \mathbf{v}^M\end{pmatrix},\\ \begin{pmatrix}\mathscr{U}_2\\ \mathscr{V}_2\end{pmatrix} &\equiv \begin{pmatrix}\mathbf{u}^{M+1}, & \cdots, & \mathbf{u}^{2M}\\ \mathbf{v}^{M+1}, & \cdots, & \mathbf{v}^{2M}\end{pmatrix},\end{aligned}$$

where the superscripts have been introduced to label the independent solutions. The independent solutions are constructed by the applied boundary conditions to be specified later on. The general solutions in the inner or outer vacuum regions can be expressed as a linear combination of the independent solutions:

$$\begin{pmatrix}\mathbf{u}\\ \mathbf{v}\end{pmatrix} = \begin{pmatrix}\mathscr{U}_1\\ \mathscr{V}_1\end{pmatrix}\mathbf{c}_v + \begin{pmatrix}\mathscr{U}_2\\ \mathscr{V}_2\end{pmatrix}\mathbf{d}_v, \tag{6.58}$$

where $\mathbf{c}_v$ and $\mathbf{d}_v$ are constant vectors in the independent solution space with $\mathbf{c}_{v1}$ and $\mathbf{d}_{v1}$ denoting the constants for inner vacuum region and $\mathbf{c}_{v2}$ and $\mathbf{d}_{v2}$ for outer vacuum region.

In the vacuum region outside the wall, the scalar potential $\mathbf{u}$ needs to satisfy the M boundary conditions at infinite ψ: $\mathbf{u} \to 0$ (or being the small solution). With these M boundary conditions at infinity imposed, there are only M independent solutions left in this vacuum region. Without loss of generality, $\mathbf{c}_{v2}$ can be zero. Consequently, equation (6.58) reduces to

$$\mathbf{u}|_{\psi_{b+}} = \mathscr{T}\mathbf{v}|_{\psi_{b+}}, \tag{6.59}$$

where the $M \times M$ matrix $\mathscr{T}$ is given by $\mathscr{T} = \mathscr{U}_2\mathscr{V}_2^{-1}|_{\psi_{b+}}$. The matrix $\mathscr{T}$ can be computed by means of the Green function method reviewed in subsection 6.1.4.2 [4]. In the vacuum region between the plasma and the wall, the independent solutions are constructed with the following boundary conditions imposed on the inner side of the wall ψ_{b-}:

$$\begin{pmatrix} \mathscr{U}_1 \\ \mathscr{V}_1 \end{pmatrix}_{\psi_{b-}} = \begin{pmatrix} \mathscr{I} \\ \mathscr{O} \end{pmatrix}, \tag{6.60}$$

$$\begin{pmatrix} \mathscr{U}_2 \\ \mathscr{V}_2 \end{pmatrix}_{\psi_{b-}} = \begin{pmatrix} \mathscr{T} \\ \mathscr{I} \end{pmatrix}, \tag{6.61}$$

where $\mathscr{O}$ is $M \times M$ zero matrix. Since the boundary conditions in equation (6.60) give rise to $\delta\mathbf{B} \cdot \nabla\psi = 0$ on the inner side of the wall, these conditions correspond to the set of independent solutions that corresponds to the perfectly conducting wall type. On the other hand, since the boundary conditions in equation (6.61) guarantee that the independent solutions to be continuous with the outer vacuum solutions in equation (6.59), these conditions correspond to the set of independent solutions that corresponds to the no-wall type. Using the general expression for the solutions in equation (6.58), the normal and parallel magnetic fields at the plasma-vacuum interface can be obtained as follows:

$$\langle\langle \mathscr{J}\nabla\psi \cdot \delta\mathbf{B}\rangle\rangle = -\mathscr{V}_1\mathbf{c}_{v1} - \mathscr{V}_2\mathbf{d}_{v1}, \tag{6.62}$$

$$-\langle\langle \mathscr{J}\mathbf{B} \cdot \delta\mathbf{B}\rangle\rangle = i\mathscr{Q}\left(\mathscr{U}_1\mathbf{c}_{v1} + \mathscr{U}_2\mathbf{d}_{v1}\right). \tag{6.63}$$

Let us first study the no-wall and the perfectly conducting wall cases. Matching the solutions on the plasma side in equations (6.56) and (6.57) and the solutions on the vacuum side in equations (6.62) and (6.63) yields

$$\mathscr{Q}\mathbf{c}_p - \mathscr{V}_1\mathbf{c}_{v1} - \mathscr{V}_2\mathbf{d}_{v1} = 0, \tag{6.64}$$

$$i\mathscr{W}_p\mathbf{c}_p - i\mathscr{Q}\left(\mathscr{U}_1\mathbf{c}_{v1} + \mathscr{U}_2\mathbf{d}_{v1}\right) = 0. \tag{6.65}$$

In the case with a perfectly conducting wall, one has $\mathbf{d}_{v1} = 0$. One can then solve equation (6.64) for $\mathbf{c}_{v1}$, yielding

$$\mathbf{c}_{v1} = \mathscr{V}_1^{-1}\mathscr{Q}\mathbf{c}_p. \tag{6.66}$$

Inserting this solution to equation (6.65) yields

$$\delta\mathscr{W}_b \mathbf{c}_p = 0, \tag{6.67}$$

where

$$\delta\mathscr{W}_b = \mathscr{W}_p - \mathscr{Q}[\mathscr{U}_1\mathscr{V}_1^{-1}]_{\psi_a}\mathscr{Q},$$

Equation (6.67) is a homogeneous equation with the solubility condition

$$\det|\delta\mathscr{W}_b| = 0. \tag{6.68}$$

The solubility condition in equation (6.68) is the dispersion relation for determining the eigenfrequency for the perfectly conducting wall case. With equation (6.68) satisfied, one can get the homogeneous solution $\mathbf{c}_p$ from equation (6.67). One can then further determine $\mathbf{c}_{v1}$ from equation (6.66). The nontrivial solutions $\mathbf{c}_p$ and $\mathbf{c}_{v1}$ can then be used to sum up the independent solutions to determine the eigenfunctions using equations (6.55) and (6.58), respectively, for the plasma and vacuum regions.

The no-wall case is similar. One only needs to set $\mathbf{c}_{v1} = 0$. One can then solve equation (6.64) for $\mathbf{d}_{v1}$, yielding

$$\mathbf{d}_{v1} = \mathscr{V}_2^{-1}\mathscr{Q}\mathbf{c}_p. \tag{6.69}$$

Inserting this solution to equation (6.65) yields

$$\delta\mathscr{W}_\infty \mathbf{c}_p = 0, \tag{6.70}$$

where

$$\delta\mathscr{W}_\infty = \mathscr{W}_p - \mathscr{Q}[\mathscr{U}_2\mathscr{V}_2^{-1}]_{\psi_a}\mathscr{Q},$$

Similarly, equation (6.70) gives rise to the dispersion relation for determining the eigenfrequency for the no-wall case

$$\det|\delta\mathscr{W}_\infty| = 0. \tag{6.71}$$

The nontrivial solutions $\mathbf{c}_p$ and $\mathbf{d}_{v1}$ from equations (6.69) and (6.70) can be then used to sum up the independent solutions to determine the eigenfunctions. Here, it is noted that in the no-wall case, one does not need to solve the vacuum solutions separately in the inner and outer vacuum regions using the continuity condition. The solution from the plasma to infinity can be obtained directly.

We now consider the resistive wall case. The thin-wall limit is considered. The thick-walled case can be found in reference [5]. There are three interfaces: the plasma and inner vacuum region, ψ_a, and the inner vacuum region and wall, ψ_{b-}, and the outer vacuum region and wall, ψ_{b+}. We first match the plasma and the inner vacuum regions. The normal magnetic field component and the combined magnetic and thermal pressures are required to be continuous at the plasma-vacuum interface. Matching the plasma solutions in equations (6.56) and (6.57) and the vacuum solutions in equations (6.62) and (6.63) at the interface ψ_a gives

$$\mathbf{d}_{v1} = \mathscr{F}_1^{-1}\delta\mathscr{W}_b\delta\mathscr{W}_\infty^{-1}\mathscr{F}_2\mathbf{c}_{v1}, \tag{6.72}$$

where $\mathscr{F}_1 = \mathscr{Q}\left[\mathscr{U}_2 - \mathscr{U}_1\mathscr{V}_1^{-1}\mathscr{V}_2\right]_{\psi_a}$, and $\mathscr{F}_2 = \mathscr{Q}\left[\mathscr{U}_1 - \mathscr{U}_2\mathscr{V}_2^{-1}\mathscr{V}_1\right]_{\psi_a}$. One can see that $\delta\mathscr{W}_\infty$ and $\delta\mathscr{W}_b$ correspond to the energy matrices without a wall and with a perfectly conducting wall at ψ_b, respectively, noting the boundary conditions in equations (6.60) and (6.61).

We next match the inner and outer vacuum solutions across the thin resistive wall. Using the thin-wall assumption, the Maxwell equation $\nabla \cdot \delta\mathbf{B} = 0$ requires that the radial magnetic field is continuous across the wall. Using equation (6.58) respectively for inner and outer vacuum regions, this yields

$$\mathbf{v}|_{\psi_{b-}} = \mathbf{v}|_{\psi_{b+}} = \mathbf{d}_{v1}.$$

According to Ampere's law, the current in the resistive wall leads to a jump in the scalar magnetic potential. Using Ohm's law to represent the current density, the Ampére law yields

$$\nabla \times \nabla \times \delta\mathbf{B} \quad = \quad -\gamma\mu_0\sigma\delta\mathbf{B}, \tag{6.73}$$

where σ is the wall conductivity. Using equation (6.58) respectively for inner and outer vacuum regions, equation (6.73) reduces to

$$\mathscr{V}\left(\mathbf{u}|_{\psi_{b+}} - \mathbf{u}|_{\psi_{b-}}\right) = \tau_w\gamma_N\mathbf{d}_{v1}, \tag{6.74}$$

where the resistive wall time $\tau_w = \mu_0\sigma db/\tau_A$, γ_N is the normalized growth rate, d is the wall thickness, b is the average wall minor radius, and

$$\begin{aligned}\mathscr{V} \quad = \quad & \mathscr{M}\left[\left[\mathscr{J}|\nabla\psi||\nabla\theta| - \mathscr{J}|\nabla\psi\cdot\nabla\theta|^2/(|\nabla\psi||\nabla\theta|)\right]\right]\mathscr{M} \\ & +n^2\left[\left[\mathscr{J}|\nabla\phi|^2|\nabla\psi|/|\nabla\theta|\right]\right].\end{aligned}$$

Since $\mathbf{c}_{v2} = 0$ in the outer vacuum region, equation (6.58) respectively for inner and outer vacuum regions yield

$$\mathbf{u}|_{\psi_{b+}} - \mathbf{u}|_{\psi_{b-}} \quad = \quad -\mathbf{c}_{v1}. \tag{6.75}$$

Using equations (6.72), (6.74), and (6.75), one can find the eigenmode equations

$$\mathscr{D}_0(\gamma_N)\mathbf{d}_{v1} \equiv \tau_w\gamma_N\mathbf{d}_{v1} + \mathscr{V}\mathscr{F}_2^{-1}\delta\mathscr{W}_\infty\delta\mathscr{W}_b^{-1}\mathscr{F}_1\mathbf{d}_{v1} = \mathscr{O}. \tag{6.76}$$

The solubility condition for the dispersion relation in equation (6.76), $\det|\mathscr{D}_0(\gamma_N)| = 0$, gives the dispersion relation for the eigenvalue of resistive wall modes. The Nyquist diagram, in general, can be used to determine the roots of this dispersion relation. Note, however, that the growth rate for resistive wall modes is much smaller than the Alfvén frequency. One can neglect the growth rate dependence of $\delta\mathscr{W}_\infty\delta\mathscr{W}_b^{-1}$ in the dispersion relation. Consequently, the eigenvalue problem is reduced to

$$-\mathscr{V}\mathscr{F}_2^{-1}\delta\mathscr{W}_\infty\delta\mathscr{W}_b^{-1}\mathscr{F}_1\mathbf{d}_{v1} = \tau_w\gamma_N\mathbf{d}_{v1}. \tag{6.77}$$

Under this simplification, the resistive wall mode growth rate γ_N on the right-hand side of equation (6.77) is used as the eigenvalue to determine the stability.

6.2 MHD MODE SPECTRUM

In this section, we discuss the MHD mode spectrum in toroidal geometry. The MHD mode spectrum in the cylinder geometry has been described in detail in Section 5.3 in Chapter 5. Readers are also referred to the author's earlier book [2] for the reviews of this topic. Here, we will focus on the relationship between singularities and spectra in the magnetohydrodynamic equations.

To study the MHD mode spectrum in tokamaks, one needs to start with the MHD equations in the Fourier space, equations (6.27), (6.29), and (6.30). Three equations describe three basic waves: the compressional Alfvén, the shear Alfvén, and the acoustic or sound waves. Therefore, one should expect three types of spectra. The compressional Alfvén wave has a high frequency and needs more energy to drive. It is usually suppressed for tokamak stability. The sound wave describes the coupling of the particle parallel motion to the waves. Because of the non-localization, the kinetic treatment is required in general for sound waves, unless in the non-resonance case. Therefore, we may focus on the shear Alfvén mode spectrum.

By solving equation (6.29) for Ξ_s and equation (6.27) for $\langle\langle \nabla \cdot \boldsymbol{\xi} \rangle\rangle$, one obtains the shear Alfvén mode equation for Ξ_ψ in equation (6.31). The highest derivative term in this equation is $(\mathscr{F}\Xi')'$. Therefore, the singularity of the equation is determined by

$$\det|\mathscr{F}| = 0. \tag{6.78}$$

Note that $\det|\mathscr{F}|$ is a function of the frequency ω and magnetic flux ψ. The plot of the solution of equation (6.78) in ω and ψ gives rise to the so-called spectrum diagram for shear Alfvén modes. Here, it should be noted that in deriving the shear Alfvén mode equation (6.31), the invertibility of the matrices $\mathscr{A}$ and $\mathscr{L}_s$ is assumed. One may plot $\det|\mathscr{A}| = 0$ and $\det|\mathscr{L}_s| = 0$ for compressional Alfvén and sound waves for the invertibility conditions. However, they are not their spectra. For spectra, one needs to derive the homogeneous equations for each of them and determine the singularities for the individual equation.

There are other works for determining the spectra based on the high n mode formalism, as described in references [6] and [7]. The vorticity equation approach based on equations (6.16)–(6.14) is used. Since high n modes are considered, the compressional Alfvén mode is suppressed. Therefore, the spectra are determined only by the vorticity equation (6.14) and sound wave equation (6.15). The typical results are plotted in figure 6.1 [7]. In Part a), the sound continuum is included; in Part b), the sound wave spectrum is suppressed.

6.3 INTERNAL KINK MODES

In this section, let us discuss the internal mode instabilities. The internal modes describe the unstable modes that developed inside the interface between the plasma and the surrounding vacuum region. The vorticity equation (6.31) can be used to study

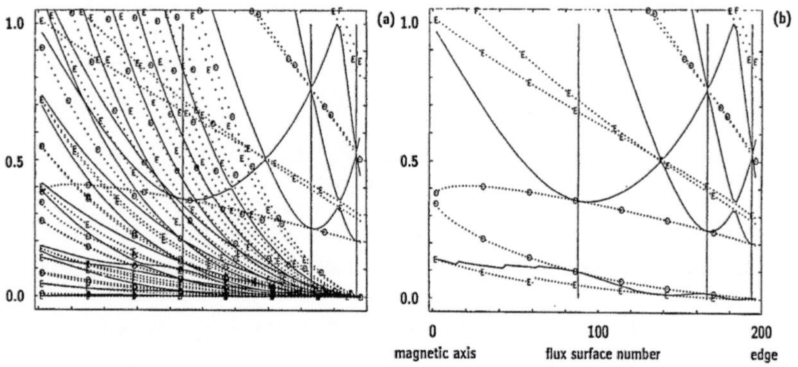

Figure 6.1 MHD spectrum: a) with the sound wave spectrum included; b) the sound waves are suppressed/ Reproduced from [7], with the permission of AIP Publishing.

the internal MHD modes

$$\frac{d}{d\psi}\left(\mathscr{F}\frac{d\Xi^{\psi}}{d\psi}+\mathscr{K}\Xi^{\psi}\right)-\mathscr{K}^{\dagger}\frac{d\Xi^{\psi}}{d\psi}-\mathscr{G}\Xi^{\psi}=0. \tag{6.79}$$

One can determine the stability from this equation either using the generalized Newcomb theorem 5 [8, 1] for marginal stability criterion or by using a direct eigenmode approach to determine the growth rate [3].

We note that equation (6.79) can be transformed into a set of differential equations of the first order

$$\mathbf{u}' = \mathscr{L}\mathbf{u}, \tag{6.80}$$

where

$$\mathbf{u} = \begin{pmatrix}\mathbf{u}_1\\ \mathbf{u}_2\end{pmatrix} = \begin{pmatrix}\Xi^{\psi}\\ \mathscr{F}\Xi^{\psi\prime}+\mathbf{K}\Xi^{\psi}\end{pmatrix}$$

and

$$\mathscr{L} = \begin{pmatrix}-\mathbf{F}^{-1}\mathbf{K} & \mathbf{F}^{-1}\\ \mathscr{G}-\mathbf{K}^{L}\mathbf{F}^{-1}\mathbf{K} & \mathbf{K}^{L}\mathbf{F}^{-1}\end{pmatrix}.$$

Equation (6.80) is a set of $2M$ linear differential equations. There are $2M$ independent solutions $\mathbf{u}^k$ with $k=1,\cdots 2M$. Its general solution can be obtained by the summation of the independent solutions

$$\mathbf{u} = \sum_{k=1}^{2M} c_k\mathbf{u}^k,$$

where the $2M$ constants c_k are determined by the boundary conditions.

One can always impose the independent solutions to satisfy the boundary conditions on one side, say the left—the magnetic axis—and obtain the remaining M independent solutions by shooting from the left as an initial value problem. The general solution with the boundary conditions at the left satisfied can be expressed as

$$\begin{pmatrix} \mathbf{u}_1 \\ \mathbf{u}_2 \end{pmatrix} = i \begin{pmatrix} \mathscr{U}_p \\ \mathscr{W}_p \mathscr{U}_p \end{pmatrix} \mathbf{c}_p, \tag{6.81}$$

where $\mathbf{c}_p$ is a constant vector with M components, $\mathscr{U}_p$ and $\mathscr{W}_p$ are $M \times M$ matrices defined by the independent solutions. This is because with the boundary conditions on the lect, $\Xi^{\psi\prime}$ and Ξ^{ψ} are no longer independent.

For internal modes, the condition for $\mathbf{u}_1 = \Xi^{\psi}$ to vanish or to be a small solution on the other end leads to the eigenmode equation

$$\mathscr{U}_p \mathbf{c}_p = 0. \tag{6.82}$$

Here, it is noted that although equation (6.82) holds, $\mathscr{W}_p \mathscr{U}_p \mathbf{c}_p \neq 0$ in general. To have the nontrivial solution, one must have

$$\det \left| \mathscr{U}_p \right| = 0. \tag{6.83}$$

There are two procedures for determining stability. One is by the generalized Newcomb theorem [1, 8]. In this procedure, only the marginal stability condition is determined. In shooting from the left to the right with the growth rate set to zero, the system is unstable if at a certain ψ equation (6.83) holds. Because the growth rate is set to be zero, the shooting has to be made between the rational surfaces or the boundaries at the magnetic axis or the plasma-vacuum interface. The other procedure is to treat equation (6.83) as the dispersion relation to determine the eigen frequency [3]. In this procedure, $\mathscr{U}_p(\omega)$ is computed. The Nyquist diagram can be used to determine if there are unstable roots. To shoot over the rational surface, a small growth rate is usually introduced. The marginal stability can be determined by exploration.

Sometimes, one desires to have the damping rate estimated in the stable case. Note that equation (6.79) is just the Euler-Lagrangian equation of the energy principle

$$\delta W_p + \delta W_k = \frac{1}{2\mu_0} \int d\psi \left(\Xi^{\psi\prime\dagger} \quad \Xi^{\psi\dagger} \right) \begin{pmatrix} \mathscr{F} & \mathscr{K} \\ \mathscr{K}^{\dagger} & \mathscr{G} \end{pmatrix} \begin{pmatrix} \Xi^{\psi\prime} \\ \Xi^{\psi} \end{pmatrix}. \tag{6.84}$$

Plugging in the solution of the Euler-Lagrangian equation (6.79), the energy integral in equation (6.84) is reduced to

$$\delta W_p + \delta W_k = \Xi^{\psi\dagger} \frac{d}{d\psi} \left(\mathscr{F} \frac{d\Xi^{\psi}}{d\psi} + \mathscr{K} \Xi^{\psi} \right) \Bigg|_{\psi_1}^{\psi_2}. \tag{6.85}$$

Suppose there is a singular point at ψ_s with $\psi_1 < \psi_s < \psi_2$, With the boundary conditions at ψ_1 and ψ_2 imposed, one can determine the independent solutions in the regions (ψ_1, ψ_s^-) and (ψ_s^+, ψ_2), respectively. With the same normalization condition imposed, one can use equation (6.85) for estimating the damping rate with the energy deficit.

6.4 EXTERNAL KINK MODES

In this section, we discuss the external modes. Comparing with the internal modes described in Section 6.3, the external modes are coupled to the perturbations in the vacuum region and the wall, which are external to the plasma torus. In particular, the right boundary, *i.e.*, the plasma-vacuum interface, gives a finite contribution to the energy integral in equation (6.85).

The configuration is described as follows. The plasma torus is surrounded by the inner vacuum region. The inner vacuum region is separated from the outer vacuum region by a wall, which can be perfectly conducting or have a finite resistivity. The outer vacuum region extends from the wall to infinity. Three cases will be considered: the no-wall, the perfectly conducting wall, and the resistive wall cases. This needs to describe the solutions in the inner and outer vacuum regions and the wall. They have been given in detail in Section 6.1.5. In the no-wall case, the dispersion relation is given by equation (6.71) as follows:

$$\det|\delta\mathscr{W}_\infty| = 0.$$

In the perfectly conducting wall case, the dispersion relation is given by equation (6.68) as follows:

$$\det|\delta\mathscr{W}_b| = 0.$$

In the resistive wall mode case, the eigenmode equation is given by equation (6.77) as follows:

$$-\mathscr{V}\mathscr{F}_2^{-1}\delta\mathscr{W}_\infty\delta\mathscr{W}_b^{-1}\mathscr{F}_1\mathbf{d}_{v1} = \tau_w\gamma_N\mathbf{d}_{v1}. \quad (6.86)$$

In studying the external modes, one usually considers the equilibria that are stable to the internal modes.

We first discuss the no-wall and perfectly conducting wall cases. One may regard the no-wall case as the case with the wall at infinity and the fixed boundary case, i.e., $\Xi_a^\psi = 0$, as the case with the perfectly conducting wall residing right at the plasma edge. The perfectly conducting wall plays a stabilizing role. The closer the perfectly conducting wall to the plasma edge, the more stable the system. There is a so-called critical wall position, as shown by the vertical dotted line in figure 6.2, beyond which the perfectly conducting wall is unable to stabilize the external kink modes.

In reality, the wall is never perfectly conducting but has resistivity. This leads to the consideration of the resistive wall modes (RWMs) [10]. From the RWM eigenmode equation (6.86), one can find the dispersion relation

$$\det|\mathscr{V}\mathscr{F}_2^{-1}\delta\mathscr{W}_\infty\delta\mathscr{W}_b^{-1}\mathscr{F}_1 + \tau_w\gamma_N| = 0.$$

In general, the Nyquist diagram or the analytical continuation method can be used to determine the unstable roots from the dispersion relation. For resistive wall modes, however, the growth rate is much smaller than the Alfvén frequency. Therefore, the

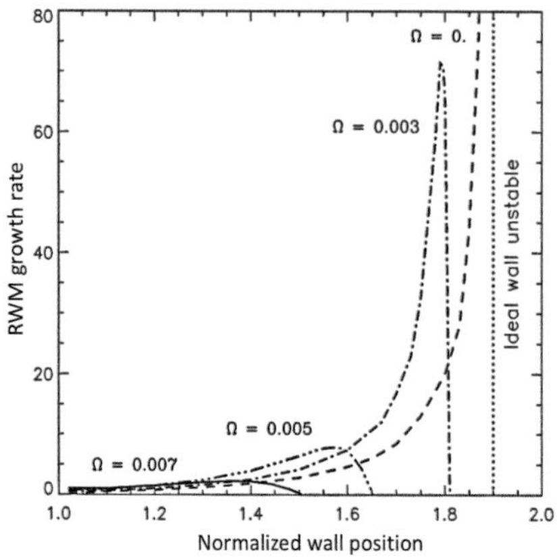

Figure 6.2 Resistive wall mode growth rate in the presence of plasma rotation. Reproduced from [9], with the permission of AIP Publishing.

growth rate dependence of $\delta\mathscr{W}_\infty\delta\mathscr{W}_b^{-1}$ can be neglected for determining the stability condition. Consequently, one obtains the reduced eigenvalue problem

$$\tau_w\gamma_N \quad = \quad -\mathscr{V}\mathscr{F}_2^{-1}\delta\mathscr{W}_\infty\delta\mathscr{W}_b^{-1}\mathscr{F}_1 \tag{6.87}$$

with the resistive wall mode growth rate γ_N on the left-hand side used as the eigenvalue to determine the stability. One can expect that the modes which are stabilized by the perfectly conducting wall can become unstable with the resistive wall modes. The RWM growth rate is determined by equation (6.76). For physics discussion, let us consider the cylinder limit. In this limit, the matrices in equation (6.86) are reduced to real numbers. One can see that RWM becomes unstable when $\delta\mathscr{W}_\infty\delta\mathscr{W}_b^{-1} < 0$, i.e., when the system is stabilized by the perfectly conducting wall $\delta\mathscr{W}_b > 0$ but unstable without the wall $\delta\mathscr{W}_\infty < 0$. Because the growth rate is inversely proportional to $\delta\mathscr{W}_b$, the RWM growth rate tends to infinity as the wall position approaches the critical wall position. This feature can be seen in the toroidal calculation in figure 6.2, the dashed curve with the rotation frequency $\Omega = 0$ [9].

Because the RWM growth rate is inversely proportional to $\delta\mathscr{W}_b$, one can expect that the non-ideal MHD effects, especially the resonance effects, can play a significant role in stabilizing RWMs in the parameter domain close to the critical wall position $\delta\mathscr{W}_b = 0$. This has been extensively examined in this field, for instance, in references [11] and [12]. For example, including the shear Alfvén resonance with the plasma toroidal rotation leads the plasma energy integrals to contain an imaginary part [12]

$$\delta W_\infty = \delta W_\infty^r + i\delta W_{rot}, \quad \delta W_b = \delta W_b^r + i\delta W_{rot}, \tag{6.88}$$

where superscript r represents the real part. Therefore, the RWM growth rate in equation (6.87) becomes

$$\tau_w \gamma_N \propto -\frac{\delta W_\infty}{\delta \mathscr{W}_b} = -\frac{\delta W_\infty^r \delta W_b^r + |\delta W_{rot}|^2 - i|\delta W_{rot}|(\delta W_b^r - \delta W_\infty^r)}{|\delta W_b^r + \delta W_{rot}|^2}. \quad (6.89)$$

Here, again, we have used the cylinder limit for physics discussion. From equation (6.89), one can see that the imaginary part induced by the shear Alfvén resonance always contributes a stabilizing term $|\delta W_{rot}|^2$ in the numerator. When the wall position is close to the critical wall position, $\delta W_b^r \to 0$. In this case, the destabilizing term $\delta W_\infty^r \delta W_b^r < 0$ in the numerator decreases, so that the stabilizing rotation effect $|\delta W_{rot}|^2$ tends to be dominant, and the shear Alfvén resonance stabilizes RWMs. The numerical results in figure 6.2 just reflect these features.

One can also investigate the kind of inverse problem for RWMs: how the error-field-induced torque can brake tokamak toroidal rotation. This is the so-called error-field amplification phenomenon [13, 14, 15]. The details are reviewed in the author's earlier book [2]. Here, let us just briefly outline the concept and results.

The torque on the error-field current structure is given as follows:

$$\tau_\phi^e = \frac{1}{2}\int_{V_w} R^2 \nabla\phi \cdot \delta \mathbf{J}^* \times \delta \mathbf{B} d\mathbf{r} + cc,$$

where "cc" represents the complex conjugate of the previous term. According to Newton's third law, this equals the torque on the rotating plasma by the error-field current structure. By a similar matching process to the derivation of RWM growth rate, one can find that [14]

$$\tau_\phi^e = \frac{in\pi}{\mu_0} \mathbf{J}_e^\dagger \mathscr{F}_1^{-1} \delta \mathscr{W}_b \delta \mathscr{W}_\infty^{-1} \mathscr{F}_2 \mathbf{J}_e + cc, \quad (6.90)$$

where $\mathbf{J}$ specifies the strength of the error field. From equation (6.90), one can see that the torque becomes infinity in the no-wall limit $\delta \mathscr{W}_\infty = 0$. In terms of plasma beta, this is lower than the perfectly conducting wall limit $\delta \mathscr{W}_b = 0$ when the rotation can play a significant role in stabilizing RWMs.

6.5 LOCALIZED INTERCHANGE MODES

We have discussed the global mode theories in the previous sections and now start to describe the analytical theory. We first describe the interchange mode theory and derive the Mercier criterion [16, 17]. Interchange mode is one of the most fundamental phenomena in the magnetically confined plasmas, which resembles the so-called Rayleigh-Taylor instability in the conventional fluid theory. To prevent the development of interchange mode in the toroidal plasma confinement, the magnetic field is configured to have finite magnetic shear. Because of the stabilization effects from the field-line bending in the system with magnetic shear, the interchange mode is usually localized at the singular layer, or the mode rational surface. The cylinder theory has been reviewed in Section 5.4.1, and one obtains the Suydam criterion. In a torus, the interchange mode experiences both bad and good curvature regions, and therefore the surface average is introduced to derive the Mercier criterion. The interchange mode theory is one of the first successful toroidal theories in this field.

There are two approaches to derive the Mercier criterion: One uses the MHD equations directly, and the other employs the energy principle. We mainly review the direct approach. The review of the energy principle approach can be found in reference [2]. Derivation of the Mercier criterion is related to the singular layer equation. Earlier derivation of the singular layer equation relies on the assumption that the modes are somewhat localized poloidally. This assumption was released in a later paper by A. Glasser et al., [18]. Here, we derive the singular layer equation using the vorticity and acoustic wave equations. Especially, the plasma compressibility effects are included in the singular layer equation.

6.5.1 SINGULAR LAYER EQUATIONS

We first derive the singular layer equation in the ideal MHD formalism, which will be used to derive the Mercier criterion in the next subsection. The plasma compressibility effects are included, which allows for obtaining a full eigenvalue problem in the ideal MHD formalism.

To derive the singular layer equation, let us first introduce the localized Hamada coordinates, which are reviewed in Section 4.1.2 in general. Suppose the modes are localized around a particular rational (or singular) magnetic surface V_0. We introduce the localized Hamada coordinates $(x,\ u,\ \theta)$, where $x = V - V_0$ and $u = m\theta - n\zeta$. In this coordinate system, the parallel derivative becomes $\mathbf{B}\cdot\boldsymbol{\nabla} = \chi'(\partial/\partial\theta) + (\Lambda x/\Xi_e)(\partial/\partial u)$, where $\Lambda = \psi'\chi'' - \chi'\psi''$ and $\Xi_e = \psi'/m = \chi'/n$. Since u describes the symmetric indirection for the equilibrium scalars in the axisymmetric system, perturbation quantities can be assumed to vary as $\exp\{ik_u u\}$ with $k_u = 2\pi n/\chi'$. It is also defined that $\Theta = \boldsymbol{\nabla}V\cdot\boldsymbol{\nabla}u/|\boldsymbol{\nabla}V|^2$.

We expand $\boldsymbol{\xi}$ and $\delta\mathbf{B}$ in three directions as follows:

$$
\begin{aligned}
\boldsymbol{\xi} &= \xi\frac{\nabla V}{|\nabla V|^2} + \mu\frac{\mathbf{B}\times\nabla V}{B^2} + v\frac{\mathbf{B}}{B^2},\\
\delta\mathbf{B} &= b\frac{\nabla V}{|\nabla V|^2} + v\frac{\mathbf{B}\times\nabla V}{B^2} + \tau\frac{\mathbf{B}}{B^2}.
\end{aligned}
$$

The radially localized modes are considered. This leads to the ordering assumption that the mode wavelength across the magnetic surface $\lambda_\perp$ is much smaller than that on the surface and perpendicular to the magnetic field line $\lambda_\wedge$. Accordingly, the following ordering scheme as in reference [18] is introduced

$$
x \sim \varepsilon,\quad \frac{\partial}{\partial V} \sim \varepsilon^{-1},\quad \frac{\partial}{\partial u} \sim \frac{\partial}{\partial \theta} \sim 1,\quad \gamma \sim x,
$$

where $\varepsilon \ll 1$ is a small parameter. We also consider the low-frequency regime

$$
|\omega/\omega_{si}| \ll 1. \tag{6.91}
$$

where ω_{si} is the parallel ion acoustic frequency. This will be further discussed in treating the acoustic wave coupling. Since the modes are assumed to vary on a slow time scale as shown in equation (6.91), the compressional Alfvén waves are decoupled. As verified *a posteriori*, we adopt the following ordering scheme:

$$\xi = \varepsilon\xi^{(1)} + \cdots, \quad \mu = \mu^{(0)} + \cdots, \quad \widetilde{\nabla\cdot\boldsymbol{\xi}} \sim \varepsilon^2, \quad \langle\nabla\cdot\boldsymbol{\xi}\rangle \sim \varepsilon, \tag{6.92}$$

$$b = \varepsilon^2 b^{(2)} + \cdots, \quad v = \varepsilon v^{(1)} + \cdots, \quad \tau = \varepsilon\tau^{(1)} + \cdots, \tag{6.93}$$

Here, the perturbed quantities are decomposed into constant and oscillatory parts along the field lines: $\xi = \bar{\xi} + \tilde{\xi}$, where $\bar{\xi} = \langle\xi\rangle \equiv \oint \xi\, dl/B / \oint dl/B$, l is arc length of magnetic field line, and $\tilde{\xi} = \xi - \langle\xi\rangle$. The ordering assumptions in equations (6.92) and (6.93) are the same as those in reference [18]. Nevertheless, it is noted that the frequency scalings in the vorticity and acoustic wave equations are different. One is normalized to the shear Alfvén frequency and the other to the parallel acoustic wave frequency.

With these ordering assumptions, we can proceed to analyze the basic set of linearized MHD equations. From the compressional Alfvén mode equation (6.13), one obtains in the leading order

$$\tau^{(1)} + \delta p^{(1)} = 0. \tag{6.94}$$

Thus, from equation (6.19) one has

$$(\nabla\cdot\boldsymbol{\xi}_{\perp})^{(0)} + 2\boldsymbol{\kappa}\cdot\boldsymbol{\xi}^{(0)} = 0. \tag{6.95}$$

We now consider $\nabla \cdot \delta\mathbf{B} = 0$. With the ordering assumptions given above, one obtain

$$\begin{aligned}&\frac{\partial b^{(2)}}{\partial x} + \frac{1}{\Xi}\frac{\partial}{\partial u}v^{(1)} + \frac{\partial v^{(1)}}{\partial\theta}\frac{\mathbf{B}\times\nabla V\cdot\nabla\theta}{B^2}\\ &+v^{(1)}\nabla\cdot\frac{\mathbf{B}\times\nabla V}{B^2} + \chi'\frac{\partial}{\partial\theta}\frac{\tau^{(1)}}{B^2} = 0.\end{aligned} \tag{6.96}$$

Noting that

$$\begin{aligned}\frac{\mathbf{B}\times\nabla V\cdot\nabla\theta}{B^2} &= \frac{1}{P'}(\mathbf{J}-\sigma\mathbf{B})\cdot\nabla\theta = \frac{J'}{P'} - \frac{\chi'}{P'}\sigma,\\ \nabla\cdot\frac{\mathbf{B}\times\nabla V}{B^2} &= \frac{2\mathbf{B}\times\boldsymbol{\kappa}}{B^2}\cdot\nabla V = -\frac{1}{P'}\chi'\frac{\partial}{\partial\theta}\sigma,\end{aligned} \tag{6.97}$$

where equation (6.18) has been used, equation (6.96) is reduced to

$$\frac{\partial b^{(2)}}{\partial x} + \frac{1}{\Xi}\frac{\partial}{\partial u}v^{(1)} + \frac{J'}{P'}\frac{\partial v^{(1)}}{\partial\theta} - \frac{\chi'}{P'}\frac{\partial\sigma v^{(1)}}{\partial\theta} + \chi'\frac{\partial}{\partial\theta}\frac{\tau^{(1)}}{B^2} = 0. \tag{6.98}$$

After the surface average, it yields

$$\frac{\partial\bar{b}^{(2)}}{\partial x} + \frac{1}{\Xi_e}\frac{\partial\bar{v}^{(1)}}{\partial u} = 0. \tag{6.99}$$

We next treat the induction equation, equation (6.2). The ∇V - projection in two significant orders yields in

$$0 = \chi' \frac{\partial \xi^{(1)}}{\partial \theta}, \tag{6.100}$$

$$b^{(2)} = \chi' \frac{\partial \xi^{(2)}}{\partial \theta} + \frac{\Lambda x}{\Xi_e} \frac{\partial \xi^{(1)}}{\partial u}, \tag{6.101}$$

The ∇u - projection in the lowest order yields

$$\chi' \frac{\partial \mu^{(0)}}{\partial \theta} = 0. \tag{6.102}$$

The projection along the magnetic field line yields

$$(\nabla \cdot \boldsymbol{\xi}_\perp)^{(0)} + 2\boldsymbol{\kappa} \cdot \boldsymbol{\xi}^{(0)} = \frac{\partial \xi^{(1)}}{\partial x} + \frac{1}{\Xi_e} \frac{\partial \mu^{(0)}}{\partial u} = 0, \tag{6.103}$$

where equations (6.16), (6.94), and (6.95) have been used.

Further on, let us work on the vorticity equation (6.14). The leading and first-order equations are as follows

$$\frac{\partial}{\partial \theta} \left(\frac{|\nabla V|^2}{B^2} \frac{\partial v^{(1)}}{\partial x} \right) + \frac{1}{P'} \frac{\partial \sigma}{\partial \theta} \frac{\partial \tau^{(1)}}{\partial x} = 0, \tag{6.104}$$

$$\begin{aligned}
& -\omega^2 \frac{\rho_m |\nabla V|^2}{B^2} \frac{\partial \mu^{(0)}}{\partial x} \\
= & -\chi' \frac{\partial}{\partial \theta} \left(\frac{|\nabla V|^2}{B^2} \frac{\partial v^{(2)}}{\partial x} - v \frac{\mathbf{B}}{B^2} \cdot \nabla \times \frac{\mathbf{B} \times \nabla V}{B^2} - \tau \frac{\mathbf{B}}{B^2} \cdot \nabla \times \frac{\mathbf{B}}{B^2} + \frac{J'}{\chi'} \tau^{(1)} \right) \\
& -v^{(1)} \left(\frac{J'}{P'} - \frac{\chi'}{P'} \sigma \right) \frac{\partial \sigma}{\partial \theta} - \tau^{(1)} \frac{\chi'}{B^2} \frac{\partial \sigma}{\partial \theta} - \Lambda x \frac{|\nabla V|^2}{\Xi_e B^2} \frac{\partial}{\partial u} \frac{\partial v^{(1)}}{\partial x} + \frac{P'}{\Xi_e B^2} \frac{\partial \tau^{(1)}}{\partial u} \\
& - \left(\frac{(P+B^2)'}{\Xi_e B^2} - \frac{1}{P'} \chi' \frac{\partial \sigma}{\partial \theta} \Theta \right) \frac{\partial \tau^{(1)}}{\partial u} \\
& - \frac{\chi'}{P'} \frac{\partial \sigma}{\partial \theta} \frac{\partial}{\partial x} \left(\Gamma P \widetilde{\nabla \cdot \boldsymbol{\xi}}^{(2)} + P' \xi^{(2)} \right).
\end{aligned} \tag{6.105}$$

The singular layer equation will be derived by averaging equation (6.105), which needs to express all unknowns in this equation in terms of $\xi^{(1)}$. It is trivial to obtain $\tau^{(1)}$ from equation (6.94) and $\mu^{(0)}$ from equations (6.102) and (6.103). The rest are derived as follows.

For $v^{(1)}$ contributions, from equations (6.100) and (6.101), one obtains $\bar{b}^{(2)} = (\Lambda x/\Xi)(\partial \xi^{(1)}/\partial u)$. With this result, one can determine $\bar{v}^{(1)}$ from equation (6.99):

$$\bar{v}^{(1)} = -\Lambda \frac{\partial}{\partial x} (x \xi^{(1)}). \tag{6.106}$$

Using equation (6.106) to determine the integration constant, one can solve equation (6.104), yielding

$$\begin{aligned}\frac{\partial v^{(1)}}{\partial x} &= -\frac{1}{P'}\left(\frac{B^2\sigma}{|\nabla V|^2}-\frac{\langle B^2\sigma/|\nabla V|^2\rangle}{\langle B^2/|\nabla V|^2\rangle}\frac{B^2}{|\nabla V|^2}\right)\frac{\partial \tau^{(1)}}{\partial x}\\ &\quad -\Lambda\frac{B^2/|\nabla V|^2}{\langle B^2/|\nabla V|^2\rangle}\frac{\partial^2}{\partial x^2}(x\xi^{(1)}).\end{aligned} \tag{6.107}$$

For $\xi^{(2)}$ contributions, from equations (6.98) and (6.101) one obtains

$$-\chi'\frac{\partial^2\xi^{(2)}}{\partial\theta\partial x} = \frac{1}{\Xi_e}\frac{\partial \tilde{v}^{(1)}}{\partial u}+\frac{J'}{P'}\frac{\partial v^{(1)}}{\partial\theta}-\frac{\chi'}{P'}\frac{\partial\sigma v^{(1)}}{\partial\theta}+\chi'\frac{\partial}{\partial\theta}\frac{\tau^{(1)}}{B^2}. \tag{6.108}$$

Here, $v^{(1)}$ can be obtained from equation (6.107) and $\tau^{(1)}$ from equation (6.94).

One also needs to solve the equation of parallel motion, equation (6.20), for $\widetilde{\nabla\cdot\boldsymbol{\xi}}$. Taking into consideration the low-frequency assumption, the second term on the left can be neglected. The equation then becomes

$$\chi'^2\frac{\partial}{\partial\theta}\left(\frac{1}{B^2}\frac{\partial\widetilde{\nabla\cdot\boldsymbol{\xi}}^{(2)}}{\partial\theta}\right)=i\frac{\rho_m\Xi_e\omega^2}{k_u\Gamma P}\frac{\mathbf{B}\times\nabla V}{B^2}\cdot\boldsymbol{\kappa}\frac{\partial\xi^{(1)}}{\partial x}, \tag{6.109}$$

where equation (6.103) has been used. Using equation (6.18), equation (6.109) can be solved

$$\chi'\frac{\partial}{\partial\theta}\widetilde{\nabla\cdot\boldsymbol{\xi}}^{(2)}=i\frac{\rho_m\omega^2}{\alpha\Gamma P}\left(B^2\sigma-\frac{\langle B^2\sigma\rangle}{\langle B^2\rangle}B^2\right)\frac{\partial\xi^{(1)}}{\partial x}. \tag{6.110}$$

Equations (6.107)—(6.110) can be used to reduce the vorticity equation (6.105). Note further that [17, 19, 20]

$$\begin{aligned}&-\sigma\chi'\frac{\partial}{\partial\theta}\frac{\nabla V\cdot\nabla u}{|\nabla V|^2}\\ &= -\sigma(\psi'\chi''-\chi'\psi'')+\frac{\sigma\mathbf{B}\times\nabla V\cdot\nabla\times(\mathbf{B}\times\nabla V)}{|\nabla V|^4},\\ &\frac{\mathbf{B}\times\nabla V\cdot\nabla\times(\mathbf{B}\times\nabla V)}{|\nabla V|^4}\\ &= \psi'\chi''-\chi'\psi''-\mathbf{B}\cdot\nabla\left(\frac{\nabla\psi\cdot\nabla\theta-\nabla\chi\cdot\nabla\zeta}{|\nabla V|^2}\right),\\ &-\frac{\nabla P\cdot\nabla(2P+B^2)}{B^2|\nabla V|^2}+\frac{\sigma\mathbf{B}\times\nabla V\cdot\nabla\times(\mathbf{B}\times\nabla V)}{|\nabla V|^4}\\ &= -\mu_0\frac{P'^2}{B^2}-P''+(\psi'J''-\chi'I'')+\mathbf{B}\cdot\nabla[(I'\nabla V\cdot\nabla\theta-J'\nabla V\cdot\nabla\zeta)/|\nabla V|^2]\\ &= \underbrace{-\mu_0\frac{P'^2}{B^2}}_{F}-\underbrace{(\psi''J'-\chi''I')}_{E}+\mathbf{B}\cdot\nabla[(I'\nabla V\cdot\nabla\theta-J'\nabla V\cdot\nabla\zeta)/|\nabla V|^2],\end{aligned} \tag{6.111}$$

where it has been noted that $P' = J'\psi' - I'\chi'$ and $P'' = J''\psi' - I''\chi' + J'\psi'' - I'\chi''$. The singular layer equation is finally obtained by averaging equation (6.105)

$$X(X\Xi)_{XX} + Q^2\Xi_{XX} + (E+F)\Upsilon + H\Xi - HX\Upsilon_{X}. + HX\Xi_X = 0, \tag{6.112}$$

where

$$\begin{aligned}
E &\equiv \frac{\langle B^2/|\nabla V|^2\rangle}{\Lambda^2}\left(J'\psi'' - I'\chi'' + \Lambda\frac{\langle\sigma B^2\rangle}{\langle B^2\rangle}\right),\\
F &\equiv \frac{\langle B^2/|\nabla V|^2\rangle}{\Lambda^2}\left(\left\langle\frac{\sigma^2 B^2}{|\nabla V|^2}\right\rangle - \frac{\langle\sigma B^2/|\nabla V|^2\rangle^2}{\langle B^2/|\nabla V|^2\rangle} + P'^2\left\langle\frac{1}{B^2}\right\rangle\right),\\
H &\equiv \frac{\langle B^2/|\nabla V|^2\rangle}{\Lambda}\left(\frac{\langle\sigma B^2/|\nabla V|^2\rangle}{\langle B^2/|\nabla V|^2\rangle} - \frac{\langle\sigma B^2\rangle}{\langle B^2\rangle}\right),\\
M &= M_c + M_t,\\
M_c &\equiv \frac{\rho_m \Xi_e^2}{k_u^2\Lambda^2}\left\langle\frac{B^2}{|\nabla V|^2}\right\rangle\left\langle\frac{|\nabla V|^2}{B^2}\right\rangle, && (6.113)\\
M_t &\equiv \frac{\rho_m \Xi_e^2}{k_u^2\Lambda^2 P'^2}\left\langle\frac{B^2}{|\nabla V|^2}\right\rangle\left(\langle\sigma^2 B^2\rangle - \frac{\langle\sigma B^2\rangle^2}{\langle B^2\rangle}\right), && (6.114)\\
G &= \frac{\langle B^2\rangle}{M\Gamma P}, \quad K = \frac{\Lambda^2}{MP'^2}\frac{\langle B^2\rangle}{\langle B^2/|\nabla V|^2\rangle},\\
Q_0 &= \left(\frac{\eta k_u^2\Lambda^2\langle B^2\rangle}{\rho M\Xi_e^2\langle B^2/|\nabla V|^2\rangle}\right)^{1/3}, \quad Q = \frac{\gamma}{Q_0}, && (6.115)\\
X_0 &= \left(\frac{\rho M\Xi_e^2\eta^2\langle B^2\rangle^2}{k_u^2\Lambda^2\langle B^2/|\nabla V|^2\rangle^2}\right)^{1/6}, \quad X = \frac{V - V_0}{X_0}, && (6.116)
\end{aligned}$$

and the perturbed quantities are given as follows:

$$\Xi = \langle\boldsymbol{\xi}\cdot\nabla V\rangle, \tag{6.117}$$

$$\Upsilon = \frac{\mathbf{B}\cdot\delta\mathbf{B}}{P'} = \Xi + \frac{\Gamma P}{P'}\langle\nabla\cdot\boldsymbol{\xi}\rangle. \tag{6.118}$$

Here, the same parameter definitions as those used in reference [18] (to be reviewed in Section 8.1.1) have been used. This allows the comparison between the current derivation and the ideal MHD limit of the resistive MHD equations in reference [18]. Although η appears in the Q_0 and X_0 definitions in equations (6.115) and (6.116), they cancel out in equation (6.112).

We still need to derive the equation for $\langle \nabla \cdot \boldsymbol{\xi} \rangle$. The surface average of the equation of parallel motion, equation (6.20), yields

$$
\begin{aligned}
&-\Gamma P\left(\frac{k_u\Lambda}{\Xi_e}\right)^2 x^2 \langle \nabla \cdot \boldsymbol{\xi}\rangle + i\Gamma P\frac{k_u\Lambda}{\Xi_e}x\left\langle \frac{1}{B^2}\widetilde{\nabla \cdot \boldsymbol{\xi}}\right\rangle \\
&+\rho_m\omega^2\left(1+\mu_0\Gamma P\left\langle\frac{1}{B^2}\right\rangle\right)\langle \nabla \cdot \boldsymbol{\xi}\rangle \\
= \; & -\rho_m\omega^2\left\langle\frac{2\mu_0 P+B^2}{B^2|\nabla V|^2}\right\rangle \xi^{(1)} - \rho_m\omega^2\frac{1}{P'}\left\langle\frac{\partial\sigma}{\partial\theta}\mu^{(1)}\right\rangle. \qquad (6.119)
\end{aligned}
$$

Here, $\widetilde{\nabla \cdot \boldsymbol{\xi}}$ has been given in equation (6.110). We only need to determine $\mu^{(1)}$. The dot product of of the induction equation (6.2) with $\mathbf{B}\times\nabla V/|\nabla V|^2$ yields

$$
\begin{aligned}
\nu^{(1)} &= \mathbf{B}\cdot\nabla\left(\frac{\boldsymbol{\xi}\cdot\mathbf{B}\times\nabla V}{|\nabla V|^2}\right) - \frac{(\mathbf{B}\times\nabla V)\cdot\nabla\times(\mathbf{B}\times\nabla V)}{|\nabla V|^4}\boldsymbol{\xi}\cdot\nabla V \\
&= \chi'\frac{\partial\mu^{(1)}}{\partial\theta} + i\frac{k_u}{\Xi_e}\Lambda x\mu^{(0)} - \frac{(\mathbf{B}\times\nabla V)\cdot\nabla\times(\mathbf{B}\times\nabla V)}{|\nabla V|^4}\xi^{(1)},
\end{aligned}
$$

Using equation (6.107) for $\nu^{(1)}$, one obtains

$$
\begin{aligned}
\chi'\frac{\partial\mu^{(1)}}{\partial\theta} &= \nu^{(1)} - i\frac{k_u}{\Xi_e}\Lambda x\mu^{(0)} + \frac{(\mathbf{B}\times\nabla V)\cdot\nabla\times(\mathbf{B}\times\nabla V)}{|\nabla V|^4}\xi^{(1)} \\
&= -\frac{\Xi_e}{ik_u}\frac{B^2/|\nabla V|^2}{\langle B^2/|\nabla V|^2\rangle}\frac{\partial\bar{b}^{(2)}}{\partial x} + \frac{1}{P'}\left(\frac{B^2/|\nabla V|^2}{\langle B^2/|\nabla V|^2\rangle}\left\langle\frac{B^2\sigma}{|\nabla V|^2}\right\rangle - \frac{B^2\sigma}{|\nabla V|^2}\right)\tau^{(1)} \\
&\quad -i\frac{k_u}{\Xi_e}\Lambda x\mu^{(0)} + \frac{(\mathbf{B}\times\nabla V)\cdot\nabla\times(\mathbf{B}\times\nabla V)}{|\nabla V|^4}\xi^{(1)} \\
&= -\Lambda\frac{B^2/|\nabla V|^2}{\langle B^2/|\nabla V|^2\rangle}\frac{\partial x\xi^{(1)}}{\partial x} + \Lambda x\frac{\partial\xi^{(1)}}{\partial x} + \frac{(\mathbf{B}\times\nabla V)\cdot\nabla\times(\mathbf{B}\times\nabla V)}{|\nabla V|^4}\xi^{(1)} \\
&\quad +\frac{1}{P'}\left(\frac{B^2/|\nabla V|^2}{\langle B^2/|\nabla V|^2\rangle}\left\langle\frac{B^2\sigma}{|\nabla V|^2}\right\rangle - \frac{B^2\sigma}{|\nabla V|^2}\right)\left(P'\xi^{(1)} + \Gamma P\langle\nabla\cdot\boldsymbol{\xi}\rangle\right) \\
&= -\Lambda\frac{B^2/|\nabla V|^2}{\langle B^2/|\nabla V|^2\rangle}\frac{\partial x\xi^{(1)}}{\partial x} + \Lambda x\frac{\partial\xi^{(1)}}{\partial x} + \frac{(\mathbf{B}\times\nabla V)\cdot\nabla\times(\mathbf{B}\times\nabla V)}{|\nabla V|^4}\xi^{(1)} \\
&\quad +\left(\frac{B^2/|\nabla V|^2}{\langle B^2/|\nabla V|^2\rangle}\left\langle\frac{B^2\sigma}{|\nabla V|^2}\right\rangle - \frac{B^2\sigma}{|\nabla V|^2}\right)\xi^{(1)} \\
&\quad +\frac{\Gamma P}{P'}\left(\frac{B^2/|\nabla V|^2}{\langle B^2/|\nabla V|^2\rangle}\left\langle\frac{B^2\sigma}{|\nabla V|^2}\right\rangle - \frac{B^2\sigma}{|\nabla V|^2}\right)\langle\nabla\cdot\boldsymbol{\xi}\rangle.
\end{aligned}
$$

Inserting this result to equation (6.119), one obtains the equation for $\langle \nabla \cdot \boldsymbol{\xi} \rangle$

$$
\begin{aligned}
&\frac{1}{P'}\Gamma P\langle\nabla\cdot\xi\rangle\left(X^2+GQ^2+KQ^2F\right) \\
&+Q^2(KE+KF)\Xi - KQ^2H(X\Xi)_X = 0.
\end{aligned}
$$

That is

$$\Gamma P \langle \nabla \cdot \xi \rangle = -\frac{K(E+F)\Xi + KH(X\Xi)_X}{X^2/Q^2 + G + KF}. \quad (6.120)$$

We have used the vorticity and acoustic wave equations (6.14) and (6.20) to derive the ideal MHD singular layer equations (6.112) and (6.120). The results agree with the ideal MHD limit of the resistive MHD equations in reference [18] as obtained in Section 8.1.2. They will be used to derive the Mercier criterion in the next subsection.

6.5.2 MERCIER CRITERION

In this subsection, we derive the Mercier criterion both using the singular layer equations and the energy principle.

In the previous subsection, we have derived the singular layer equations (6.112) and (6.120) with plasma compressibility effects included. In the outer limit, one can find from equation (6.120) that $\langle \nabla \cdot \boldsymbol{\xi} \rangle = 0$. Therefore, one has $\Upsilon = \Xi$ and the vorticity equation (6.112) is reduced to

$$\frac{\partial}{\partial x}\left(x^2 - M\omega^2\right)\frac{\partial \xi^{(1)}}{\partial x} + \left(\frac{1}{4} + D_I\right)\xi^{(1)} = 0, \quad (6.121)$$

where

$$D_I \equiv E + F + H - \frac{1}{4}. \quad (6.122)$$

One can use the singular layer equation (6.121) to derive the stability criterion for localized interchange modes in the toroidal geometry, *i.e.*, the Mercier criterion. In the marginal stability $\omega^2 = 0$, equation (6.121) reduces to the Euler differential equation with the solution

$$\xi = \xi_0 x^{-\frac{1}{2} \pm \sqrt{-D_I}}. \quad (6.123)$$

One can use Newcomb's theorem 5 [8] to determine the system stability in this case: the system is unstable if and only if the solution of equation (6.121) vanishes at two or more regular points or is small if they are singular points. One can see from the solution in equation (6.123) that ξ becomes oscillating when $-D_I < 0$, i.e., crossing the x-axis multiple times. Therefore, the interchange mode stability condition, or Mercier's criterion, is

$$-D_I > 0. \quad (6.124)$$

One can also derive the singular layer equation and the Mercier criterion from the MHD energy principle in equation (6.7). This has been detailed in references [19] and [2]. Here we just describe the results. Using the ordering assumptions in equations (6.92) and (6.93), one can find that the plasma energy in equation (6.7) is reduced to

$$2\delta W_p = c_0 \int \left[x^2 \left(\frac{d\xi}{dx}\right)^2 - \left(D_I + \frac{1}{4}\right)\xi^2 \right] dx + c_0 \left(\Delta + \frac{1}{2}\right) x\xi^2 \Big|_a^b, \quad (6.125)$$

where

$$\Delta \equiv \frac{1}{2} - \frac{1}{\Lambda}\left\langle \frac{\sigma B^2}{|\nabla Z|^2} \right\rangle,$$

$$c_0 = \frac{\Lambda^2}{\mu_0} \frac{[\oint (dl/B)]^2}{\oint (B^2/|\nabla V|^2)(dl/B)}.$$

One can also reduce the kinetic energy and the plasma compressional energy. The results are as follows

$$\begin{aligned} -2(\delta W_k + \delta W_c) &= 2\omega^2 \int \rho |\boldsymbol{\xi}|^2 d\mathbf{r} - 2\int \Gamma P |\nabla \cdot \boldsymbol{\xi}|^2 d\mathbf{r}. \\ &= c_0 \int M\omega^2 \left|\frac{d\xi}{dx}\right|^2 dx. \end{aligned} \quad (6.126)$$

Using equations (6.125) and (6.126), one can obtain the singular layer equation (6.121) by the minimization of the total energy.

One can derive the Mercier criterion directly from the energy integral in equation (6.125). For internal modes, the last term in equation (6.125) can be neglected. Using Schwartz's inequality [19]

$$\int \left|\frac{\partial}{\partial x} x\xi\right|^2 dx \geq \frac{1}{4}\int |\xi|^2 dx,$$

the field line bending term, i.e., the first term on the right-hand side of equation (6.125) can be reduced to

$$\int x^2 \left|\frac{\partial \xi}{\partial x}\right|^2 dx + x|\xi|^2\Big|_a^b = \int \left|\frac{\partial}{\partial x} x\xi\right|^2 dx \geq \frac{1}{4}\int |\xi|^2 dx.$$

The energy principle in equation (6.125) is reduced to

$$\frac{1}{2} M\omega^2 \int |\xi|^2 dx \geq -\frac{1}{2} D_I \int |\xi|^2 dx.$$

One can thus find that the stability condition in equation (6.124). The energy principle in equation (6.125) can also be used to study the peeling modes, which will be described in Section 6.7.

Here, let us further discuss the compressibility effect. The toroidal and cylindrical descriptions of this effect differ. The appearance of the apparent mass effect is the main factor. From the singular layer equation (6.121), one can see that the inertia effect appears when $x \sim \omega$, noting that ω is normalized by the shear Alfvén frequency. In solving the plasma compressibility effect in equation (6.20), we show that $\langle \boldsymbol{\nabla} \cdot \boldsymbol{\xi} \rangle$ is actually $\mathscr{O}(\varepsilon)$. Its effect needs to be included if one wishes to keep the internal effect in the singular layer equation (6.112). The Mecier criterion is for the marginal stability condition. To obtain the actual growth rate of ideal MHD interchange modes, the singular layer equations with the $\langle \boldsymbol{\nabla} \cdot \boldsymbol{\xi} \rangle$ term included need to match the outer region solutions in equation (6.123). Here, it is noted that in the outer limit $x \to \infty$, the $\langle \boldsymbol{\nabla} \cdot \boldsymbol{\xi} \rangle$ term becomes negligible. Therefore, the matching can be facilitated smoothly. Note that the sound wave frequency appearing in the parallel acoustic wave equation (6.20) is smaller than the Alfvén frequency in the vorticity equation or singular layer equation (6.121). There is also a regime in which $\widetilde{\boldsymbol{\nabla} \cdot \boldsymbol{\xi}} \sim \mathscr{O}(\varepsilon)$. This remains open to be exploited. All of these are related to the low-frequency gap in figure 6.1.

The mass factor M in equation (6.113) contains two parts. M_c is from the perpendicular motion, which has the counterpart in the cylinder description, and M_t from the parallel motion resulting from the toroidal coupling. M_t is often referred to as the apparent mass. In the kinetic description, the apparent mass, M_t, is enhanced by the so-called small parallel ion speed effect. In the large aspect ratio configurations, the enhancement factor is of order $\sqrt{R/a}$, where a and R are, respectively, minor and major radii [21, 22].

6.6 BALLOONING REPRESENTATION AND MODES

In this section, the high-n ballooning mode theory is reviewed [23, 24, 25]. Magnetically confined plasmas in toroidal configurations have good and bad curvature regions, depending on whether the pressure gradient and the magnetic field line curvature point in the same direction or not. In Section 6.5, the average effects of the good and bad curvature have been considered. The criterion is given by the Mercier criterion. Due to the field line rotational transform and the Shafraov shift, tokamaks usually have average good curvature with the magnetic shear stabilization taken into account, *i.e.*, it is Mercier-interchange-mode stable. However, when plasma beta becomes high, the field line bending stabilization becomes insufficient to overcome the pressure-driven energy in the bad curvature region. Plasma turns to balloon out on the low-field side of the plasma torus. To minimize the field line bending and the magnetic shear stabilization, the modes tend to be featured by $k_\perp \gg k_\|$, *i.e.*, the perpendicular wavelength is much smaller than the parallel one. In general, the ballooning modes allow normal and geodesic wave lengths to be of the same order $\lambda_\perp \sim \lambda_\wedge$. This leads to considering the case with the high toroidal mode number n. This is different from the interchange mode theory. The ballooning modes have a high toroidal mode number n, while the interchange modes can be either of low or high n. But the interchange theory described in Section 6.5 requires $\lambda_\perp \ll \lambda_\wedge$.

In the mathematical aspect, the invention of ballooning-mode representation represents a mathematical breakthrough in dealing with the high n modes. The discovery of ballooning invariance and its application to reduce the two-dimensional problem into one-dimensional one shows that, even in a field in which one often has to cope with lengthy or sometimes "ugly" modelings, there can still exist simplicity and beauty. Unlike the interchange mode theory, the ballooning mode theory is not just a localized theory for a single resonance surface but a global mode theory.

In this section, we first derive the high-n mode equations and then describe the ballooning mode representation. The ballooning mode analyses in the so-called s-α model will be given in the end. Besides the conventional ballooning representation, the representation in configuration space is also described [26].

6.6.1 HIGH N MODE EQUATIONS

In this subsection, let us discuss how to reduce the ideal MHD equations in the high n limit by introducing the so-called stream function [24]. The linearized MHD equation (6.5) contains three components. Three types of waves are described in this vector equation: the compressional Alfvén, shear Alfvén, and the acoustic waves. For high n modes, the high-frequency compressional Alfvén waves are usually suppressed, and only the shear Alfvén and the acoustic waves remain.

For ballooning modes, one can assume the compressional Alfvén modes are suppressed due to the high toroidal mode number. Therefore, one can use equation (6.19) for further analyses. In the lowest order, this equation indicates that $\nabla \cdot \boldsymbol{\xi}_\perp \sim \xi/R$, which is satisfied by introducing the so-called stream function $\delta\varphi$ [24], such that

$$\boldsymbol{\xi}_\perp = \mathbf{B} \times \nabla\delta\varphi/B^2. \tag{6.127}$$

With this representation, two components $\boldsymbol{\xi}_\perp$ are reduced to a single variable $\delta\varphi$. The equation for compressional Alfvén waves is satisfied by this representation.

Using the representation in equation (6.127), one can reduce the so-called vorticity equation (6.14) to describe the shear Alfvén waves in the case with $k_\perp \gg k_\parallel$

$$\begin{aligned}\mathbf{B} \cdot \nabla\frac{1}{B^2}\nabla \cdot \left(B^2\nabla_\perp \frac{\mathbf{B} \cdot \nabla\delta\varphi}{B^2}\right) + \nabla \cdot \left(\rho\omega^2\frac{\nabla_\perp\delta\varphi}{B^2}\right)\\ +P'_\psi\nabla \times \frac{\mathbf{B}}{B^2} \cdot \nabla\left(\frac{\mathbf{B} \times \nabla\psi}{B^2} \cdot \nabla\delta\varphi\right) + \Gamma P\nabla \times \frac{\mathbf{B}}{B^2} \cdot \nabla\nabla \cdot \boldsymbol{\xi} = 0.\end{aligned} \tag{6.128}$$

Note here that every term in this equation is of order n^2 or $\nabla_\perp^2$. Using the representation in equation (6.127), the parallel equation of motion, equation (6.15), can be reduced to

$$\begin{aligned}&\Gamma P\mathbf{B} \cdot \nabla\left(\frac{1}{B^2}\mathbf{B} \cdot \nabla\nabla \cdot \boldsymbol{\xi}\right) + \rho_m\omega^2\frac{B^2 + \mu_0\Gamma P}{B^2}\nabla \cdot \boldsymbol{\xi}\\ = &\ \rho_m\omega^2\frac{2\mathbf{B} \times \boldsymbol{\kappa}}{B^2} \cdot \nabla\delta\varphi,\end{aligned} \tag{6.129}$$

where $\nabla \cdot \boldsymbol{\xi}$ is used as a new unknown [27]. Noting that the perturbed density $\delta n \propto \nabla \cdot \boldsymbol{\xi}$, one can see that equation (6.129) describes the coupling of the acoustic or sound waves. Equations (6.128) and (6.129) constitute the set of equations for studying the high n modes.

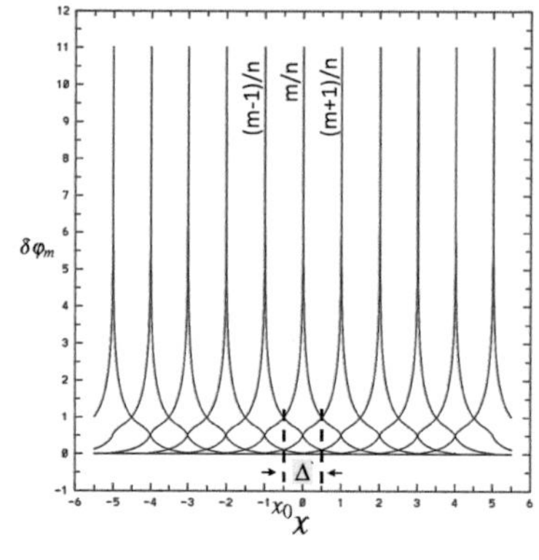

Figure 6.3 High n ballooning mode invariance.

6.6.2 BALLOONING MODE REPRESENTATION

The set of equations (6.128) and (6.129) decouples the compressional Alfvén modes. However, it remains a two-dimensional problem. One has to solve the radial and poloidal dependences of the perturbations $\delta\varphi$ and $\nabla \cdot \boldsymbol{\xi}$. The success of high n ballooning mode theory lies in the invention of the ballooning representation, which reduces the two-dimensional problem into a one-dimensional one. There are two types of formulations: one is based on the Fourier space [25, 23]. and the other is based on the configuration space [26]. The latter can be used for the free boundary modes. The inversion of ballooning representation is described in reference [28].

Let us first describe the ballooning representation in the Fourier space [25, 23]. In tokamak geometry with toroidal symmetry, one can introduce the following Fourier decomposition for the perturbed field:

$$\delta\varphi(nq,\theta,\zeta) \quad = \quad \sum_{m=-\infty}^{+\infty} \delta\varphi_m(nq)\exp\{-in\zeta + m\theta\}. \tag{6.130}$$

Here, nq has been used to represent the radial coordinate. This is based on the assumption that the q profile is monotonic, which is the applicability condition for the ballooning representation.

For high n modes, the distance between nearby mode rational surfaces, for example, $q = m/n$ and $q = (m+1)/n$ surfaces, is of order $1/n$. This is much smaller than the equilibrium scale length, for example, minor radius a. Therefore, in the lowest order with $1/n$, the spatial variance of equilibrium quantities can be neglected and the poloidal Fourier harmonics remain radially invariant as shown in figure 6.3, referred to as the ballooning invariance:

$$\delta\varphi_m(nq) = \delta\varphi(nq-m). \tag{6.131}$$

This representation taps the similarity of Fourier components in the lowest order by assuming that each poloidal Fourier harmonic has the same shape in this order and considering the overall envelope in the next order. We only describe the lowest-order theory and refer to the next-order theories in references [23] and [26].

Using the ballooning invariance in equation (6.131), one can express the Fourier decomposition in equation (6.130) as follows:

$$\delta\varphi(nq,\theta,\zeta) = \sum_{m=-\infty}^{+\infty} \delta\varphi(nq-m)\exp\{-in\zeta+m\theta\}. \tag{6.132}$$

Note that $\delta\varphi_m(nq)$ is two-dimensional: radial nq and poloidal m, while $\delta\varphi(nq-m)$ is one-dimensional with the dependencies of m and nq mixed. The original two-dimensional problem is transformed into a one-dimensional one. There are two methods to further apply the representation in equation (6.132). The mixed argument $nq-m$ for $\delta\varphi$ needs special treatment. There are two approaches.

Let us first describe the conventional approach based on the integral transformation

$$\delta\varphi(nq) = \frac{1}{2\pi}\int_{-\infty}^{+\infty} \delta\varphi(\eta)\exp\{inq\eta\}d\eta. \tag{6.133}$$

For brevity, here the same notations have been used for the original and transformed functions. Inserting it into equation (6.132) one obtains

$$\delta\varphi(nq,\theta,\phi) = \frac{1}{2\pi}\exp\{-in\zeta\}\int_{-\infty}^{+\infty} \delta\varphi(\eta)\sum_{m}\exp\{im(\theta-\eta)\}d\eta. \tag{6.134}$$

Using the identity

$$\frac{1}{2\pi}\sum_{m=-\infty}^{+\infty} \exp\{im(\theta-\eta)\} = \sum_{j=-\infty}^{+\infty} \delta(\eta-\theta-j2\pi),$$

equation (6.134) is further reduced to

$$\begin{aligned}\delta\varphi(nq,\theta,\zeta) &= \sum_{j=-\infty}^{+\infty} \delta\varphi(\theta+j2\pi)\exp\{-in(\zeta-q(\theta+j2\pi))\} \\ &= \sum_{j=-\infty}^{+\infty} \delta\hat{\varphi}(\theta+j2\pi),\end{aligned} \tag{6.135}$$

where

$$\delta\hat{\varphi}(\theta) = \delta\varphi(\theta)\exp\{-in\beta\} \tag{6.136}$$

with $\beta = \zeta - q(\theta+\theta_k)$ and θ_k being an initial value. This indicates that one can use the ekonal representation in equation (6.136) with the boundary condition $\delta\varphi(\theta) \to 0$ as $\theta \to \pm\infty$ without concerning the periodic boundary condition. The boundary conditions are required to assure the convergence of the Laplace transform in equation (6.133). The periodic boundary condition in θ can be met afterward with the summation in equation (6.135).

We now apply equation (6.136) to equations (6.128) and (6.129). The flux coordinates (χ, θ, ζ) as described in Chapter 4 are employed. One thus has $\mathbf{\nabla}_\perp \gg \mathbf{\nabla}_\parallel$, $\mathbf{\nabla}_\perp = in\mathbf{\nabla}\beta$, with $\beta = n\zeta - q\theta$, and $\mathbf{B}\cdot\mathbf{\nabla} = \chi'\partial/\partial\theta$. One obtains the coupled ballooning mode equations [27]

$$\frac{\partial}{\partial\theta}\left(\frac{|\nabla\beta|^2}{B^2}\frac{\partial}{\partial\theta}\delta\varphi\right)+\frac{2P'}{\chi'^4}(\kappa_n+q\theta\kappa_g)\delta\varphi-\frac{\rho_m\gamma^2}{\chi'^2B^2}|\nabla\beta|^2\delta\varphi$$

$$=-\frac{2P'}{\chi'^4}(\kappa_n+q\theta\kappa_g)D, \tag{6.137}$$

$$\frac{\partial}{\partial\theta}\left(\frac{1}{B^2}\frac{\partial D}{\partial\theta}\right)-\frac{\rho_m\gamma^2}{\chi'^2}\frac{B^2+\mu_0P}{B^2\Gamma P}D=\frac{2\rho_m}{P'\chi'^2}\gamma^2(\kappa_n+q\theta\kappa_g)\delta\varphi, \tag{6.138}$$

where $D=\Gamma P\mathbf{\nabla}\cdot\boldsymbol{\xi}$, $\kappa_n=\boldsymbol{\kappa}\cdot\mathbf{\nabla}\theta\times\mathbf{\nabla}\zeta$, and $\kappa_g=\boldsymbol{\kappa}\cdot\mathbf{\nabla}V\times\mathbf{\nabla}\theta=-(\chi'/2P')\mathbf{B}\cdot\mathbf{\nabla}\sigma$, and hats for unknowns have been dropped for brevity. Note also $|\nabla\beta|^2=|\nabla\zeta|^2-2q'\theta\nabla V\cdot\nabla\zeta+q'^2\theta^2|\nabla V|^2$. The derivatives here are along a reference magnetic field line labeled by χ and β in the so-called Celbsch coordinates $\mathbf{B}=\mathbf{\nabla}\beta\times\mathbf{\nabla}\chi$. The boundary conditions are $\delta\varphi, D\to 0$ at $\theta\to\pm\infty$. They are the coupled second-order differential equations. Equation (6.137) describes the shear Alfvén wave, and equation (6.138) governs the parallel acoustic wave. The perpendicular WKB representation has been formally introduced: $\mathbf{k}_\perp\sim n\mathbf{\nabla}\beta$, which is not periodic. Nonetheless, the periodic boundary condition can always be achieved by the summation in equation (6.135).

Next, we describe an alternative ballooning mode representation [26]. The conventional ballooning formulation in equation (6.135) relies on the integral transformation, which leads to the differential equations for a single Fourier component in the infinite θ domain. Reference [26] gave another ballooning mode formulation by considering all the Fourier components but only in a single resonance interval: $x_0\to x_0+\Delta$, where Δ is the width of the interval, and x_0 denotes its left end, as shown in figure 6.3. The ballooning invariance in equation (6.131) is then used to determine the boundary conditions to connect the nearby Fourier components

$$\delta\varphi_m(x_0+\Delta) = e^{-i\theta_k}\delta\varphi_{m+1}(x_0), \tag{6.139}$$

$$\frac{d}{dx}\delta\varphi_m(x_0+\Delta) = e^{-i\theta_k}\frac{d}{dx}\delta\varphi_{m+1}(x_0). \tag{6.140}$$

Suppose the farthest sidebands are respectively $m_{\min}$ and $m_{\max}$. Their boundary conditions are given as follows

$$\delta\varphi_{m_{\max}}(x_0+\Delta) = 0, \tag{6.141}$$

$$\delta\varphi_{m_{\min}}(x_0) = 0, \tag{6.142}$$

or to be small solutions. The boundary conditions in equations (6.139)–(6.142) fully determine the eigenvalue problem for the set of differential equations of the Fourier components in the interval $[x_0, x_0+\Delta]$.

Apparently, two ballooning formulations, computing multiple Fourier components in a single resonance interval in the configurational space and computing a single Fourier component in the transformed infinite θ space, use equivalent computation resources. The single interval representation can be used to develop a layer code, which has a more apparent poloidal transport picture than conventional flux tube code based on the conventional ballooning mode representation. Besides, the single interval representation in the configuration space can be generalized to treat the free boundary ballooning mode problem, and even to include the pedestal physics [26].

6.6.3 ASYMPTOTIC BEHAVIOR

In this subsection, the asymptotic analyses of the MHD ballooning mode equations (6.137) and (6.138) are described. This is needed to fully determine the stability condition from the equations. It is noted that the inertia effects are only important in the singular layer. In the ballooning mode representation space, this corresponds to $\theta \sim 1/\varepsilon_\theta \to \infty$ and is therefore referred to as the outer region. In the inner region with $\theta \sim 1$, the ideal MHD theory applies. Here, it is noted that the outer region in the ballooning formalism corresponds to the inner layer in the configuration description as treated in subsection 6.5. We first lay out the ordering scheme. The singular layer equations will be derived next.

To proceed with asymptotic analyses in the outer inertia region, let us first lay out the ordering scheme. The low-frequency regime is considered. The balance between the field line bending term and the inertia term determines the width of singular layer, *i.e.*, the small parameter ε_θ, as can be seen in the ordering analyses later on in equation (6.146). These lead to the following ordering assumptions:

$$\omega \ll \omega_{si} \quad \text{and} \quad \varepsilon_\theta \sim \beta_t^{1/2}(\omega/\omega_{si}), \tag{6.143}$$

where $\omega_{si} \sim v_{thi}/Rq$ is the ion acoustic frequency, q is the safety factor, v_{thi} is the ion thermal velocity, and β_t is the plasma beta. Noting that the plasma beta is usually small, it is assumed in addition that

$$\beta_t \lesssim a/R, \tag{6.144}$$

where the minor radius a is introduced for dimensionless processing.

We need to analyze the orderings for each equation, equation (6.137) and equation (6.138). Let us first examine the sound wave equation (6.138). Using the orderings in equations (6.143) and (6.144), one can find the relative orderings of two terms on the left-hand side of equation (6.138) as follows:

$$\omega_{si}^2,\ \omega^2. \tag{6.145}$$

Next, let us consider equation (6.137). Using the orderings in equations (6.143) and (6.144), one can also estimate the orderings for the first, second, and last terms in equation (6.137), respectively, as follows:

$$\frac{1}{\varepsilon_\theta^2},\ \beta_t\frac{R}{a}\frac{1}{\varepsilon_\theta},\ \beta_t\frac{\omega^2}{\omega_{si}^2}\frac{1}{\varepsilon_\theta^2}. \tag{6.146}$$

The term on the right of equation (6.137), which gives the so-called apparent mass effects, can be proved *a posteriori* to be of the same order as the last term on the left, i.e., the perpendicular inertia term.

Here, it should be pointed out that in these ordering considerations, the orderings for beta and aspect ratio are not essential. In the ordering sequence in equation (8.141) for the sound wave, the only necessary ordering is the low frequency regime, *i.e.*, $\omega_{si} \gg \omega$. In the ordering sequence in equation (8.143) for the vorticity equation or shear Alfvén wave, the only necessary ordering is the mode localization, *i.e.*, $\varepsilon_\theta \ll 1$.

With the ordering scheme given, we can proceed to analyze the ballooning mode equations. As in the ideal MHD case, two length scale analyses are performed with $(\theta, \varepsilon_\theta z)$, where z corresponds to the slowly varying scale. We decompose $\delta\varphi$ as follows

$$\delta\varphi = \delta\varphi_0 + \varepsilon_\theta \delta\varphi_1 + \cdots .$$

Instead, D is decomposed into the oscillating and average parts

$$D = \tilde{D} + \bar{D},$$

where $\bar{D} = \langle D \rangle == \oint JBd\theta\, D / \oint JBd\theta$ and $\tilde{D} = D - \bar{D}$.

We first treat the acoustic wave equation (6.138). Since $\omega_{si} \gg \omega$, the equation for $\tilde{D}$ in leading order becomes

$$\frac{\partial}{\partial\theta}\left(\frac{1}{B^2}\frac{\partial\tilde{D}}{\partial\theta}\right) = \frac{2\rho_m}{P'\chi'^2}\gamma^2 qz\kappa_g \delta\varphi_0.$$

Noting that $\kappa_g = -(\chi'/2P')\partial\sigma/\partial\theta$, this equation can be solved, yielding

$$\frac{\partial\tilde{D}}{\partial\theta} = \frac{\rho_m}{P'^2}\gamma^2\left(B^2\sigma - \frac{B^2\left\langle B^2\sigma\right\rangle}{\langle B^2\rangle}\right) qz\delta\varphi_0. \tag{6.147}$$

The equation for the average part is as follows

$$\begin{aligned}&\left\langle\frac{1}{B^2}\right\rangle\frac{\partial^2\bar{D}}{\partial z^2} + \rho\omega^2\frac{1}{\Gamma P}\bar{D}\\ = {} & \frac{\partial}{\partial z}\left\langle\frac{1}{B^2}\frac{\partial\tilde{D}}{\partial\theta}\right\rangle + \frac{2\rho_m}{P'\chi'^2}\gamma^2\left(qz\left\langle\kappa_g\delta\varphi_1\right\rangle + \langle\kappa_n\rangle\,\delta\varphi_0\right).\end{aligned} \tag{6.148}$$

Here, $\tilde{D}$ is given in equation (6.147), and $\delta\varphi_1$ will be derived later in equation (6.150).

Let us now treat the vorticity equation (8.133). In the lowest order, one has

$$\frac{\partial\delta\varphi_0}{\partial\theta} = 0.$$

This yields

$$\delta\varphi_0 = \delta\varphi_0(z). \tag{6.149}$$

Using the solution in equation (8.147), equation (8.133) in the first order can be solved

$$\delta\varphi_1(\theta,z) = f_1(z)+\frac{d\delta\varphi_0}{dz}\int\widetilde{\left(\frac{B^2}{|\nabla V|^2}\right)}d\theta + \frac{\delta\varphi_0}{\chi'^2 q' z}\int\left[\widetilde{\left(\frac{B^2}{|\nabla V|^2}\right)}+\widetilde{\left(\frac{B^2\sigma}{|\nabla V|^2}\right)}\right]d\theta, \tag{6.150}$$

In the second order, the average of equation (6.137) over the fast variable θ gives the solubility condition for $\delta\varphi_2$:

$$\frac{\partial}{\partial z}z^2\frac{\partial\delta\varphi_0}{\partial z}-\left(\frac{1}{4}+D_I\right)\delta\varphi_0+M\omega^2z^2\delta\varphi_0 + \frac{2P'}{\chi'^4|\nabla q|^2\langle 1/B^2\rangle}\langle\kappa_n\rangle\bar{D}=0, \tag{6.151}$$

where $M = M_c + M_t$ and

$$D_I = \frac{\langle B^2/|\nabla\chi|^2\rangle}{\chi'^2\langle d\Lambda/d\theta\rangle^2}\left[\mu_0\left\langle\frac{d\Lambda}{d\theta}\left(\sigma-\frac{\langle B^2\sigma/|\nabla\chi|^2\rangle}{\langle B^2/|\nabla\chi|^2\rangle}\right)\right\rangle - \frac{\mu_0^2}{\chi'^2}\left(\left\langle\frac{B^2\sigma}{|\nabla\chi|^2}\right\rangle^2-\frac{\langle\sigma B^2/|\nabla\chi|^2\rangle^2}{\langle B^2/|\nabla\chi|^2\rangle}\right)-\mu_0P'\frac{2\boldsymbol{\kappa}\cdot\nabla\chi}{\chi'|\nabla\chi|^2}\right]-\frac{1}{4},$$

$$M_c = \frac{\mu_0\rho}{\chi'^2}\left\langle\frac{B^2}{|\nabla\chi|^2}\right\rangle\left\langle\frac{|\nabla\chi|^2}{B^2}\right\rangle,$$

$$M_t = \frac{\mu_0\rho}{\chi'^2}\frac{\langle B^2/|\nabla\chi|^2\rangle}{P'^2}\left(\langle B^2\sigma^2\rangle-\frac{\langle B^2\sigma\rangle^2}{\langle B^2\rangle}\right). \tag{6.152}$$

Here, M_t represents the so-called apparent mass effect [17].

In the inner region limit $|z|\to 0$, equation (6.148) yields $\bar{D}=0$, and also the inertia term, the third term, in equation (6.151) can be neglected. The asymptotic solution of equation (6.151) becomes

$$\delta\varphi = a_1|z|^s + a_2|z|^{-1-s}, \tag{6.153}$$

where a_1 and a_2 are constants and $s=-1/2+\sqrt{-D_I}$. The solution $a_1|\theta|^s$ is the so-called large solution, while the solution $a_2|\theta|^{-1-s}$ is the small solution. Only the small solution is acceptable in the ideal MHD description since the large solution causes the energy integral to diverge.

Note that equation (6.153) is in the ballooning transformation space using the Laplace transformation in equation (6.133). A backward transform reduces equation (6.153) exactly to the singular layer equation in configuration space in equation (6.121).

6.6.4 BALLOONING STABILITY CONDITIONS

With the ballooning mode equations (6.137) and (6.138) derived, using the asymptotic behavior in equation (6.153) for boundary conditions, one can study the ballooning stability conditions. This needs the equilibrium parameters. In this field, one often uses the so-called steep-pressure-gradient equilibrium model to draw the stability conditions [29, 30, 31]. Let us first review this equilibrium model and then describe how to determine the stability conditions.

In the steep-pressure-gradient equilibrium model, the finite beta modification is only taken into account for magnetic shear, while other equilibrium parameters remain at their low beta values. In the ballooning mode equations, the magnetic shear appears in the perpendicular wave vector

$$\nabla\beta = \Lambda\nabla\chi + \frac{\mathbf{B}\times\nabla\chi}{|\nabla\chi|^2}. \tag{6.154}$$

in which Λ reflects the magnetic shear effects. In the literature, references [31] and [32] gave a systematic description of this equilibrium model. They are also reviewed in the author's earlier books [20, 2]. Here, let us briefly outline the model as described in reference [32].

Using equation (6.154), one can derive the equation for Λ as follows

$$\begin{aligned} 0 &= \nabla\cdot(\nabla\phi\times\nabla\beta) \\ &= \nabla\cdot\left(\nabla\phi\times\frac{\mathbf{B}\times\nabla\chi}{|\nabla\chi|^2}\right)+\nabla\cdot(\Lambda\nabla\phi\times\nabla\chi) \\ &= \nabla\cdot\left[\frac{(\nabla\phi\cdot\nabla\chi)\mathbf{B}-(\mathbf{B}\cdot\nabla\phi)\nabla\chi}{|\nabla\chi|^2}\right]+\nabla\phi\times\nabla\chi\cdot\nabla\Lambda, \\ &= -\nabla\cdot\left[\frac{(\mathbf{B}\cdot\nabla\phi)\nabla\chi}{|\nabla\chi|^2}\right]+\nabla\phi\times\nabla\chi\cdot\nabla\Lambda, \end{aligned} \tag{6.155}$$

where ϕ is the axisymmetric toroidal angle, and it has been noted that $\nabla\phi\cdot\nabla\chi = 0$. Equation (6.155) then yields

$$\nabla\phi\times\nabla\chi\cdot\nabla\Lambda = \nabla\cdot\left[\frac{(\mathbf{B}\cdot\nabla\phi)\nabla\chi}{|\nabla\chi|^2}\right]. \tag{6.156}$$

In the steep-pressure-gradient equilibrium model, the intention is to count in the Shafranov shift effects on the magnetic shear, while keeping other equilibrium parameters intact. Expanding the poloidal flux: $\chi = \chi_0+\chi_1+\cdots$, with $\chi_0 \gg \chi_1$, equation (6.156) is reduced to

$$\nabla\phi\times\nabla\chi\cdot\nabla\Lambda = \nabla\cdot\left[\frac{(\mathbf{B}\cdot\nabla\phi)\nabla\chi}{|\nabla\chi|^2}\right]^{(0)} - \left[\frac{(\mathbf{B}\cdot\nabla\phi)}{|\nabla\chi|^2}\right]^{(0)}\nabla\cdot\nabla\chi_1. \tag{6.157}$$

Here, it has been noted that in the steep-pressure-gradient case, the radial derivative is dominant. Under this assumption, the Grad-Shafranov equation in the first order becomes

$$\nabla \cdot \nabla \chi_1 = -(R^2)^{(1)} \frac{\partial P}{\partial \chi_0}. \tag{6.158}$$

Therefore, equation (6.157) becomes

$$\nabla \phi \times \nabla \chi \cdot \nabla \Lambda = \nabla \cdot \left[\frac{(\mathbf{B} \cdot \nabla \phi) \nabla \chi}{|\nabla \chi|^2}\right]^{(0)} + \left[\frac{(\mathbf{B} \cdot \nabla \phi)}{|\nabla \chi|^2}\right]^{(0)} (R^2)^{(1)} \frac{\partial P}{\partial \chi_0}. \tag{6.159}$$

For simplicity, it is limited to consider the tokamak configuration with a circular cross-section and a large aspect ratio. In the (r, θ, ϕ) coordinates, one has $\nabla \phi \times \nabla \chi = \mathbf{B}_\theta / R$, $\mathbf{B} \cdot \nabla \phi = B_\phi / R$, and $\mathbf{B}_\theta \cdot \nabla = (B_\theta / r)(\partial / \partial \theta)$, where θ is the poloidal angle, R and r are major and minor radii, respectively. Note that magnetic-field-line curvature $\boldsymbol{\kappa} = -\mathbf{R}/R^2$, one further has $\kappa_n \propto \cos\theta$ and $\kappa_g \propto \sin\theta$. In this case, equation (6.159) is reduced to

$$\frac{\partial \Lambda}{\partial \theta} = \frac{1}{R B_\theta} \frac{dq}{dr} + 2 \frac{q^2 r}{R B_\phi B_\theta^2} \cos\theta \frac{dP}{dr}. \tag{6.160}$$

Finally, introducing $\Lambda \equiv \Lambda_s B_\phi / |\nabla \chi|^2$, one has

$$\Lambda_s = s(\theta - \theta_k) - \alpha(\sin\theta - \sin\theta_k). \tag{6.161}$$

Here, $\alpha = -(2\mu_0 R q^2 / B^2)(dP/dr)$, $s = d\ln q / d\ln r$, and θ_k is the integration constant, as mentioned earlier. The ballooning mode equations (6.137) and (6.138) are then reduced to

$$\frac{d}{d\theta}\left((1+\Lambda_s^2)\frac{d\delta\varphi}{d\theta}\right) + \alpha(\cos\theta + \Lambda_s \sin\theta)\delta\varphi + \frac{2\Gamma R r q P}{B}(\cos\theta + \Lambda_s \sin\theta)\delta\Xi$$
$$+ \frac{\omega^2}{\omega_A^2}(1+\Lambda_s^2)\delta\varphi = 0, \tag{6.162}$$

$$\frac{1}{R^2 q^2}\frac{\partial^2 D}{\partial \theta^2} + \frac{1}{\Gamma P}\rho_m \omega^2 D = -\frac{2\rho_m \omega^2}{R^2 B_\theta}(\cos\theta + \Lambda_s \sin\theta)\delta\varphi. \tag{6.163}$$

To further analyze this set of equations, we note that the acoustic wave primarily propagates along the magnetic field lines, which is different from the shear Alfvén wave. One can then first solve the acoustic wave equation (6.163) and insert the solution to the shear Alfvén equation (6.162). The orderings of two terms on the left-hand side of equation (6.163) are respectively ω_{si}^2 and ω^2, where ω_{si} is the ion acoustic frequency. As originally proposed in reference [33] and further developed in references [34] – [40], one can consider three frequency regimes: the low frequency ($\omega \ll \omega_{si}$), the comparable frequency ($\omega \sim \omega_{si}$), and intermediate frequency regimes ($\omega_{si} \ll \omega \ll \omega_{se}$), where ω_{se} is the electron acoustic frequency. In principle, kinetic treatments are generally required for treating the acoustic waves. Charged particles are not localized along the field lines. Also, ions and electrons have different parallel velocities, and their resonance conditions with waves are different. These lead to the excitation of the parallel electric field. In the low-frequency regime, there is the so-called small parallel ion speed effects [34]; In the comparable frequency regime, the ion-wave resonance and the excitation of the parallel electric field need to be considered [36]; and in the intermediate regime, the parallel electric field can also be excited [33, 36, 40]. Nevertheless, discussing the ideal MHD case helps to understand the eigenmode structure.

In the low frequency regime, we assume that $\bar{D}$ effects can be neglected and keep only the apparent mass effect from $\tilde{D}$ as discussed in Section 6.6.3. The second term in equation (6.163) can be neglected in this case. Note that the acoustic wave coupling only gives rise to an enhanced inertia effect, and the inertia term is only important in the outer region $\theta \to \infty$. Since the slow variables $s\theta$ and $\delta\varphi$ can be regarded as constant in the acoustic wave scale, equation (6.163) can be solved in the outer region limit, yielding

$$D = \frac{2\rho_m q^2 \omega^2}{B_\theta} s(\theta - \theta_k) \sin\theta \delta\varphi. \tag{6.164}$$

Inserting equation (6.164) into equation (6.162) results in the following single ballooning mode equation

$$\begin{aligned} &\frac{d}{d\theta}\left((1+\Lambda_s^2)\frac{d\delta\varphi}{d\theta}\right) + \alpha(\cos\theta + \Lambda_s \sin\theta)\delta\varphi \\ &+ \frac{\omega^2}{\omega_A^2}(1+2q^2)s^2(\theta-\theta_k)^2\delta\varphi = 0. \end{aligned} \tag{6.165}$$

The acoustic wave coupling gives rise to the so-called apparent mass effect: i.e., the inertia term (the last term in equation (6.165)) is enhanced by a factor $(1+2q^2)$ as shown generally in subsection 6.6.3 [17]. In the kinetic description, the $2q^2$ apparent mass term is further enhanced by the so-called small parallel ion speed effect by a factor $1/\sqrt{r/R}$ in the large aspect ratio case [21, 34].

In the comparable frequency regime, the second term in equation (6.163) is kept. Since the acoustic wave coupling still gives rise to an enhanced inertia effect, similar to the low frequency case, one obtains [35]

$$D = \frac{2\rho_m q^2 \omega^2}{B_\theta} \frac{1}{1-\omega^2/\omega_{si}^2} s(\theta - \theta_k) \sin\theta \delta\varphi. \tag{6.166}$$

Inserting equation (6.166) into equation (6.162) results in the following single ballooning mode equation

$$\begin{aligned} &\frac{d}{d\theta}\left((1+\Lambda_s^2)\frac{d\delta\varphi}{d\theta}\right) + \alpha(\cos\theta + \Lambda_s \sin\theta)\delta\varphi \\ &+ \frac{\omega^2}{\omega_A^2}\left(1 + \frac{2q^2}{1-\omega^2/\omega_{si}^2}\right)s^2(\theta-\theta_k)^2\delta\varphi = 0. \end{aligned} \tag{6.167}$$

In the intermediate frequency regime, the first term in equation (6.163) can be neglected, and one obtains the solution

$$D = -\frac{2\Gamma P}{R^2 B_\theta}(\cos\theta + \Lambda_s \sin\theta)\delta\varphi. \tag{6.168}$$

Inserting equation (6.168) into equation (6.162), one obtains [33]

$$\frac{d}{d\theta}\left((1+\Lambda_s^2)\frac{d\delta\varphi}{d\theta}\right) + \alpha(\cos\theta + \Lambda_s \sin\theta)\delta\varphi$$
$$-\frac{4\mu_0\Gamma q^2 P}{B^2}(\cos\theta + \Lambda_s \sin\theta)^2\delta\varphi + \frac{\omega^2}{\omega_A^2}(1+\Lambda_s^2)\delta\varphi = 0. \tag{6.169}$$

The acoustic wave coupling (3rd term) can give rise to the so-called second harmonic toroidal Alfvén eigenmodes [36].

The marginal stability condition ($\omega^2 = 0$) for ballooning modes can be determined by Newcomb's theorem 5 [42]: the system is unstable if and only if the solution $\delta\varphi$ of equation (6.137) with $\omega = \delta\Xi = 0$ vanishes at two or more points (or is the small solution at infinity). In the large aspect ratio case with circular cross section, using the steep-pressure-gradient equilibrium model in equation (6.165), references [29] and [43] obtained the stability boundaries. The θ_k effects were added later by reference [41]. The results are summarized in figure 6.4 in the $\alpha - s$ coordinates. From figure 6.4, one can see that there are two stability regimes: the first and the second stability regimes. The θ_k effects reduce the parameter domain in the second stability regime.

6.7 PEELING AND PEELING-BALLOONING MODES

In this section, we discuss the peeling [44, 45] and peeling-ballooning modes [46, 47]. We only outline the underlying physics and their mathematical treatments. One may find a detailed review from the author's earlier book [2].

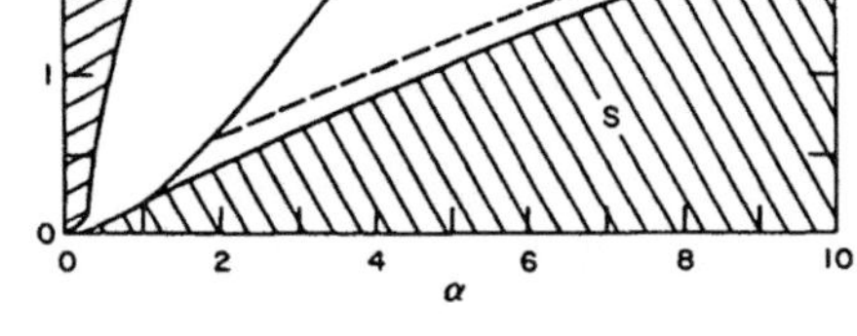

Figure 6.4 Ballooning mode stability boundaries in the s-α equilibrium model. Reproduced from [41], with the permission of IAEA.

Peeling mode is closely related to the interchange mode reviewed in Section 6.5. It is a kind of free boundary interchange modes. We have derived the energy principle for the radially localized modes in equation (6.125) in Section 6.5

$$\frac{1}{2}M\omega^2\int|\xi|^2dx = c_0\int\left[x^2\left|\frac{d\xi}{dx}\right|^2-\left(D_I+\frac{1}{4}\right)|\xi|^2\right]dx + c_0\left(\Delta+\frac{1}{2}\right)x|\xi|^2\Big|_{x_a}^{x_b}, \quad (6.170)$$

with its Euler-Lagrange equation given by equation (6.121) in the same section

$$\frac{d}{dx}x^2\frac{d\xi}{dx}-\left(\frac{1}{4}+D_I\right)\xi=0. \quad (6.171)$$

Its solution is

$$\xi=\xi_0|x|^{-\frac{1}{2}\pm\sqrt{-D_I}}. \quad (6.172)$$

Before moving further, let us discuss the Euler-Lagrange equation (6.171) and its solution in equation (6.172). The Euler-Lagrange equation (6.171) has two singular points: $x=0$ and $+\infty$. Its solution in equation (6.172) can be large or small depending on the sign in the exponent and the location of the singular point. At $x=+\infty$, the negative sign corresponds to the small solution and the positive sign to the large solution, and vice versa at $x=0$. If the singular point is inside plasmas, only the small solution is acceptable, while the large solution causes the energy integral to diverge. At the plasma edge, however, one of the two singular points can appear in the vacuum. The large solution becomes acceptable if the corresponding singular point is in the vacuum region. This gives extra room for minimization. There are two types of modes: Type I, the singular point $x=0$ is located in the vacuum, and Type II, the singular point $x=+\infty$ is located in the vacuum.

Carrying out integration by parts and using equation (6.171), the energy integral in equation (6.170) becomes

$$\frac{1}{2}M\omega^2\int|\xi|^2dx \geq \left[\frac{x^2}{2}\left(\xi^*\frac{d\xi}{dx}+\xi\frac{d\xi^*}{dx}\right)+\left(\Delta+\frac{1}{2}\right)x|\xi|^2\right]_{x_a}^{x_b}.$$

Then, inserting the solution in equation (6.172), one obtains the stability criterion

$$\sqrt{-D_I}+\sigma_T\Delta>0. \quad (6.173)$$

where $\sigma_T=\mp1$ respectively for Type I and II modes.

From the definition of D_I in equation (6.122), it can be proved that

$$-D_I = \Delta^2-\Lambda_p, \quad (6.174)$$

where

$$\begin{aligned}\Lambda_p &\equiv \frac{\langle B^2/|\nabla V|^2\rangle}{\Lambda^2}\left(J'\psi'' - I'\chi'' + \left\langle\frac{\sigma^2 B^2}{|\nabla V|^2}\right\rangle + p'^2\left\langle\frac{1}{B^2}\right\rangle\right)\\ &= \frac{\langle B^2/|\nabla V|^2\rangle}{\Lambda^2}\left(J'\psi'' - I'\chi'' + \left\langle\frac{\mathbf{J}^2}{|\nabla V|^2}\right\rangle\right).\end{aligned}$$

Therefore, the stability criterion for peeling modes in equation (6.173) becomes

$$\sqrt{\Delta^2 - \Lambda_p} + \sigma_T\Delta > 0. \tag{6.175}$$

To analyze the criterion in equation (6.175), one can make the following choices: for $\Delta > 0$ consider Type I peeling modes, whereas for $\Delta < 0$, consider Type II. In this way, one always has $\sigma_T\Delta < 0$. Therefore, from equation (6.175), the stability criterion for both types of peeling modes can be obtained identically as follows:

$$\Lambda_P < 0. \tag{6.176}$$

From equation (6.174) one can see that the peeling stability criterion in equation (6.176) is more stringent than the Mercier criterion in equation (6.124).

Next, let us discuss the peeling-ballooning modes. The peeling-ballooning modes are actually the free boundary ballooning modes. They can be computed by the two-dimensional codes, such as the ELITE [46] and AEGIS [3] codes. Here, we just explain the one-dimensional treatment, i.e., the free boundary ballooning mode representation [26], in order to explore the underlying physics.

The peeling mode involves only a single resonant Fourier component with coupling to the two nearby sidebands, while the free boundary ballooning modes involve multiple mode couplings and are nonlocal. The free boundary ballooning mode representation is intended to reduce the two-dimensional problem to one-dimensional one using the high n ordering.

As discussed at the beginning of this section, the conventional ballooning invariance in equation (6.131) becomes invalid at the plasma edge since the harmonic with the singular point lying in the vacuum region is different from the harmonic with the singular point in the plasma region. However, the ordering separation for high n modes, *i.e.,* the distance between the nearby mode rational surfaces and the equilibrium scale length, remains unchanged. The cause leading the conventional ballooning mode representation to become inapplicable is due to the boundary conditions. The solution of differential equations can be divided into two steps: determining the independent solutions and using the independent solutions to satisfy the boundary conditions. Therefore, one can still employ the ballooning invariance for the independent solutions, which is more fundamental, i.e.,

$$^j\delta\varphi_{m\mp 2k}(x \pm k) = {}^j\delta\varphi_m(x), \tag{6.177}$$

where the index j labels the independent solutions and k denotes the resonant interval. The difference from the conventional ballooning invariance in equation (6.131) lies in the fact that the invariance here is for the independent solutions, instead of the Fourier components of the perturbation. With the set of independent solutions in one resonant interval determined, using the representation in equation (6.177), one obtains the set of independent solutions in all other intervals. The independent solutions in all intervals can then be used to satisfy the boundary conditions either with the singular surface inside the plasma or in the vacuum region. This is the principle for the free boundary ballooning mode representation discovered in reference [47].

6.8 TOROIDAL ALFVEN EIGENMODES

In this section, we review the toroidal Alfvén eigenmode (TAE) theories [37, 38, 39, 36]. Unlike the MHD stability theory, which is related to the low-frequency regime and a single resonant Fourier harmonic on the rational surface, TAE has a frequency comparable to the shear Alfvén frequency and involves the coupling of two Fourier components. Because the frequency is high, the resonance with the energetic particles, for example, the α particles, can cause their losses. This makes the TAE physics particularly relevant to the burning plasma physics in magnetic fusion research.

To explain the TEA theory, let us consider the equilibrium with a large aspect ratio and circular cross section, as will also be discussed later in Section 6.9. In this equilibrium model, one has $B = B_0(1 - \varepsilon \cos\theta)$, $\mathbf{B} \cdot \nabla = i(B_\phi/R)(m/q - n)$, and

$$\frac{2\mathbf{B} \times \boldsymbol{\kappa}}{B^2} \cdot \nabla r = -\frac{2}{R} \sin\theta.$$

For simplicity, the toroidal mode number n is assumed to be high. In this case, we can start from equations (6.128) and (6.129), which are recited as follows:

$$\mathbf{B} \cdot \nabla \frac{1}{B^2} \nabla \cdot \left(B^2 \nabla_\perp \frac{\mathbf{B} \cdot \nabla \delta\varphi}{B^2} \right) + \nabla \cdot \left(\rho\omega^2 \frac{\nabla_\perp \delta\varphi}{B^2} \right)$$
$$+ P'_\psi \nabla \times \frac{\mathbf{B}}{B^2} \cdot \nabla \left(\frac{\mathbf{B} \times \nabla\psi}{B^2} \cdot \nabla\delta\varphi \right) + \Gamma P \nabla \times \frac{\mathbf{B}}{B^2} \cdot \nabla\nabla \cdot \boldsymbol{\xi} = 0, \quad (6.178)$$

and

$$\Gamma P \mathbf{B} \cdot \nabla \left(\frac{1}{B^2} \mathbf{B} \cdot \nabla\nabla \cdot \boldsymbol{\xi} \right) + \rho_m \omega^2 \frac{B^2 + \mu_0 \Gamma P}{B^2} \nabla \cdot \boldsymbol{\xi}$$
$$= \rho_m \omega^2 \frac{2\mathbf{B} \times \boldsymbol{\kappa}}{B^2} \cdot \nabla\delta\varphi, \quad (6.179)$$

We first solve equation (6.179). Noting that TAE frequency is higher than the acoustic wave frequency, one can ignore the first term in equation (6.179). Consequently, equation (6.179) can be solved, yielding

$$i(\nabla \cdot \boldsymbol{\xi})_m = \frac{1}{BR} \left(\frac{d\delta\varphi_{m+1}}{dr} - \frac{d\delta\varphi_{m-1}}{dr} \right). \quad (6.180)$$

Inserting this solution, equation (6.178) is reduced to [48] (see also Section 6.9)

$$\underbrace{\frac{d}{dr}\left[r^3\left(\frac{1}{q}-\frac{n}{m}\right)^2\frac{d}{dr}E_m\right]}_{field\ line\ bending}$$
$$\underbrace{-\frac{d}{dr}\left(r^3\frac{R^2\omega^2}{m^2v_A^2}\frac{d}{dr}E_m\right)-\varepsilon\left(r^3\frac{R^2\omega^2}{m^2v_A^2}\frac{d^2}{dr^2}E_{m+1}\right)-\varepsilon\left(r^3\frac{R^2\omega^2}{m^2v_A^2}\frac{d^2}{dr^2}E_{m-1}\right)}_{inertia}$$
$$\underbrace{+2\frac{\Gamma Pr^3}{B^2m^2}\frac{d^2E_m}{dr^2}-\frac{\Gamma Pr^3}{B^2m^2}\frac{d^2E_{m+2}}{dr^2}-\frac{\Gamma Pr^3}{B^2m^2}\frac{d^2E_{m-2}}{dr^2}}_{plasma\ compressibiliy}$$
$$\underbrace{-wE_m-\frac{\alpha r^2}{2mq^2}\frac{dE_{m+1}}{dr}+\frac{\alpha r^2}{2mq^2}\frac{dE_{m-1}}{dr}-\frac{\alpha r}{2q^2}E_{m+1}+\frac{\alpha r}{2q^2}E_{m-1}}_{pressure\ driven\ effcts}=0, \quad (6.181)$$

where $E_m = \delta\varphi_m/r$, $v_A^2 = B_0^2/\rho_m$, B_0 denotes the magnetic field at the magnetic axis, and w is introduced to denote the magnetic well effects. Here, the two sideband contributions in the inertia effects result from the large aspect ratio expansion of magnetic field strength B tied to the Alfvén frequency. The plasma compressibility terms also give sideband contributions. It applies only for the the case with a frequency higher than that of the ion acoustic wave. In the low-frequency case with the mode frequency ω much smaller than the sound wave one ω_s, it mainly gives rise to the so-called apparent effects. The sideband contributions are of order $\omega^2/\omega_s^2 \ll 1$ as compared to the case given in equation (6.180).

As shown in figure 6.5, both the inertia and plasma compressibility can cause the TAE gaps. As will be seen, the inertia-effect-generated gap is located at half of Alfvén frequency, and the plasma-compressibility-generated gap is located at the Alfvén frequency. They are referred to respectively as the first and second TAE gaps. The related Alfvén eigenmodes are referred to as the TAEs of the first and second kinds. The second kind of TAEs is often referred to as the elliptic TAEs (EAEs), since the elliptic cross-section can lead to the gap in this frequency range. Since this kind of TAEs can occur in the circular cross-section case [36]. We therefore refer to it as the second kind of TAEs.

Here, we also need to discuss the low-frequency spectrum, which is not included in figure 6.5. In sections 6.5.1 and 6.6.3, we have derived the singular layer equations both in the configuration and ballooning representation spaces. From the singular layer equations, equation (6.112) and equation (6.120), in the configuration space, for example, one can see that the average plasma compressibility effects from $\langle \nabla \cdot \boldsymbol{\xi} \rangle$ are coupled to the shear Alfvén mode equation, equation (6.112). The denominator of the solution of $\langle \nabla \cdot \boldsymbol{\xi} \rangle$ in equation (6.120) can vanish. This affects the low-frequency spectrum. The numerically computed spectrum shown in figure 6.1 somehow reflects this coupling feature.

We first consider that equation (6.181) can be used to study both the first and second kinds of TAEs. To be specific, we use the first kind of TAEs to explain the theory. In this case, the coupling of two harmonics: $m = m$ and $m+1$ is considered. We focus on the resonce layer. Using equation (6.181), the governing equations for these two harmonics are as follows:

$$\frac{d}{dr}\left[(m-nq)^2\frac{dE_m}{dr}\right]-\frac{d}{dr}\left(\omega_N^2\frac{dE_m}{dr}\right)+2\Gamma q^2\beta\frac{d^2E_m}{dr^2}=\varepsilon\omega_N^2\frac{d^2E_{m+1}}{dr^2}, \tag{6.182}$$

$$\frac{d}{dr}\left[(m+1-nq)^2\frac{dE_{m+1}}{dr}\right]-\frac{d}{dr}\left(\omega_N^2\frac{dE_{m+1}}{dr}\right)+2\Gamma q^2\beta\frac{d^2E_{m+1}}{dr^2}=\varepsilon\omega_N^2\frac{d^2E_m}{dr^2}, \tag{6.183}$$

where $\omega_N = \omega/\omega_A$ is the normalized frequency.

We first consider equations (6.182) and (6.183) in lowest order. The equations have an important feature. In the field line bending terms, $(m-nq)^2 = [(m+1)-nq]^2 = (1/2)^2$ on the surface $q_0 = (m+1/2)/n$. This defines the TAE resonance layer. In this layer, the set of equations (6.182) and (6.183) has a solution with the eigenfrequency

$$\omega_{N0} \quad = \quad \sqrt{\frac{1}{4}+2\Gamma q^2\beta}. \tag{6.184}$$

The β modification is kept here, since q^2 may be big.

Moving to the next order, we define that $\delta\omega_N = \omega_N - \omega_{N0}$ and $\delta q = q - q_0$. Here, it has been assumed that q is monoyonic so that one can use it as the radial coordinate. The set of equations (6.182) and (6.183) become

$$\begin{aligned}
\frac{\partial}{\partial\delta q}(2\omega_N\delta\omega_N - n\delta q)\frac{\partial}{\partial\delta q}E_m &= -\varepsilon\omega_N^2\frac{\partial^2}{\partial\delta q^2}E_{m+1},\\
\frac{\partial}{\partial\delta q}(2\omega_N\delta\omega_N + n\delta q)\frac{\partial}{\partial\delta q}E_{m+1} &= -\varepsilon\omega_N^2\frac{\partial^2}{\partial\delta q^2}E_m.
\end{aligned}$$

Introducing the even and odd modes: $E_\pm = E_m \pm E_{m+1}$, this set of equations becomes

$$\begin{aligned}
\frac{\partial}{\partial\delta q}2\omega_N\delta\omega_N\frac{\partial}{\partial\delta q}E_+ - \frac{\partial}{\partial\delta q}n\delta q\frac{\partial}{\partial\delta q}E_- &= -\varepsilon\omega_N^2\frac{\partial^2}{\partial\delta q^2}E_+,\\
\frac{\partial}{\partial\delta q}2\omega_N\delta\omega_N\frac{\partial}{\partial\delta q}E_- - \frac{\partial}{\partial\delta q}n\delta q\frac{\partial}{\partial\delta q}E_+ &= \varepsilon\omega_N^2\frac{\partial^2}{\partial\delta q^2}E_-.
\end{aligned}$$

We can integrate this set of equations once over δq, yielding

$$\begin{aligned}
&\mathscr{D}\begin{pmatrix}\frac{\partial E_+}{\partial\delta q}\\ \frac{\partial E_-}{\partial\delta q}\end{pmatrix}\\
&\equiv \begin{pmatrix}2\omega_N\delta\omega_N+\varepsilon\omega_N^2 & -n\delta q\\ -n\delta q & 2\omega_N\delta\omega_N-\varepsilon\omega_N^2\end{pmatrix}\begin{pmatrix}\frac{\partial E_+}{\partial\delta q}\\ \frac{\partial E_-}{\partial\delta q}\end{pmatrix} = \begin{pmatrix}A_+\\ A_-\end{pmatrix},
\end{aligned} \tag{6.185}$$

where $\mathscr{D}$ is a 2×2 matrix and $A_\pm$ are the integration constants. The TAE dispersion relation can be obtained by integrating equation (6.185) across the TAE singular layer (i.e., from δq^- to δq^+) once more

$$\left(\Delta_+ - \int_{\delta q^-}^{\delta q^+} \frac{\mathscr{D}_{22} d\delta q}{\det|\mathscr{D}|}\right)\left(\Delta_- - \int_{\delta q^-}^{\delta q^+} \frac{\mathscr{D}_{11} d\delta q}{\det|\mathscr{D}|}\right) = \left(\int_{\delta q^-}^{\delta q^+} \frac{\mathscr{D}_{12} d\delta q}{\det|\mathscr{D}|}\right)^2, \quad (6.186)$$

where $\Delta_\pm \equiv E_\pm|_{\delta q^-}^{\delta q^+}/A_\pm$ are determined by the outer solutions to the left and right of the TAE singular layer and $\mathscr{D}_{ij}$ represent the elements of matrix $\mathscr{D}$. With $\Delta_\pm$ being determined by the outer-region solutions, equation (6.186) determines the TAE eigenfrequency to the first order.

Let us now analyze the dispersion relation in equation (6.186). We first note that the denominators of integranda in this dispersion relation involve the determinant $\det|\mathscr{D}|$. The singularity emerges as soon as $\det|\mathscr{D}| = 0$. This leads to the so-called continuum damping as reviewed in Chapter 5. To find eigenmodes that are not damped by the continuum, one needs to find the condition $\det|\mathscr{D}| \neq 0$. Using the definition of $\mathscr{D}$ in equation (6.185), one can find that the condition $\det|\mathscr{D}| = 0$ is given by

$$\left(2\omega_{N0}\delta\omega_N + \varepsilon\omega_{N0}^2\right)\left(2\omega_{N0}\delta\omega_N - \varepsilon\omega_{N0}^2\right) = n^2\delta q^2.$$

Its solution is simply

$$n\delta q = \pm\omega_{N0}\sqrt{4\delta\omega_N^2 - \varepsilon^2\omega_{N0}^2}.$$

Therefore, the nonexistence condition for $\det|\mathscr{D}| = 0$ for a real δq becomes $4\delta\omega_N^2 < \varepsilon^2\omega_{N0}^2$. This indicates the mode frequency must fall in the gap, as shown in the first TAE gap in figure 6.5:

$$\omega_{N-} < \omega_N < \omega_{N+}, \quad (6.187)$$

where

$$\omega_{N\pm} = \omega_{N0} \pm \frac{\varepsilon}{2}\omega_{N0}. \quad (6.188)$$

This indicates that the gap width is $\Delta\omega_N = \omega_{N+} - \omega_{N-} = \varepsilon\omega_{N0}$, which is proportional to ε. The gap exists only in the toroidal geometry with finite ε. That is why it is referred to as "toroidal" Alfvén eigenmodes.

Similar analyses can be made for the harmonics $(m+1) - nq$ and $-[(m-1) - nq]$ to show the existence of the second kind of TAEs in the second TAE gap as shown in figure 6.5 due to the plasma compressibility. One may consider the coupling of two harmonics $\pm(m - nq)$. However, as pointed out earlier after equation (6.180), the sideband contributions are of order $\omega^2/\omega_s^2 \ll 1$ as compared to the 2nd TAE case. The gap width is negligible in this case.

If the finite Larmor radius effects are taken into consideration, the Alfvén types of singularities can be resolved, so that the discrete modes can emerge in the continuum. These types of modes are referred to as kinetic TAEs (i.e., KTAEs). Due to the correction of gyrokinetic theory [49], several missing FLR effects are recovered. Consequently, KTAE theories need to be reevaluated.

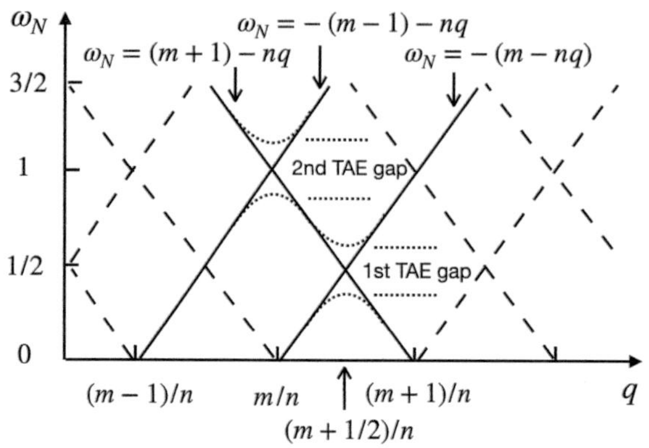

Figure 6.5 TAE gaps.

6.9 LIMIT OF LARGE ASPECT RATIO AND CIRCULAR CROSS SECTION

We have described the general theoretical framework of linear ideal MHD analyses. In this section, we use the analytical equilibrium with a large aspect ratio and circular cross section as described in Section 4.7 to show the formalism and the applications [50]. The basic set of equations (6.13)–(6.15) is reduced to the limit case of a large aspect ratio and circular cross-section. We first describe the global formulation and then the localized mode description. In this way, the treatments of multi-mode couplings in the interchange and ballooning mode formalisms are explained. This is fundamental to understanding the toroidal theory.

6.9.1 GENERAL FORMULATION

First, we treat the compressional Alfvén mode equation (6.13). Since the compressional Alfvén modes have high frequency, for most cases of interest, they are assumed to be suppressed. This is particularly the case when $\delta_L \equiv \lambda_\perp / L \ll 1$, where L is the equilibrium scalelength. In this case, equation (6.13) together with equation (6.16) yields

$$(B^2 + \mu_0 P)\nabla \cdot \boldsymbol{\xi} - B^2 \mathbf{B} \cdot \nabla \left(\frac{\mathbf{B} \cdot \boldsymbol{\xi}}{B^2} \right) = \mathscr{O}\left(\delta_L^2\right).$$

This allows for the following representation

$$\boldsymbol{\xi} = \frac{\mathbf{B} \times \nabla \phi}{B^2} + \frac{\xi_\parallel}{\rho} \mathbf{B} + \mathscr{O}\left(\max\{\delta_L^2, \beta \delta_L\}\right), \qquad (6.189)$$

where φ is the stream function, and $\xi_\parallel$ describes the parallel perturbation. With the representation in equation (6.189), we can reduce the shear Alfvén and acoustic wave equations. The voticity equation (6.14) becomes

$$\begin{aligned}
&\mathbf{B}\cdot\boldsymbol{\nabla}\frac{1}{B^2}\boldsymbol{\nabla}\cdot\left(B^2\boldsymbol{\nabla}_\perp\frac{\mathbf{B}\cdot\boldsymbol{\nabla}\varphi}{B^2}\right)+\mathbf{B}\times\boldsymbol{\nabla}\frac{\mathbf{B}\cdot\boldsymbol{\nabla}\varphi}{B^2}\cdot\boldsymbol{\nabla}\left(\frac{R^2}{I}\boldsymbol{\nabla}\cdot\frac{\boldsymbol{\nabla}\psi}{R^2}\right)\\
&+\frac{2\mathbf{B}\times\boldsymbol{\kappa}}{B^2}\cdot\boldsymbol{\nabla}\left(\frac{\partial P}{\partial\psi}\frac{\mathbf{B}\times\boldsymbol{\nabla}\psi}{B^2}\cdot\boldsymbol{\nabla}\varphi\right)+\boldsymbol{\nabla}\cdot\left(\rho\omega^2\frac{\boldsymbol{\nabla}_\perp\varphi}{B^2}\right)\\
&-\frac{2\mathbf{B}\times\boldsymbol{\kappa}}{B^2}\cdot\boldsymbol{\nabla}(\rho D)-\rho\left(\frac{2\mathbf{J}_\parallel}{B^2}\cdot\nabla+\mathbf{J}\cdot\boldsymbol{\nabla}\frac{1}{B^2}\right)D=0,
\end{aligned}\tag{6.190}$$

and the acoustic wave equation (6.15) is reduced to

$$\begin{aligned}
&\mathbf{B}\cdot\boldsymbol{\nabla}\left(\frac{\rho}{B^2}\mathbf{B}\cdot\boldsymbol{\nabla}D\right)+\omega^2\frac{(B^2+\mu_0P)\rho^2}{B^2\Gamma P}D\\
=\;&\rho\omega^2\frac{2\mathbf{B}\times\boldsymbol{\kappa}}{B^2}\cdot\boldsymbol{\nabla}\varphi,
\end{aligned}\tag{6.191}$$

where

$$D=\frac{B^2\Gamma P}{\rho^2(B^2+\mu_0P)}\left(\mathbf{B}\cdot\boldsymbol{\nabla}\xi_\parallel+\rho\frac{2\mathbf{B}\times\boldsymbol{\kappa}}{B^2}\cdot\boldsymbol{\nabla}\varphi\right).\tag{6.192}$$

We now use the equilibrium solutions for the case with a large aspect ratio and circular cross section in Section 4.7 to express metric parameters, such as $\boldsymbol{\nabla}r\cdot\boldsymbol{\nabla}r$, $\boldsymbol{\nabla}r\cdot\boldsymbol{\nabla}\theta$, and etc., and introduce the Fourier decomposition

$$\varphi(r,\theta,\zeta)\;=\;\sum_{m'}\varphi^{(m')}(r)\exp\left[i(m\theta-n\zeta)\right].$$

Here, the magnetic flux coordinates have been used with r labelling the magnetic surfaces. It is further assumed that the main harmonic is denoted by m and the ordering sequence $\varphi^{(m)}$, $\varepsilon\varphi^{(m\pm1)},\cdots$ applies for the Fourier components. It is also assumed that the magnetic shear $s\lesssim\varepsilon^{1/2}$, which is consistent with the Mercier criterion to be derived in this limit.

The vorticity equation (6.190) can be reduced term by term as follows. The field-line-bending term is reduced to

$$\begin{aligned}
&\left(-\frac{\mathscr{I}}{R_0r}\right)\mathbf{B}\cdot\boldsymbol{\nabla}\frac{1}{B^2}\boldsymbol{\nabla}\cdot\left(B^2\boldsymbol{\nabla}_\perp\frac{\mathbf{B}\cdot\boldsymbol{\nabla}\varphi}{B^2}\right)\\
=\;&\left(\frac{m^2}{R_0^2r^2}\right)\left[\frac{d}{dr}\left(r^3k_{\parallel m}^2\frac{d}{dr}E_0\right)\right.\\
&\left.-r(m^2-1)k_{\parallel m}^2E_0+k_{\parallel m}\frac{d}{dr}\left(r^3\frac{dk_{\parallel m}}{dr}\right)E_0\right],
\end{aligned}$$

where, $k_{\|m} = (1/q - n/m)$ and $E_0(r) \equiv \varphi_0(r)/r$. The kink term is reduced to

$$\begin{aligned}
&\left(-\frac{\mathscr{I}}{R_0 r}\right)\mathbf{B}\times\nabla\frac{\mathbf{B}\cdot\nabla\varphi}{B^2}\cdot\nabla\left(\frac{R^2}{I}\nabla\cdot\frac{\nabla\psi}{R^2}\right)\\
&= \left(\frac{m^2}{R_0^2 r^2}\right)\left[-k_{\|m}\frac{d}{dr}\left(r^3\frac{dk_{\|m}}{dr}\right)E_0 + \frac{(m+1)r^2}{2m^2q^2}\frac{\partial\alpha_p}{\partial r}E_1^{(m+1)}\right.\\
&\quad +\frac{r^2\alpha_p}{2m^2q^2}\frac{1}{r}\frac{d}{dr}\left(rE_1^{(m+1)}\right) - \frac{(m-1)r^2}{2m^2q^2}\frac{\partial\alpha_p}{\partial r}E_1^{(m-1)}\\
&\quad \left. +\frac{r^2\alpha_p}{2m^2q^2}\frac{1}{r}\frac{d}{dr}\left(rE_1^{(m-1)}\right)\right].
\end{aligned}$$

The pressure driving term, which contributes to the main harmonic, is reduced to

$$\begin{aligned}
&\left(-\frac{\mathscr{I}}{R_0 r}\right)\frac{2\mathbf{B}\times\boldsymbol{\kappa}}{B^2}\cdot\nabla\left(\frac{\partial P}{\partial\psi}\frac{\mathbf{B}\times\nabla\psi}{B^2}\cdot\nabla\varphi\right)\\
&= \left(\frac{m^2}{R_0^2 r^2}\right)\left[-\frac{r}{2q^2}\alpha_p^2 E_0 - \frac{r^2}{Rq^2}\alpha_p\left(1-\frac{1}{q^2}\right)E_0\right.\\
&\quad -\frac{m+1}{2q^2m^2}r^{-m+1}\frac{d}{dr}\left(\alpha_p r^{m+1}E_1^{(m+1)}\right)\\
&\quad \left. +\frac{m-1}{2q^2m^2}r^{m+1}\frac{d}{dr}\left(\alpha_p r^{-m+1}E_1^{(m-1)}\right)\right].
\end{aligned}$$

In order to reduce the plasma compressibility term, we need to solve the parallel acoustic wave equation (6.191), In the low-frequency regime, the second term on the left can be neglected. Consequently, one has

$$\begin{aligned}
D^{(m+1)} &= -i\frac{Rq^2}{B}\omega^2 r^m\frac{d}{dr}\left(r^{-m+1}E_0\right),\\
D^{(m-1)} &= i\frac{Rq^2}{B}\omega^2 r^{-m}\frac{d}{dr}\left(r^{m+1}E_0\right).
\end{aligned}$$

Using these results, the plasma compressibility term can be reduced to

$$\begin{aligned}
&\frac{\mathscr{I}}{R_0 r}\frac{2\mathbf{B}\times\boldsymbol{\kappa}}{B^2}\cdot\nabla(\rho D)\\
&= -\left(\frac{\partial}{\partial r}+\frac{m+1}{r}\right)\frac{q^2\rho\omega^2}{B^2}r^m\frac{\partial}{\partial r}\left(r^{-m+1}E_0\right)\\
&\quad -\left(\frac{\partial}{\partial r}-\frac{m-1}{r}\right)\frac{q^2\rho\omega^2}{B^2}r^{-m}\frac{\partial}{\partial r}\left(r^{m+1}E_0\right)\\
&= \left(\frac{m^2}{R_0^2 r^2}\right)\left\{-\frac{d}{dr}\left[r^3\frac{R^2\omega^2}{m^2v_A^2}\mathscr{M}_\|\frac{d}{dr}E_0\right] + r(m^2-1)\frac{R^2\omega^2}{m^2v_A^2}\mathscr{M}_\| E_0\right.\\
&\quad \left. -r^2\frac{R^2}{m^2}\frac{d}{dr}\left(\frac{\omega}{v_A}\right)^2\mathscr{M}_\| E_0\right\}.
\end{aligned}$$

Finally, the inertia term is reduced to

$$
\begin{aligned}
&\left(-\frac{\mathscr{I}}{R_0 r}\right)\boldsymbol{\nabla}\cdot\left(\rho\omega^2\frac{\boldsymbol{\nabla}_\perp\varphi}{B^2}\right)\\
=&\left(\frac{m^2}{R_0^2 r^2}\right)\left[-\frac{d}{dr}\left(r^3\frac{R^2\omega^2}{m^2 v_A^2}\frac{d}{dr}E_0\right)+r(m^2-1)\frac{R^2\omega^2}{m^2 v_A^2}E_0\right.\\
&\left.-r^2\frac{R^2}{m^2}\frac{d}{dr}\left(\frac{\omega}{v_A}\right)^2 E_0\right],
\end{aligned}
$$

where $v_A^2 = B^2/\rho$.

Using these reductions, the vorticity equation (6.190) for the main harmonic becomes

$$
\begin{aligned}
&\frac{d}{dr}\left[r^3\left(\frac{1}{q}-\frac{n}{m}\right)^2\frac{d}{dr}E_0\right]-r\left[(m^2-1)\left(\frac{1}{q}-\frac{n}{m}\right)^2\right]E_0\\
&-r\frac{\alpha_p^2}{2q^2}E_0-\frac{r^2}{Rq^2}\alpha_p\left(1-\frac{1}{q^2}\right)E_0\\
&-\frac{\alpha_p r^{-m}}{2mq^2}\frac{d}{dr}\left(r^{m+2}E_1^{(m+1)}\right)+\frac{\alpha_p r^m}{2mq^2}\frac{d}{dr}\left(r^{-m+2}E_1^{(m-1)}\right)\\
&-\frac{d}{dr}\left(r^3\mathscr{M}\frac{R^2\omega^2}{m^2 v_A^2}\frac{d}{dr}E_0\right)\\
&+r(m^2-1)\mathscr{M}\frac{R^2\omega^2}{m^2 v_A^2}E_0-r^2\mathscr{M}\frac{R^2}{m^2}\frac{d}{dr}\left(\frac{\omega}{v_A}\right)^2 E_0=0,
\end{aligned}
\tag{6.193}
$$

$\mathscr{M} = 1+2q^2$. Similarly, the vorticity equation (6.190) for the side bands (with the poloidal mode number $m\pm 1$) becomes

$$
\begin{aligned}
&\frac{d}{dr}\left[r^3\left(\frac{1}{q}-\frac{n}{m+1}\right)^2\frac{dE_1^{(m+1)}}{dr}\right]-m(m+2)r\left(\frac{1}{q}-\frac{n}{m+1}\right)^2 E_1^{(m+1)}\\
=&-\frac{n^2 r^{m+2}}{m^2(m+1)^2}\frac{m}{2}\frac{d}{dr}\left(\alpha_p r^{-m}E_0\right),
\end{aligned}
\tag{6.194}
$$

$$
\begin{aligned}
&\frac{d}{dr}\left[r^3\left(\frac{1}{q}-\frac{n}{m-1}\right)^2\frac{dE_1^{(m-1)}}{dr}\right]-m(m-2)r\left(\frac{1}{q}-\frac{n}{m-1}\right)^2 E_1^{(m-1)}\\
=&\frac{n^2 r^{-m+2}}{m^2(m-1)^2}\frac{m}{2}\frac{d}{dr}\left(\alpha_p r^{m}E_0\right).
\end{aligned}
\tag{6.195}
$$

Equations (6.193)–(6.195) constitute a complete set of MHD equations for low-beta, large aspect-ratio circular tokamak, with a small magnetic shear. Note that this set of equations is valid for general ideal MHD modes, including both global and localized modes.

6.9.2 ANALYSIS OF THE INTERCHANGE AND BALLOONING MODES

In this subsection, we apply equations (6.193)–(6.195) to analyze the interchange and ballooning modes to explain how the localized mode and high n mode analyses are related to the global Fourier mode description.

For the interchange mode, the following auxiliary orderings are introduced

$$s \sim \frac{r}{R} \sim \alpha_p, \ \ x \equiv \frac{r-r_s}{r_s} \sim \frac{\omega}{s\gamma_A} \ll 1, \tag{6.196}$$

where r_s labels the reference rational surface. In this case, the sideband equations (6.194) and (6.195) can be solved

$$\frac{d}{dx}E_1^{(m+1)} = -\frac{m}{2r_s}\alpha_p E_0, \tag{6.197}$$

$$\frac{d}{dx}E_1^{(m-1)} = \frac{m}{2r_s}\alpha_p E_0. \tag{6.198}$$

With the solution in equations (6.197) and (6.198), the main harmonic equation (6.193) is reduced to

$$\begin{aligned}&\frac{d}{dx}\left(s^2x^2\frac{d}{dx}E_0\right) - \frac{r}{R}\alpha_p\left(1-\frac{1}{q^2}\right)E_0\\ &-\frac{d}{dx}\left(\frac{q^2\mathscr{M}}{m^2}\frac{\omega^2}{\gamma_A^2}\frac{d}{dx}E_0\right) = 0,\end{aligned} \tag{6.199}$$

Here, it has been noted that the second and the last two terms in equation (6.193) are negligible under the ordering assumptions in equation (6.196). Equation (6.199) is the singular layer equation for low-beta, large aspect-ratio circular tokamaks. The Mercier criterion can be readily obtained as

$$\frac{s^2}{4} + \frac{r}{R}\alpha_p\left(1-\frac{1}{q^2}\right) > 0. \tag{6.200}$$

Equation (6.200) recovers results for the Mercier criterion in references [51] and [52]. From the derivation of the criterion, one can see that the driving term, the second term in equation (6.200), results from both the magnetic well effects of the main harmonic and the sideband couplings.

For ballooning modes, we can neglect the small terms in the high-n ordering. Equation (6.193) is then reduced to

$$\begin{aligned}&\frac{d}{dr}\left[r^3\left(\frac{1}{q}-\frac{n}{m}\right)^2\frac{d}{dr}E_0\right] - r\frac{\alpha_p^2}{2q^2}E_0 - \frac{r^2}{Rq^2}\alpha_p\left(1-\frac{1}{q^2}\right)E_0\\ &-\frac{\alpha_p r^{-m}}{2mq^2}\frac{d}{dr}\left(r^{m+2}E_1^{(m+1)}\right) + \frac{\alpha_p r^m}{2mq^2}\frac{d}{dr}\left(r^{-m+2}E_1^{(m-1)}\right)\\ &-\frac{d}{dr}\left(r^3\mathscr{M}\frac{R^2\omega^2}{m^2v_A^2}\frac{d}{dr}E_0\right) = 0.\end{aligned} \tag{6.201}$$

Applying the ballooning invariance in equation (6.131) in Section 6.193 and noting that $q = m/n + sx + \cdots$, where $s = r_s(d\ln q/dr)$ and $x = (r - r_s)/r_s$, equation (6.201) becomes

$$\frac{d}{dx}\left(s^2x^2 - \omega_N^2\right)\frac{dE_0}{dx} - \frac{\alpha_p^2}{2}E_0 - \frac{r}{R}\alpha_p\left(1 - \frac{1}{q^2}\right)E_0$$
$$-\frac{\alpha}{2}\left(E_0(x+1) + E_0(x-1)\right)$$
$$-\frac{\alpha}{2m}\left(\frac{dE_0(x+1)}{dx} - \frac{dE_0(x-1)}{dx}\right) = 0.$$

Here, we see that the ballooning mode invariance can be used to include the sideband couplings with the shifted representations of the main harmonic. Introducing the Fourier transform as in equation (6.133), one obtains

$$\frac{d}{d\theta}\left(s^2\theta^2 + \omega_N^2\right)\frac{dE_0}{d\theta} + \frac{\alpha_p^2}{2}E_0 + \frac{r}{R}\alpha_p\left(1 - \frac{1}{q^2}\right)E_0$$
$$+\alpha\left(\cos\theta + s\theta\sin\theta\right)E_0 = 0.$$

Noting that when the steep-pressure-gradient equilibrium model is introduced ($s\theta \rightarrow \Lambda_s$), this equation becomes the ballooning equation analyzed in Section 6.6.4 with the magnetic well effect kept.

The analyses of TAEs in the equilibrium with a large aspect ratio and circular cross section have been described in Section 6.8. We do not further elaborate on them here.

6.10 CONCLUSIONS AND DISCUSSION

In this chapter, we have described the linear ideal MHD theory in the toroidal geometry. First, the global description in the Fourier space is presented. This helps not only the numerical computation, but also helps clarify the spectra for three major waves: the compressional Alfvén, shear Alfvén, and acoustic waves. The MHD modes in toroidal geometry can now be computed numerically. Nevertheless, the refined treatments of the singular layer are still needed. On the one hand, the numerical treatment of the singular layers often needs analytical assistance. On the other hand, many of the subtle physics mechanisms beyond the ideal MHD description enter through the singular layer. This leads to an extensive discussion of the theories of the interchange, ballooning, and toroidal Alfvén modes next. We have also used the equilibrium with a large aspect ratio and circular cross-section to further explain these theories.

The ideal MHD stability theories reviewed in this chapter help to extend the ideal MHD description to non-ideal MHD ones. The basic formalisms to include non-ideal MHD effects have been outlined in Chapter 2. Here, we particularly point out that in the current formulation of the interchange and ballooning mode theories, the vorticity and acoustic wave equations (6.14) and (6.20) are employed. This makes it straightforward to include the nonideal MHD effects in the direction parallel to the magnetic field.

References

1. Glasser A 2016 The direct criterion of Newcomb for the stability of an axisymmetric toroidal plasma, *Phys. Plasmas 23*, 072505. https://doi.org/10.1063/1.4958328
2. Zheng L 2020 Magnetically Confined Fusion Plasma Physics, ideal MHD theory (London: IOP Publishing). DOI: 10.1088/978-1-64327-138-5.
3. Zheng L-J, Kotschenreuther M 2006 AEGIS: An adaptive ideal-magnetohydrodynamics shooting code for axisymmetric plasma stability *J. Comput. Phys. 211*, 748–66. https://doi.org/10.1016/j.jcp.2005.06.009
4. Chance M S 1997 Vacuum calculations in azimuthally symmetric geometry *Phys. Plasmas 4*, 2161–180. https://doi.org/10.1063/1.872380
5. Zheng L-J, Kotschenreuther M 2005 Wall thickness effect on the resistive wall mode stability in toroidal plasmas *Phys. Plasmas 12*, 072504. https://doi.org/10.1063/1.1943347
6. Cheng C Z, Chance M S 1986 Low-n shear Alfvén spectra in axisymmetric toroidal plasmas, *Phys. Fluids 29*, 3695–701. https://doi.org/10.1063/1.865801
7. Chu M S, Greene J M, Lao L L, Turnbull A D, Chance M S 1992 A numerical study of the high-n shear Alfvén spectrum gap and the high-n gap mode, *Phys. Fluids B 4*, 3713–721. https://doi.org/10.1063/1.860327
8. Newcomb W A 1960 Hydromagnetic stability of a diffuse linear pinch, *Ann. Phys. NY 10*, 232–67. https://doi.org/10.1016/0003-4916(60)90023-3
9. Zheng L-J, Kotschenreuther M, Van Dam J W 2009 Kinetic analysis of the resistive wall modes in the ITER advanced tokamak scenario *Nucl. Fusion 49*, 075021. DOI 10.1088/0029-5515/49/7/075021
10. Freidberg J P 1987 Ideal Magnetohydrodynamics (New York: Plenum).
11. Hu B, Betti R 2004 Resistive Wall Mode in Collisionless Quasistationary Plasmas *Phys. Rev. Lett. 93*, 105002. https://doi.org/10.1103/PhysRevLett.93.105002
12. Zheng L-J, Kotschenreuther M, Chu M S 2005 Rotational stabilization of resistive wall modes by the shear Alfvén resonance, *Phys. Rev. Lett. 95*, 255003, https://doi.org/10.1103/PhysRevLett.95.255003
13. Boozer A. H. 2001 Error field amplification and rotation damping in tokamak plasmas *Phys. Rev. Lett. 86*, 5059. https://doi.org/10.1103/PhysRevLett.86.5059
14. Zheng L-J, Kotschenreuther M 2005 Rotation, error field, and wall thickness effects on resistive wall modes in ITER, *International Sherwood Fusion Theory Conference, Stateline, Nevada, USA* O4.
15. Zheng L-J, Kotschenreuther M, Waelbroeck F L 2006 Shielding of the error field by a liquid metal wall in tokamaks *Nucl. Fusion* 46, L9-12. DOI 10.1088/0029-5515/46/8/L03
16. Mercier C 1960 A necessary condition for hydromagnetic stability of plasma with axial symmetry *Nucl. Fusion 1*, 47. DOI 10.1088/0029-5515/1/1/004
17. Greene J M, Johnson J L 1962 Stability criterion for arbitrary hydromagnetic equilibria *Phys. Fluids 5*, 510–7. https://doi.org/10.1063/1.1706651
18. Glasser A H, Greene J M, Johnson J L (1975) Resistive instabilities in general toroidal plasma configurations, *Phys. Fluids 18*, 875–88. https://doi.org/10.1063/1.861224
19. Greene J M, Johnson J L 1968 Interchange instabilities in ideal hydromagnetic theory *Plasma Phys. 10*, 729. DOI 10.1088/0032-1028/10/8/301

20. Zheng L 2015 IOP Concise Physics: Advanced Tokamak Stability Theory (San Rafael, CA: Morgan & Claypool and London: IOP Publishing).
21. Mikhailovsky A 1974 Theory of Plasma Instabilities (New York: Consultants Bureau).
22. Zheng L-J, Tessarotto M 1994 Collisionless kinetic ballooning mode equation in the low-frequency regime *Phys. Plasmas 1*, 3928–35, https://doi.org/10.1063/1.870865
23. Connor J W, Hastie R J, Taylor J B 1979 High mode number stability of an axisym- metric toroidal plasma *Proc. R. Soc. A 365*, 1–17. https://doi.org/10.1098/rspa.1979.0001
24. Chance M S *et al.* 1979 MHD stability limits on high-beta tokamaks Proc. 7th Int. Conf. Plasma Physics and Controlled Fusion Research, (Innsbruck, Austria, 23–30 August, 1978) vol 1 (Vienna: International Atomic Energy Agency) p 677.
25. Lee Y C, Van Dam J W 1977 Kinetic theory of ballooning instabilities, Proc. Finite Beta Theory Workshop vol CONF-7709167, ed B Coppi and W Sadowski (Washington, D.C.: US Department of Energy) p 93.
26. Zheng L J 2012 Free boundary ballooning mode representation *Phys. Plasmas 19*, 102506. https://doi.org/10.1063/1.4759012
27. Correa-Restrepo D 1982 Resistive ballooning modes in three-dimensional configurations *Z. Naturforsch. A 37*, 848. https://doi.org/10.1515/zna-1982-0818
28. Hazeltine R D, Hitchcock D A and Mahajan S M 1981 Uniqueness and inversion of the ballooning representation *Phys. Fluids 24*, 180–1. https://doi.org/10.1063/1.863238
29. Connor J W, Hastie R J and Taylor J B 1978 Shear, periodicity, and plasma ballooning modes *Phys. Rev. Lett. 40*, 396–9. https://doi.org/10.1103/PhysRevLett.40.396
30. Greene J, Chance M 1981 The second region of stability against ballooning modes *Nucl. Fusion 21*, 453. DOI 10.1088/0029-5515/21/4/002
31. Berk H L, Rosenbluth M N, Shohet J L 1983 Ballooning mode calculations in stellarators *Phys. Fluids 26*, 2616–20. https://doi.org/10.1063/1.864453
32. Vergara N 1986 Stability of ballooning modes in tokamaks with energetic particles *PhD dissertation: DOE/ET-53088-245, IFSR #245.*
33. Tang W, Connor J, Hastie R 1980 Kinetic-ballooning-mode theory in general geometry *Nucl. Fusion 20*, 1439. DOI 10.1088/0029-5515/20/11/011
34. Zheng L-J, Tessarotto M 1994 Collisionless kinetic ballooning mode equation in the low-frequency regime *Phys. Plasmas 1*, 3928–35. DOI:https://doi.org/10.1063/1.870865.
35. Zheng L-J, Tessarotto M 1994 Collisionless kinetic ballooning equations in the comparable frequency regime *Phys. Plasmas 1*, 2956-62.
36. Zheng L-J, Chen L 1998 Plasma compressibility induced toroidal Alfvén eigenmode *Phys. Plasmas 5*, 444–9. https://doi.org/10.1063/1.872728
37. Cheng C Z, Chance M S 1986 Low-n shear Alfvén spectra in axisymmetric toroidal plasmas *Phys. Fluids 29*, 3695–701. https://doi.org/10.1063/1.865801
38. Rosenbluth M N, Berk H L, Van Dam J W, Lindberg D M 1992 Mode structure and continuum damping of high-n toroidal Alfvén eigenmodes *Phys. Fluids B 4*, 2189–202. DOI: 10.1103/PhysRevLett.68.596
39. Chen L, Zonca F 2016 Physics of Alfvén waves and energetic particles in burning plasmas *Rev. Mod. Phys. 88*, 015008. DOI: https://doi.org/10.1103/RevModPhys.88.015008
40. Zheng L-J, Chen L 1998 Kinetic toroidal Alfvén eigenmodes in finite-β tokamak plasmas *Phys. Plasmas 5*, 1056–61. https://doi.org/10.1063/1.872635
41. Chen L, Bondeson A, Chance M S 1987 Asymptotic stability boundaries of ballooning modes in circular tokamaks *Nucl. Fusion 27*, 1918. DOI 10.1088/0029-5515/27/11/015
42. Newcomb W A 1960 Hydromagnetic stability of a diffuse linear pinch *Ann. Phys. NY 10*, 232–67. https://doi.org/10.1016/0003-4916(60)90023-3

43. Lortz D and Nührenberg J 1978 Ballooning stability boundaries for the large-aspect-ratio tokamak *Phys. Lett. A 68*, 49–50. https://doi.org/10.1016/0375-9601(78)90753-3
44. Lortz D 1975 The general 'peeling' instability *Nucl. Fusion 15*, 49-54. DOI 10.1088/0029-5515/15/1/007
45. Hastie R J private communication; published in Wesson J 1978 Hydromagnetic stability of tokamaks *Nucl. Fusion 18*, 87–132.
46. Snyder P B, Wilson H R, Ferron J R, Lao L L, Leonard A W, Osborne T H, Turnbull A D, Mossessian D, Murakami M, Xu X Q 2002 Edge localized modes and the pedestal: a model based on coupled peeling–ballooning modes *Phys. Plasmas 9*, 2037–43.
47. Zheng L J 2012 Free boundary ballooning mode representation *Phys. Plasmas 19*, 102506.
48. Zheng L-J, Chu M S, Chen L 1999 Effect of toroidal rotation on the localized modes in low beta circular tokamaks *Phys. Plasmas 6*, 1217–26.
49. Zheng L 2022 Magnetically Confined Fusion Plasma Physics, Volume 3, Kinetic theory (London: IOP Publishing). https://doi.org/10.1088/978-0-7503-3779-3.
50. Zheng L J, Chu M S, and Chen L 1999 Effect of toroidal rotation on the localized modes in low beta circular tokamaks *Phys. Plasmas 6*, 1217–26. https://doi.org/10.1063/1.873365
51. Ware A A, Haas F A 1966 Stability of a Circular Toroidal Plasma under Average Magnetic Well Conditions *Phys. Fluids 9*, 956–64. https://doi.org/10.1063/1.1761797
52. Shafranov V D, Yurchenko E I 1967 Zh. Eksp. Teor. Fiz. 53 1157 (in Russian); 1968 Condition for flute instability of a toroidal-geometry plasma *Sov. Phys.-JETP 26* 682–6 (Engl. transl.).

7 Linear resistive MHD stability theory in cylindrical geometry

In the previous chapters, the ideal MHD theory has been presented. In this and the following chapters, the resistive MHD theory will be presented. The linear resistive MHD theories in the cylindrical and toroidal geometries will be described in this and the next chapters. The nonlinear theory will be given in Chapter 9.

Resistivity effects break the frozen-in law, which requires plasma and magnetic field lines to move together in the direction perpendicular to the magnetic field line. Nevertheless, it should be pointed out that the resistivity is usually small, especially in the magnetic fusion plasmas. It, therefore, only plays a significant role in the singular layer in the ideal MHD description. This leads the treatment of resistive MHD to often invoke the matching technique. The resistive MHD solution in the singular layer is matched to the external resolutions to each side of the singular layer. Here, it should be noted that the asymptotical matching is different from solving the same differential equations with multiple region matching. The asymptotical matching matches the solutions of different differential equations in two nearby regions, which only become identical in a certain limit in their common region. It is not a point matching, but a smooth asymptotic overlap of two solutions in a common area. It is usually not a simple task. Fortunately, after decades of effort, a fairly comprehensive theoretical framework has been established. We especially describe the theories given in references [1, 2, 3].

To clarify further, let us refer to figure 7.1 for a schematic explanation of the resistive MHD modes in comparison with the ideal MHD ones. This is an extension of a similar explanation given in reference [4]. The ideal MHD stability theory has been described in Chapter 5 and 6. In figure 7.1, the solid curves represent the solutions of the ideal MHD equations with vanishing frequency/growth rate. According to Newcomb's theorem 5 [5], if ξ has two zeros (or a small solution at the singular layer) inside the plasma, the system is unstable for internal modes. The ideal MHD unstable modes are depicted in figure 7.1 by the curves E and E': Curve E has two zeros, while curve E' has one zero, and the order is small at the singular layer.

In the ideal MHD stable cases, there can still exist the resistive MHD instabilities in figure 7.1. The ideal MHD stable modes are depicted by the curves $E = F$, F', and

DOI: 10.1201/9781003522133-7

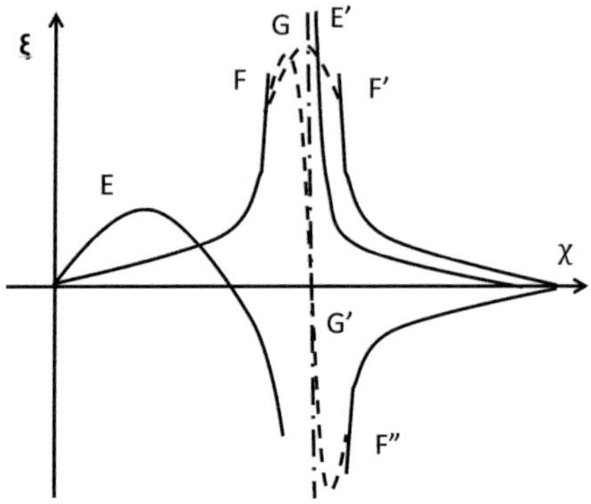

Figure 7.1 Sketch of the ideal and resistive MHD modes.

F''. Each of them contains only one zero, while the other side is large at the singular layer. The dashed curves G and G' represent the resistive MHD solutions in the vicinity of the singlar layer. They connect the stable MHD modes across the singular layers to form unstable resistive MHD modes. There are two types of resistive MHD modes: the even and odd modes according to the parities of the perturbed radial magnetic field δB_χ (or ψ). The even one corresponds to the tearing modes, while the odd one is referred to as the resistive interchange modes. Note that δB_χ and the radial displacement ξ have opposite parity. Therefore, in figure 7.1, the curve F-G-F' represents the resistive interchange modes, while the curve F-G'-F'' describes the tearing modes. For tearing modes, there is a finite normal perturbed magnetic field at the singular layer. This causes the magnetic field lines to reconnect at the singular layer, so that the magnetic islands and stochastic field line regions emerge as explained in Section 9.1. Note that the interchange modes in tokamak interchange not only the plasma and magnetic energies, but also the equilibrium current. This makes the interchange and tearing modes become coupled as explained in Section 9.5 [6].

The chapter arrangement is as follows. We first lay out the linear resistive MHD equations with the basic ordering considerations pointed out. The singular layer equations are then derived. The external limits of the singular layer solutions are analyzed, which help classify the resistive MHD modes. The solutions of the singular layer equations in the Fourier space are given afterward. After that, the solutions in the real space are derived under the so-called constant-ψ assumption to obtain the tearing mode solution. We mainly focus on the incompressible case. The plasma compressibility will be included in the end in Section 7.6. The conclusions and discussion are given in the end.

7.1 BASIC SET OF EQUATIONS

In this section, we describe the basic set of equations to describe the resistivity effects on the MHD modes. The starting basic set of equations is as follows:

$$
\begin{aligned}
\frac{\partial \rho}{\partial t} + \nabla \cdot (\rho \mathbf{v}) &= 0, \\
\frac{\partial}{\partial t}(\rho \mathbf{v}) + \nabla \cdot (\rho \mathbf{v}\mathbf{v}) &= \mathbf{J} \times \mathbf{B}, \\
\frac{\partial \mathbf{B}}{\partial t} &= \nabla \times (\mathbf{v} \times \mathbf{B} - \eta \mathbf{J}), \\
\frac{\partial}{\partial t}(P/\rho^{\Gamma-1}) + \nabla \cdot (\mathbf{v}P/\rho^{\Gamma-1}) &= 0, \\
\nabla \cdot \mathbf{B} &= 0, \\
\mu_0 \mathbf{J} &= \nabla \times \mathbf{B}.
\end{aligned}
$$

They can be linearized to obtain the set of linearized resistive MHD equations, constituted by the momentum equation, Ohm's law, Ampère's law, and the law of adiabatic:

$$\rho \gamma^2 \boldsymbol{\xi} = \mathbf{j} \times \mathbf{B} + \mathbf{J} \times \mathbf{b} + \nabla p, \tag{7.1}$$

$$\mathbf{b} - \frac{\eta}{\mu_0 \gamma} \nabla \times \nabla \times \mathbf{b} = \nabla \times (\boldsymbol{\xi} \times \mathbf{B}), \tag{7.2}$$

$$\mu_0 \mathbf{j} = \nabla \times \mathbf{b}, \tag{7.3}$$

$$p = -\boldsymbol{\xi} \cdot \nabla P - \Gamma P \nabla \cdot \boldsymbol{\xi}. \tag{7.4}$$

Here, we have used the upper and lower cases to denote the unperturbed and perturbed quantities.

As in the ideal MHD case, we use the vorticity equation to describe the perpendicular momentum balance as given in equation (6.14)

$$
\begin{aligned}
& -\mathbf{B} \cdot \nabla \frac{\mathbf{B} \cdot \mathbf{j}}{B^2} + \nabla \cdot \frac{\mathbf{B}}{B^2} \times \rho_m \omega^2 \boldsymbol{\xi} \\
= \; & \mathbf{b} \cdot \nabla \sigma - \mathbf{J} \cdot \nabla \frac{\mathbf{B} \cdot \mathbf{b}}{B^2} + \nabla \times \frac{\mathbf{B}}{B^2} \cdot \nabla p,
\end{aligned} \tag{7.5}
$$

with $\sigma = \mathbf{J} \cdot \mathbf{B}/B^2$. Besides, from equation (3.118), one has

$$\frac{1}{B^2}\left(\nabla \times \boldsymbol{\xi} \times \mathbf{B} - \mathbf{B}\frac{\mu_0 \boldsymbol{\xi} \cdot \nabla p}{B^2}\right) \cdot \mathbf{B} = -\nabla \cdot \boldsymbol{\xi} + \mathbf{B} \cdot \nabla \left(\frac{\boldsymbol{\xi} \cdot \mathbf{B}}{B^2}\right) - 2\boldsymbol{\xi} \cdot \boldsymbol{\kappa}. \tag{7.6}$$

Since the compressional Alfvén modes of high frequency are decoupled from the low frequency modes, equations (7.5) and (7.6) help simplify the analyses.

Note that the resistivity occurs only in Ohm's law in equation (7.2). Depending on the significance of resistivity, three cases are usually investigated. Here, let us first classify them. The detailed analyses will be given in the following sections. The first is the ideal MHD case with

$$|\mathbf{b}| \gg \left| \frac{\eta}{\mu_0 \gamma} \nabla \times \nabla \times \mathbf{b} \right|. \tag{7.7}$$

The ideal MHD case has been described in Chapter 5. The second is with the optimal ordering

$$|\mathbf{b}| \sim \left| \frac{\eta}{\mu_0 \gamma} \nabla \times \nabla \times \mathbf{b} \right|, \tag{7.8}$$

which is usually related to the resistive interchange modes with the growth rate $\gamma \sim \eta^{1/3}$. And the third is with the strongest resistivity effects

$$|\mathbf{b}| \ll \left| \frac{\eta}{\mu_0 \gamma} \nabla \times \nabla \times \mathbf{b} \right|. \tag{7.9}$$

which is usually related to the tearing modes with the growth rate $\gamma \sim \eta^{3/5}$.

7.2 RESISTIVE SINGULAR LAYER EQUATIONS

Since the resistivity is small, it plays a significant role only in the singular layer or mode resonance layer, as we discussed in the ideal MHD theory in Chapters 5 and 6. Using the basic set of equations and the ordering scheme described in the last subsection, we now proceed to derive the resistive singular layer equations in the cylinder geometry in this section. In Chapter 6, we have derived the ideal MHD singular layer equation from the vorticity equation approach. The corresponding parts can be simplified to the current cylinder geometry with the resistivity effects added. Nevertheless, to be self-consistent, the derivation process is outlined.

In the cylinder equilibrium with circular cross-section, the magnetic field can be expressed as

$$\mathbf{B} = B_z(r)\mathbf{e}_z + B_\theta(r)\mathbf{e}_\theta \tag{7.10}$$

in the cylindrical coordinates (r, θ, z). Because of poloidal and longitudinal symmetry, one can adopt the Fourier decomposition for perturbed quantities as follows

$$\xi = \xi(r)\exp\{im\theta - inkz - i\omega t\}. \tag{7.11}$$

Therefore, one has

$$\mathbf{B} \cdot \nabla \xi = i\left(\frac{m}{r} - nkB_z\right)\xi. \tag{7.12}$$

To derive the singular layer equation, it is assumed that the radial scale length of the perturbation is of order ε, while the parallel one $\mathbf{B} \cdot \nabla \sim \varepsilon$. We first decompose the field line displacement and the perturbed magnetic field as follows

$$\begin{aligned}\boldsymbol{\xi} &= \xi\mathbf{e}_r + \mu\frac{\mathbf{B}\times\mathbf{e}_r}{B^2} + \nu\frac{\mathbf{B}}{B^2},\\ \mathbf{b} &= b\mathbf{e}_r + \nu\frac{\mathbf{B}\times\mathbf{e}_r}{B^2} + \tau\frac{\mathbf{B}}{B^2}.\end{aligned}$$

As approved *a posteriori*, one can assume the following orderings:

$$\begin{aligned}\xi &= \varepsilon\xi^{(1)} + \cdots, \quad \mu = \mu^{(0)} + \cdots, \quad \nu^{(0)} = \nu^{(0)} + \cdots,\\ b &= \varepsilon^2 b^{(2)} + \cdots, \quad \nu = \varepsilon\nu^{(1)} + \cdots, \quad \tau = \varepsilon\tau^{(1)} + \cdots.\end{aligned}$$

Here, ξ and b are ordered to be one order smaller than μ and ν respectively since one has $\nabla\cdot\boldsymbol{\xi}$ is small due to the suppression of compressional Alfvén modes and $\nabla\cdot\mathbf{b} = 0$. These are the same as the ideal MHD case described in Chapters 5 and 6. These ordering assumptions are confirmed in the derivation of resistive singular layer equations.

We first consider $\nabla\cdot\mathbf{b} = 0$. Note that

$$\begin{aligned}\nabla\cdot\mathbf{b} &= \frac{\partial b^{(2)}}{\partial r} + \frac{i}{r_s B^2}\left(nB_z + nkr_s B_\theta\right)\nu^{(1)}\\ &\quad + \nu^{(1)}\nabla\cdot\frac{\mathbf{B}\times\mathbf{e}_r}{B^2} + \frac{\mathbf{B}}{B^2}\cdot\nabla\tau^{(1)} = 0. \qquad (7.13)\end{aligned}$$

This can be further reduced as follows. Noting that in the cylinder equilibrium, one has

$$\begin{aligned}\nabla\cdot\frac{\mathbf{B}\times\mathbf{e}_r}{B^2} &= \nabla\cdot\frac{B_z\mathbf{e}_\theta - B_\theta\mathbf{e}_z}{B^2} = 0,\\ \frac{\mathbf{B}}{B^2}\cdot\nabla\tau^{(1)} &\sim \varepsilon,\end{aligned}$$

equation (7.13) is reduced to

$$\frac{\partial b^{(2)}}{\partial r} + \frac{i}{r_s B^2}\left(nB_z + nkr_s B_\theta\right)\nu^{(1)} = 0. \qquad (7.14)$$

Next, noting that $\mathbf{j}\times\mathbf{B} + \mathbf{J}\times\mathbf{Q} + \nabla p = \nabla\left(\mathbf{B}\cdot\mathbf{Q}/(\mu_0 B^2) + p\right)$ and the localization assumption, one has $\mathbf{B}\cdot\mathbf{Q}/(\mu_0 B^2) + p = 0$ in the lowest order, *i.e.*,

$$\tau^{(1)} = \mu_0 p'\xi^{(1)} + \mu_0\Gamma p(\nabla\cdot\boldsymbol{\xi})^{(1)}. \qquad (7.15)$$

Using this equation, equation (7.6) is reduced to

$$\begin{aligned}\left(1 + \frac{\mu_0\Gamma P}{B^2}\right)\nabla_\perp\cdot\boldsymbol{\xi} + 2\boldsymbol{\kappa}\cdot\xi &= \frac{\partial\xi^{(1)}}{\partial r} + \frac{i}{r_s B^2}\left(nB_z + nkr_s B_\theta\right)\mu^{(0)}\\ &\quad + 2\boldsymbol{\kappa}\cdot\xi = 0. \qquad (7.16)\end{aligned}$$

Since $\boldsymbol{\kappa} \cdot \xi = \boldsymbol{\kappa} \cdot \mathbf{e}_r \xi^{(1)}$, which is one order smaller than $\partial \xi^{(1)}/\partial r$, equation (7.16) is reduced to

$$\frac{\partial \xi^{(1)}}{\partial r} + \frac{i}{r_s B^2} (nB_z + nkr_s B_\theta)\, \mu^{(0)} = 0. \tag{7.17}$$

We now reduce the vorticity equation (7.5). The field line bending term is reduced as follows:

$$\begin{aligned} \mathbf{B} \cdot \nabla \left(\frac{\mathbf{B} \cdot \nabla \times \mathbf{b}}{B^2} \right) &= \mathbf{B} \cdot \nabla \left[\frac{\varepsilon}{B^2} \mathbf{B} \cdot \left(\mathbf{e}_r \times \frac{\mathbf{B} \times \mathbf{e}_r}{B^2} \right) \frac{\partial v^{(1)}}{\partial r} \right] \\ &= \mathbf{B} \cdot \nabla \left[-\frac{i\varepsilon r_s}{mB_z + nkr_s B_\theta} \frac{\partial^2 b^{(2)}}{\partial r^2} \right], \end{aligned} \tag{7.18}$$

where equation (7.14) has been used. The inertia term is reduced as follows:

$$\begin{aligned} \nabla \cdot \frac{\mathbf{B}}{B^2} \times (\rho \gamma^2 \boldsymbol{\xi}) &= -\frac{1}{r} \frac{\partial}{\partial r} \left(\frac{\rho \gamma^2 \mu}{B^2} \right) \\ &= -\frac{\varepsilon \rho \gamma^2}{B^2} \frac{1}{r} \frac{\partial \mu^{(0)}}{\partial r} \\ &= -\frac{i\varepsilon r_s}{mBz + nkr_s B_\theta} \rho \gamma^2 \frac{\partial^2 \xi^{(1)}}{\partial r^2}, \end{aligned} \tag{7.19}$$

where equation (7.17) has been used. The pressure-driven term becomes

$$\nabla \cdot \left[\frac{1}{B^2} \mathbf{B} \times \nabla (\boldsymbol{\xi} \cdot \nabla p) \right] = \frac{1}{B^2} \mathbf{j} \cdot \nabla (\boldsymbol{\xi} \cdot \nabla p) + \frac{\mathbf{B} \times \nabla B^2}{B^4} (\boldsymbol{\xi} \cdot \nabla p). \tag{7.20}$$

Adding the second term on the right-hand side of equation (7.5) and using equation (7.6) for $\tau^{(1)}$, one obtains

$$\frac{1}{B^2} \left(2\mathbf{j} + \frac{\mathbf{B} \times \nabla B^2}{B^2} \right) \cdot \nabla \tau^{(1)} = \frac{1}{B^4} \mathbf{B} \times \nabla \left(2\mu_0 p + B^2 \right) \cdot \nabla \tau^{(1)}, \tag{7.21}$$

where the following equation has been used

$$\mathbf{j} = \frac{\mathbf{B} \times \nabla \mu_0 p}{B^2} + \frac{\mathbf{j} \cdot \mathbf{B}}{B^2} \mathbf{B}$$

and the term $\mathbf{B} \cdot \nabla \tau^{(1)} \sim \varepsilon^2$ has been neglected. Using equation (3.116), equation (7.21) can be further reduced to

$$\begin{aligned} \frac{1}{B^4} \mathbf{B} \times \nabla \left(2\mu_0 p + B^2 \right) \cdot \nabla \tau^{(1)} &= -\frac{2}{B^2} \mathbf{B} \times \nabla \tau^{(1)} \cdot \boldsymbol{\kappa} \\ &= -2 \frac{i}{r_s^3 B^2} (mB_z + nkr_s B_\theta)\, \tau^{(1)}. \end{aligned} \tag{7.22}$$

Finally, using equations (7.18), (7.19), and (7.22), the vorticity equation (7.5) becomes

$$\begin{aligned}\rho\gamma^2\frac{\partial^2\xi^{(1)}}{\partial r^2} &= 2\frac{B_\theta^2}{r_s^3B^4}(mB_z+nkr_sB_\theta)^2\tau^{(1)}-i\frac{B_\theta}{r_s}mq'(r-r_s)\frac{\partial^2 b^{(2)}}{\partial r^2}\\ &= \frac{2m^2B_\theta^2}{r_s^2B_z^2}\tau^{(1)}-i\frac{B_\theta}{r_s}mq'(r-r_s)\frac{\partial^2 b^{(2)}}{\partial r^2}.\end{aligned}\tag{7.23}$$

One also needs to derive the momentum equation (7.1) in the parallel direction, yielding

$$\rho\gamma^2 v^{(0)} = -\frac{dp}{dr}b^{(2)}-i\frac{B_\theta}{r_s}nq'(r-r_s)\tau^{(1)}.\tag{7.24}$$

Next, let us reduce the induction equation (7.2). In the radial direction, it becomes

$$\begin{aligned}b^{(2)}-\frac{\eta}{\gamma}\frac{\partial^2 b^{(2)}}{\partial r^2} &= \mathbf{B}\cdot\nabla\xi^{(1)}\\ &= -i\frac{B_\theta}{r_s}nq'(r-r_s)\xi^{(1)}.\end{aligned}\tag{7.25}$$

One can also project the induction equation (7.2) in the parallel direction and obtain

$$\begin{aligned}&\tau^{(1)}-\frac{\eta}{\gamma}\frac{\partial^2\tau^{(1)}}{\partial r^2}\\ &= -i\frac{B_\theta}{r_s}nq'(r-r_s)v^{(0)}-B^2(\nabla\cdot\boldsymbol{\xi})^{(1)}-\frac{d}{dr}(\mu_0 p+B^2)\xi^{(1)},\end{aligned}\tag{7.26}$$

where equation (3.116) has been used to obtain $\boldsymbol{\kappa}$. Using equation (7.24) for $v^{(0)}$, equation (7.15) for $(\nabla\cdot\boldsymbol{\xi})^{(1)}$, and the equilibrium equation (4.106) for reducing the last term, equation (7.26) is reduced to

$$\begin{aligned}&\tau^{(1)}-\frac{\eta}{\gamma}\frac{\partial^2\tau^{(1)}}{\partial r^2}\\ &= -i\frac{B_\theta}{r_s}nq'(r-r_s)\frac{1}{\rho\gamma^2}\left(-\frac{dp}{dr}b^{(2)}-i\frac{B_\theta}{r_s}nq'(r-r_s)\tau^{(1)}\right)\\ &\quad -B^2\frac{1}{\Gamma p}\left(\tau^{(1)}-\mu_0 p'\xi^{(1)}\right)+\left(\frac{2B_\theta^2}{r}+\mu_0\frac{dp}{dr}\right)\xi^{(1)}.\end{aligned}\tag{7.27}$$

Equations (7.23), (7.25), and (7.27) constitute the complete set of equations in the singular layer orderings. We further introduce the notations

$$b^{(2)} = -i\frac{L_R nq'B_\theta}{r_s}\psi\tag{7.28}$$

$$\tau^{(1)} = \frac{dp}{dr}\Upsilon,\tag{7.29}$$

$$\gamma = \left(\frac{\eta n^2q'^2B_\theta^2}{\mu_0^2\rho r_s^2}\right)^{1/3}Q,,\tag{7.30}$$

and

$$r - r_s = L_R x, \quad L_R \equiv \left(\frac{\eta^2 \rho r_s^2}{\mu_0 n^2 q'^2 B_\theta^2} \right)^{1/6}. \tag{7.31}$$

The complete set of singular layer equations becomes

$$\begin{aligned} Q^2 \xi'' &= -D_s \Upsilon - x\psi'' \\ &= -D_s \Upsilon + Qx^2 \xi - Qx\psi, &(7.32) \\ \psi'' &= Q(\psi - x\xi), &(7.33) \\ \Upsilon'' &= Q\left(1 + \frac{x^2}{Q^2} + \frac{B^2}{\Gamma P}\right)\Upsilon + Q\left(\frac{u}{D_s} - 1 - \frac{B^2}{\Gamma P}\right)\xi - \frac{1}{Q} x\psi, &(7.34) \end{aligned}$$

where primes denote the derivative with respect with x, $u = 4k/q'^2$, and D_s is given in equation (5.67).

As usual, the dimensionless parameters: the resistive diffusion time and the Lundquist number are introduced respectively as follows.

$$\tau_R = \frac{\mu_0 r_s}{\eta} \quad \text{and} \quad S = \frac{\tau_R}{\tau_A},$$

where

$$\tau_A = \left(\frac{B_\theta^2 q'^2}{\mu_0 \rho r_s^2} \right)^{-1}.$$

Noting that $q' \sim r_s B_T / R B_\theta$, τ_A here represents the toroidal Alfvén time. With these parameters, one can find that

$$\begin{aligned} \gamma \tau_A &= n^{2/3} S^{-1/3} Q, &(7.35) \\ L_R / r_s &= n^{1/3} S^{-1/3}. &(7.36) \end{aligned}$$

In the incompressibility limit, $\Gamma P \to 0$, from equation (7.34) one obtains $\Upsilon = \xi$. Therefore, (7.32) - (7.34) is reduced to

$$\begin{aligned} Q^2 \xi'' &= -D_s \xi - x\psi'' \\ &= -D_s \xi + Qx^2 \xi - Qx\psi, &(7.37) \\ \psi'' &= Q(\psi - x\xi). &(7.38) \end{aligned}$$

For simplicity, in the following sections we mainly limit ourselves to study the incompressible case. The compressibility effects will be included in Section 7.6.

7.3 EXTERNAL LIMITS OF SINGULAR LAYER SOLUTIONS

In Section 7.2, we have obtained the set of singular layer equations (7.37) and (7.38) in the incompressibility limit. These equations need to be solved and matched to the solutions in outer regions on the left- and right-hand sides of the layer. This needs to study the external or outer limits of singular layer equations, which the current section is going to focus on. These studies help classify the resistive MHD modes.

First, we note that in the outer regions the inertia and resistivity effects are unimportant so that the equation describing the outer region becomes simply the ideal MHD one, i.e., equation (5.65), in subsection 5.4.1 in deriving the Suydan criterion.

The resistivity effects originate from equation (7.38). This leads us to consider its outer limits. There are two terms in this equation related ψ. This leads to two possibilities: One is $\psi''/Q \ll \psi \sim x\xi$, and the other is $\psi''/Q \sim x\xi \gg \psi$. Let us consider them one by one.

In the case with $\psi''/Q \ll \psi \sim x\xi$, equation (7.38) becomes $\psi - x\xi = 0$. Equation (7.37) simply becomes

$$Q^2\xi'' = -D_s\xi - x(x\xi)''. \tag{7.39}$$

In the outer limit $x \to \infty$, the inertia term $Q^2\xi''$ on the left becomes negligible while two terms on the right become of the same order. Noting further that $x(x\boldsymbol{\xi})'' = (x^2\xi')'$, equation (7.39) is simply equation (5.66) used to derive the Suydan criterion in subsection 5.4.1. The two asymptotic solutions are

$$\xi = x^s \quad \text{and} \quad \xi = x^{-1-s} \tag{7.40}$$

and the corresponding ψ solutions are respectively $\psi = x^{s+1}$ and $\psi = x^{-s}$, where

$$s = 1/2 + (1-4D_s)^{1/4}/2. \tag{7.41}$$

These two outer limits of the inner layer solution match exactly with the solutions in the outer regions.

Next, we consider the case with $\psi''/Q \sim x\xi \gg \psi$. In this case, equation (7.38) becomes $\psi'' = -Qx\boldsymbol{\xi}$ and equation (7.37) in the large x limit reduces to

$$Q\xi'' = x^2\xi, \tag{7.42}$$

where the term $D_s\xi$ is assumed to be negligible as compared with $Qx^2\xi$. One can directly verifies that in the large x limit, the solution of equation (7.42) is

$$\xi = \exp\left\{\pm x^2/2Q^{1/2}\right\}. \tag{7.43}$$

Let us now discuss the determination of the system of differential equations, (7.37) and (7.38). As shown in figure 7.2, there are three regions: the left outer region, the singular layer, and the right outer region. There are 4 boundaries that delineate these regions: 0, c^-, c^+, and a. In these regions, there are two second-order differential equations in the left and right outer regions and one fourth-order differential equation in the singular layer. According to the differential equation theory, there are 8 constants remaining to be determined in this system: $2+4+2$. However, there are only 6 boundary and matching conditions: 2 for both ends and 2 by 2 for

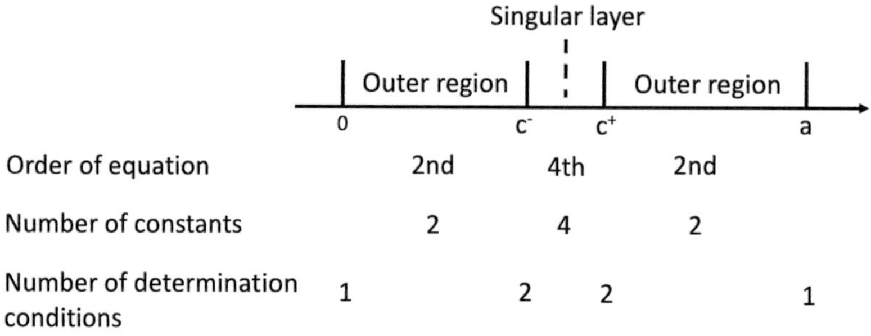

Figure 7.2 Determination conditions.

internal matchings between the layer and outer regions. There are only 2 matching conditions on each interface because the outer differential equations are of the second order. Upon these countings, one immediately sees that there are two more constants that need to be determined. Examining the outer limits of inner layer solutions in equation (7.43), one can see that one of the solutions is exponentially growing and the other vanishes exponentially as $x \to \pm\infty$. The exponentially growing solution should be excluded, and the exponentially decaying solutions can be ignored in the matching. Consequently, two independent solutions in the singular layer are left out and the system becomes completely determined.

Here, it is pointed out that, although the matchings for the solutions of differential equations are discussed, the matching in the perturbation theory needs to have an overlapping range. Since the solutions in equation (7.40) are the same type of solution as the one in the outer regions in the case described for Suydam criterion in subsection 5.4.1. Therefore, one can expect the matching over a range can be well facilitated.

The numerical solution of the singular layer equations (7.37) and (7.38) shows there are even and odd modes according to the parity of ψ as shown in figure 7.3. As will be discussed later on, the even modes are related to the tearing type of modes, which causes the field line reconnection, and the odd modes are related to the interchange type of modes. In the following sections, the analytical solutions of the singular layer equations will be reviewed.

7.4 SOLUTION OF THE SINGULAR LAYER EQUATIONS IN THE FOURIER SPACE

In this section, we discuss the solution of the singular layer equations in the incompressibility case, equations (7.37) and (7.38), using the Fourier transformation method. It has long been realized that the set of singular layer equations (7.37) and (7.38) can be reduced by the Fourier transformation [2, 3]. This is because the set of equations is a fourth-order differential equation system, while its dependence on

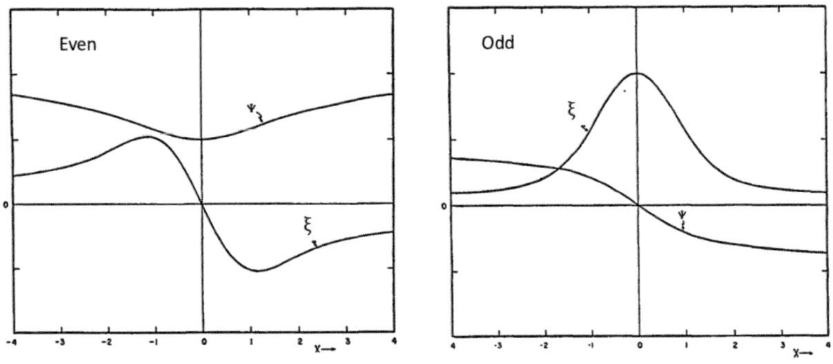

Figure 7.3 Even and odd modes, Reproduced from reference [2] with the permission of IAEA.

x is just quadratic. The Fourier transformation can reduce it to the second-order differential equation system. However, this procedure has a convergence issue. For the applicability of the Fourier transform, the function to be transformed is required to approach to zero at the limits $x \to \pm\infty$. From the external limits of ξ described in Section 7.3, one can see that in most cases, only the exponentially converged boundary solution in equation (7.43) can satisfy this requirement. For the boundary conditions in equation (7.40), the Laplace transform is usually required since the requirement in the so-called "s-doman" in the Laplace transform is much less stringent than that in the Fourier transform. In the later discussion, one can see that for the resistive instabilities, one can focus on the even and odd branches. In this case, one can use the one-sided Laplace transform to treat the problem with the boundary conditions in equation (7.40). Anyway, here we only review the theory with the Fourier transform. The Laplace transform theory can be found in Chapter 8 for the resistive MHD stability theory in toroidal geometry.

We introduce the Fourier transformation

$$\hat{\xi} = \int_{-\infty}^{+\infty} dx \xi(x) \exp\{-i\lambda x\} \tag{7.44}$$

with the inverse transformation

$$\xi = \frac{1}{2\pi}\int_{-\infty}^{+\infty} d\lambda \hat{\xi}(\lambda) \exp\{i\lambda x\}. \tag{7.45}$$

To assure the convergence of the transformation in equation (7.44), $\xi(x) \to 0$ as $x \to \pm\infty$ is required. With the transformation, the singular layer equations (7.37) and (7.38) become, respectively,

$$-\lambda^2 \hat{\xi} = -\frac{D_s}{Q^2}\hat{\xi} + \frac{1}{Q}\frac{d^2\hat{\xi}}{d\lambda^2} - \frac{1}{Q}\frac{d\hat{\psi}}{d\lambda}$$

and

$$-\lambda^2\hat{\psi} = Q\hat{\psi} - Q\frac{d\hat{\xi}}{d\lambda}.$$

Combining them together, one obtains

$$-\lambda^2\hat{\xi} = -\frac{D_s}{Q^2}\hat{\xi} + \frac{1}{Q}\frac{d^2\hat{\xi}}{d\lambda^2} - \frac{1/Q}{\lambda^2+Q}\frac{d}{d\lambda}\left(Q\frac{d\hat{\xi}}{d\lambda}\right).$$

This is further reduced to

$$\frac{d}{d\lambda}\frac{\lambda^2}{\lambda^2+Q}\frac{d\hat{\xi}}{d\lambda} - Q\left(\lambda^2 - \frac{D_s}{Q^2}\right)\hat{\xi} = 0. \tag{7.46}$$

Let us first examine the asymptotic behaviors of the transformed equation (7.46). In the small λ limit, it reduces to

$$\frac{d}{d\lambda}\lambda^2\frac{d\hat{\xi}}{d\lambda} + D_s\hat{\xi} = 0.$$

One simply has the solutions in this limit

$$\hat{\xi} = \lambda^s \text{ and } \lambda^{-1-s}. \tag{7.47}$$

In the large λ limit, equation (7.46) is reduced to

$$\frac{d^2\hat{\xi}}{d\lambda^2} - Q\lambda^2\hat{\xi} = 0.$$

One therefore has the asymptotic solutions

$$\hat{\xi} = \exp\left\{\pm\frac{1}{2}Q^{1/2}\lambda^2\right\}. \tag{7.48}$$

Here, the exponentially growing solution should be excluded. It is interesting to point out that the asymptotic solutions in equations (7.47) and (7.48) resembles the boundary conditions in (7.40) and (7.43).

The eigenmode solution of equation (7.46) can be directly verified to be

$$\hat{\xi} = \lambda^s \exp\left\{-\frac{1}{2}Q^{1/2}\lambda^2\right\}\sum_{i=0}^{k} a_i\lambda^{2i}$$

with a_i being constants determined at each order by the recurrence relations and the eigenvalue

$$Q^{3/2} = \frac{D_s}{s+2k+1/2+[4sk+(2k+1/2)^2]^{1/2}}, \tag{7.49}$$

where k is an integer that truncates the series. Note that Q is the normalized growth rate. Equation (7.49) shows that the system is unstable if $D_s > 0$ and there are multiple unstable modes. Comparing with the ideal MHD case in subsection 5.4.1, one can see that the stability criterion is more stringent than the Suydam criterion. In the Suydam stability criterion, one has $D_s > 1/4$, while in both cases, there are multiple unstable modes. Noting that $1/4$ in the Suydam criterion represents the magnetic shear stabilization, one can see that the resistivity eliminates the stabilization effects. From the definition of Q in equation (7.35) one can obtain the growth rate scaling with respect to the Lundquist number or resistivity $\gamma\tau_A \sim S^{-1/3} \sim \eta^{1/3}$, as mentioned in Section 7.1 after equation (7.8).

7.5 CONSTANT-ψ ASSUMPTION, TEARING MODES

In Section 7.4, the general solution of the singular layer equations in the Fourier space has been described. In this section, the solution in the configuration space will be discussed with the matching technique. It particularly focuses on the tearing mode branch to show how the exponentially decaying solution, as shown in the boundary condition in equation (7.43) can be matched to the outer solutions to form a tearing mode. We consider the incompressible case as described by equations (7.37) and (7.38).

In constructing the eigenvalue problem by matching, the overall framework is laid out in figure 7.2. Three regions are considered: the left outer region, the singular layer or inner region, and the right outer region. The outer regions are treated as the ideal MHD region with the inertia effects dropped. The resistivity and inertia effects are taken into account in the singular layer. Since the outer regions are ideal MHD-like, which have been investigated in Chapter 5, we focus on the singular layer description.

As usual, let us start with the ordering consideration of the singular layer so that perturbation analyses can be carried out. As discussed in Section 7.3 for the external limits of singular layer solutions, two ordering scenarios are considered for the Ohm's law in equation (7.38): $\psi'' \ll Q\psi$ and $\psi'' \gg Q\psi$. In the first scenario, the Ohm's law in equation (7.38) yields in the lowest order $\psi = x\xi$. Its further analyses resemble to the ideal MHD case in studying the accumulating points in Section 5.3.2. We therefore focus on the second scenario.

We consider the small Q and D_s case with the ordering scheme

$$\psi'' \sim Qx\xi \gg Q\psi. \tag{7.50}$$

In equations (7.37) and (7.38) for singular layer, the localized coordinate is defined as $x = (r - r_s)/L_R$ with L_R given in equation (7.31). Instead, for the outer solutions as discussed in equation (7.40) are expressed with the relative scale r_s

$$\xi_{L,R} = A^{L,R} \left| \frac{r - r_s}{r_s} \right|^{s} + B^{lL,R} \left| \frac{r - r_s}{r_s} \right|^{-1-s}, \tag{7.51}$$

where the indices L, R indicate the left and right outer regions and s is defined in equation (7.41). The difference between the relative scales in the singular layer L_B

and the outer regions r_s implies that an auxiliary ordering assumption, $L_B \ll r_s$, is introduced, which is similar to the ideal MHD case in discussing the accumulating point in Section 5.3.2. It is further assumed that L_B is specified by equation (7.9), which is narrower than that in the optimal ordering scheme in equation (7.8) for resistive interchange modes.

Consider the case with D_s being small. One then has $s \to 0$. Using equation (7.51) and noting that $\psi = x\xi$ in the ideal MHD, the solutions in the outer regions can be expressed as

$$\psi^{L,R} = B^{L,R} + A^{L,R} x,$$

as approaching the singular layer. One can expect that the outer limits of the inner solution assume the same structure,

$$\psi_s^{L,R} = B_s^{L,R} + A_s^{L,R} x. \tag{7.52}$$

Taking into consideration the scale difference between the inner and outer solutions, matching equations (7.51) and (7.52) yields

$$\frac{A^L}{B^L} = \left(\frac{r_s}{L_R}\right)^{2s+1} \frac{A_s^L}{B_s^L} \quad \text{and} \quad \frac{A^R}{B^R} = \left(\frac{r_s}{L_R}\right)^{2s+1} \frac{A_s^R}{B_s^R}. \tag{7.53}$$

This formally yields the dispersion relation

$$\Delta' = \Delta, \tag{7.54}$$

where

$$\Delta' = \frac{A^R}{B^R} - \frac{A^L}{B^L}, \tag{7.55}$$

$$\Delta = \left(\frac{a}{L_R}\right)^{2s+1} \left(\frac{A_s^R}{B_s^R} - \frac{A_s^L}{B_s^L}\right). \tag{7.56}$$

Here, Δ' is given by the outer solutions, and Δ can be obtained by solving the equations (7.37) and (7.38) in the inner region.

Before solving for Δ, let us establish the ordering scheme. As shown in equation (7.43) in Section 7.3, under the ordering assumption in equation (7.50) one has

$$\xi \sim \exp\left\{\pm x^2/2Q^{1/2}\right\}. \tag{7.57}$$

As Q is small, equation (7.57) requires the length scale becoming small as well

$$x \sim Q^{1/4} \sim \varepsilon. \tag{7.58}$$

In solving equation (7.38), it is assumed in the leading order

$$\psi^{(0)''} = 0.$$

Its solution is

$$\psi^{(0)} = B_s^{(0)} + A_s^{(0)} x \approx B_s^{(0)}. \tag{7.59}$$

Here, we have noted that $A^{L,R}/B_{L,R}$ for outer regions are usually of order unity, which excludes the internal kink mode case, and $(r_s/L_R)^{2s+1} \gg 1$ since $s \to 0$ in the small D_s case. Equation (7.53) therefore shows that $A_s^{(0)}$ is negligible in this order. This is related to the so-called constan-ψ approximation.

In the expansion $(\xi,\psi) = (\xi^{(0)},\psi^{(0)}) + \varepsilon(\xi^{(1)},\psi^{(1)}) + \cdots$ with equation (7.59), equation (7.37) becomes

$$Q^2\xi^{(0)''} + D_s\xi^{(0)} - Qx^2\xi^{(0)} = -Qx\psi^{(0)} = -QxB_s^{(0)}, \tag{7.60}$$

where equation (7.59) has been used. Note here that each term on the left-hand side of this equation is of the same order as $D_s \sim Q^{3/2}$.

We now solve equation (7.60) to construct the dispersion relation. Expand $\xi^{(0)}$ with the Hermit polynomials H_n [7] as follows

$$\xi^{(0)} = e^{-x^2/2Q^{1/2}} \sum_{n=0}^{\infty} c_n H_n(x/Q^{1/4}). \tag{7.61}$$

One can find that the left-hand side of equation (7.60) becomes

$$\begin{aligned} & Q^2\xi^{(0)''} + D_s\xi^{(0)} - Qx^2\xi^{(0)} \\ = \; & Q^{3/2} e^{-x^2/2Q^{1/2}} \sum \left(\frac{D_s}{Q^{3/2}} - 2n - 1 \right) c_n H_n. \end{aligned} \tag{7.62}$$

The right-hand side of equation (7.60) can also be represented as summation of Hermit polynomials using the following generating function

$$x = 2^{1/2} e^{-x^2/2Q^{1/2}} \sum \frac{14^n \Gamma(n+1) H_{2n+1}(x)}{4^n \Gamma(n+1)}. \tag{7.63}$$

Using equations (7.62) and (7.63), the lowest-order equation (7.60) can be solved. The solution can be used to calculate the jump

$$\begin{aligned} \delta &= \frac{\psi^{(1)'}|_{-\infty}^{+\infty}}{B_s^{(0)}} = \frac{1}{B_s^{(0)}} Q \int_{-\infty}^{+\infty} (\psi^{(0)} - x\boldsymbol{\xi}^{(0)}) dx \\ &= 2\pi Q^{5/4} \frac{\Gamma\left(\frac{1}{4}(3 - D_s/Q^{3/2})\right)}{\Gamma\left(\frac{1}{4}(1 - D_s/Q^{3/2})\right)}. \end{aligned}$$

Using equation (7.56), one obtains

$$\Delta = \left(\frac{a}{L_R}\right)^{2s+1} \delta = \left(\frac{r_s}{L_R}\right)^{2s+1} 2\pi Q^{5/4} \frac{\Gamma\left(\frac{1}{4}(3 - D_s/Q^{3/2})\right)}{\Gamma\left(\frac{1}{4}(1 - D_s/Q^{3/2})\right)}. \tag{7.64}$$

Inserting equation (7.64) back to equation (7.54), one obtains the dispersion relation.

Noting that Δ' as given in equation (7.55) is given by the outer solutions and Δ is given in equation (7.64), the dispersion relation in equation (7.54) can be used to determine the normalized eigenfrequency Q. When the term $D_s/Q^{3/2}$ in the Gamma function in equation (7.64) is small, one can see that the stability condition is $\Delta' < 0$. In this case, one has the ordering estimate

$$\Delta' = \Delta \sim \frac{r_s}{L_R} Q^{5/4}. \tag{7.65}$$

Assuming $\Delta' \sim 1$, equation (7.65) yields

$$Q \sim L_R^{4/5}. \tag{7.66}$$

Noting the definitions of Q and L_R respectively in equations (7.35) and (7.36), one obtains the growth rate scaling with respect to the Lundquist number or resistivity, as follows:

$$\gamma \tau_A \sim S^{-3/5} \sim \eta^{3/5}.$$

This is the growth rate ordering mentioned in Section 7.1 after equation (7.9). Since $\psi^{(0)}$ is even, this is the tearing mode. Using equations (7.58) and (7.66), one can also determine the x scaling with respect to the Lundquist number or resistivity.

7.6 PLASMA COMPRESSIBILITY EFFECTS

In this section, let us briefly discuss how to include the plasma compressibility effects [3]. The basic set of singular layer equations is given by equations (7.32) - (7.34) in this case.

We keep the ordering consideration in equation (7.58) and the constant-ψ assumption in equation (7.59). Instead of treating equation (7.60) in the incompressible case, the basic set of equations (7.32)–(7.34) is used to derive the equations for $\xi^{(0)}$ and $\Upsilon^{(0)}$

$$Q^2\xi^{(0)''} - Qx^2\xi^{(0)} + D_s\Upsilon^{(0)} = -QxB_s^{(0)}, \tag{7.67}$$

$$Q^2\Upsilon^{(0)''} - Qx^2\Upsilon^{(0)} + Q^3\frac{B^2}{\Gamma p}\Upsilon^{(0)} - Q^3\left(\frac{\lambda}{D_s} - \frac{B^2}{\Gamma p}\right)\xi^{(0)} = -QxB_s^{(0)}. \tag{7.68}$$

Similarly, equations (7.67) and (7.68) can be solved using the expansion with the Hermit polynomials H_n. The results can be used to compute Δ in the singular layer, yielding

$$\Delta = \frac{a}{L_B}\frac{\pi D_s}{4Q^{1/4}}F(\alpha, \nu, u), \tag{7.69}$$

where

$$\begin{aligned} F \equiv & \left(\alpha - 1 + \frac{1+\alpha^2\nu - \alpha\nu + \alpha u}{\lambda}\right)\frac{\Gamma\left[\frac{1}{4}(3+\alpha\nu-\lambda)\right]}{\Gamma\left[\frac{1}{4}(5+\alpha\nu-\lambda)\right]} \\ & + \left(\alpha - 1 + \frac{1+\alpha^2\nu - \alpha\nu + \alpha u}{\lambda}\right)\frac{\Gamma\left[\frac{1}{4}(3+\alpha\nu+\lambda)\right]}{\Gamma\left[\frac{1}{4}(5+\alpha\nu+\lambda)\right]}, \end{aligned}$$

and

$$\begin{aligned}\alpha &= Q^{3/2}/D_s,\\ \nu &= D_s B^2/\Gamma p,\\ \lambda^2 &= \alpha^2\nu^2+2\nu-u.\end{aligned}$$

Inserting equation (7.69) back into equation (7.54), one obtains the dispersion relation, which can be used to determine the normalized eigenfrequency Q.

7.7 CONCLUSIONS AND DISCUSSION

In this chapter, the resistive MHD stability theory in the cylinder geometry is reviewed. This extends the ideal MHD stability theory in the cylinder geometry in Chapter 5, which will be referred in the next chapter, Chapter 8, for resistive MHD stability theory in the toroidal geometry.

The resistivity effect breaks the restriction of the ideal MHD frozen-in-law. It weakens the stabilizing effects of the magnetic field line bending term and also allows the magnetic field reconnection to occur. Consequently, the resistive interchange and tearing modes appear. In the resistive interchange mode case, the Suydam criterion $1-4D_s >$ in the ideal MHD case becomes the more stringent criterion $-D_s > 0$ for stability in the resistive interchange mode case. The tearing mode is a new type of mode in the resistive MHD description. Its criterion is simply $\Delta' \leq 0$. We find two types of scalings: $\gamma\tau_A \sim S^{-1/3} \sim \eta^{1/3}$ and $\gamma\tau_A \sim S^{-3/5} \sim \eta^{3/5}$, respectively for resistive interchange and tearing modes.

We mainly review the incompressible case. The plasma compressibility was included only in the end in Section 7.6. This is because the toroidal and cylindrical descriptions of the plasma compressibility effects differ vastly. There is the so-called apparent mass effect, which becomes resistivity-dependent in the toroidal description. It makes the basic growth rate scalings inapplicable in the toroidal description. This will be discussed in the next chapter.

References

1. Furth H P, Killeen J, Rosenbluth M N 1963 Finite-Resistivity Instabilities of a Sheet Pinch. *Phys. Fluids 6*, 459–484. https://doi.org/10.1063/1.1706761
2. Coppi B, Greene J M, Johnson J L 1966 Resistive instabilities in a diffuse linear pinch. *Nuclear Fusion 6*, 101-117. DOI 10.1088/0029-5515/6/2/003
3. Greene J M 1976 Introduction to resistive instabilities. Report by Centre de Recherches en Physique des Plasmas, Ecole Polytechnique de Lausanne, LRP 114/76.
4. Johnson J L, Greene J M 1967 Resistive interchanges and the negative v'' criterion, *Plasma Phys. 9*, 611. DOI 10.1088/0032-1028/9/5/311
5. Newcomb W A 1960 Hydromagnetic stability of a diffuse linear pinch *Ann. Phys. NY 10*, 232-67 https://doi.org/10.1016/0003-4916(60)90023-3

6. Zheng L J, Furukawa M 2010 Current-interchange tearing modes: Conversion of interchange-type modes to tearing modes *Phys. Plasmas 17*, 052508. https://doi.org/10.1063/1.3425857
7. Abramowitz M, Stegun I A (ed) 1972 Handbook of Mathematical Functions (Washington, D.C.: National Bureau of Standards).

8 Linear resistive MHD stability theory in toroidal geometry

In this chapter, the cylinder theory in Chapter 7 is extended to the toroidal geometry. This includes the well-known resistive MHD theory by Glasser, Greene, and Johnson [1, 2] and the resistive ballooning theory in reference [3, 4]. The underlying physics has been discussed in Chapter 7 with figure 7.1. The singular layer treatments in the cylindrical and toroidal geometries share some similarities. However, since the poloidal direction is no longer symmetric in the toroidal system, one more dimension needs to be considered. The technique of flux surface average is introduced to reduce the extra dimension. The theoretical framework is hard to understand without mathematical details. The chapter is aimed at addressing this issue. First, the derivation of the resistive MHD singular layer equations will be given. The solution of the equations and the matching techniques will be followed. Subsequently, the resistive interchange modes and tearing modes theory will be presented. We will also describe the resistive ballooning theory [4] and compare them with the singular layer theory by Glasser et al. In the end, conclusions and discussion are given.

For reference, let us list here the linear resistive MHD equations given in section 7.1:

$$\rho_m \gamma^2 \boldsymbol{\xi} = \delta\mathbf{J} \times \mathbf{B} + \mathbf{J} \times \delta\mathbf{B} - \boldsymbol{\nabla}\delta P, \tag{8.1}$$

$$\delta\mathbf{B} + \frac{\eta}{\gamma}\boldsymbol{\nabla} \times (\boldsymbol{\nabla} \times \delta\mathbf{B}) = \boldsymbol{\nabla} \times \boldsymbol{\xi} \times \mathbf{B}, \tag{8.2}$$

$$\mu_0 \delta\mathbf{J} = \boldsymbol{\nabla} \times \delta\mathbf{B}, \tag{8.3}$$

$$\delta P = -\boldsymbol{\xi} \cdot \boldsymbol{\nabla} P - \Gamma P \boldsymbol{\nabla} \cdot \boldsymbol{\xi}. \tag{8.4}$$

The vorticity equation is still applicable for resistive MHD:

$$-\boldsymbol{\nabla} \cdot \frac{\mathbf{B}}{B^2} \times \rho_m \gamma^2 \boldsymbol{\xi} = \mathbf{B} \cdot \boldsymbol{\nabla}\frac{\mathbf{B} \cdot \delta\mathbf{J}}{B^2} + \delta\mathbf{B} \cdot \boldsymbol{\nabla}\sigma - \mathbf{J} \cdot \boldsymbol{\nabla}\frac{\mathbf{B} \cdot \delta\mathbf{B}}{B^2} + \boldsymbol{\nabla} \times \frac{\mathbf{B}}{B^2} \cdot \boldsymbol{\nabla}\delta P. \tag{8.5}$$

However, a modification of the parallel acoustic equation in the ideal MHD case, equation (6.15), is required. Note that the parallel momentum equation is

$$\rho_m \gamma^2 \mathbf{B} \cdot \boldsymbol{\xi} = -\boldsymbol{\nabla} P \cdot \delta\mathbf{B} - \mathbf{B} \cdot \boldsymbol{\nabla}\delta P,$$

DOI: 10.1201/9781003522133-8

Using the induction equation (8.2), it is reduced to

$$\rho_m \gamma^2 \mathbf{B}\cdot\boldsymbol{\xi} = -\frac{\eta}{\gamma} P' \nabla V \cdot (\nabla^2 \delta \mathbf{B}) + \Gamma P \mathbf{B}\cdot\nabla\nabla\cdot\boldsymbol{\xi}.$$

Operating $\mathbf{B}\cdot\nabla\frac{1}{B^2}$ on it, one obtains

$$\begin{aligned} &\rho_m \gamma^2 \mathbf{B}\cdot\nabla\frac{\mathbf{B}\cdot\boldsymbol{\xi}}{B^2} \\ = &-\frac{\eta}{\gamma} P' \mathbf{B}\cdot\nabla\left\{\frac{1}{B^2}\nabla V\cdot\nabla^2(\delta\mathbf{B})\right\} + \Gamma P \mathbf{B}\cdot\nabla\left[\frac{1}{B^2}\mathbf{B}\cdot\nabla\nabla\cdot\boldsymbol{\xi}.\right]. \end{aligned} \tag{8.6}$$

Besides, noting equation (3.118), one has

$$\begin{aligned} &\frac{1}{B^2}\left(\nabla\times(\boldsymbol{\xi}\times\mathbf{B}) - \mathbf{B}\frac{\mu_0\boldsymbol{\xi}\cdot\nabla p}{B^2}\right)\cdot\mathbf{B} \\ = &-\nabla\cdot\boldsymbol{\xi} + \mathbf{B}\cdot\nabla\left(\frac{\boldsymbol{\xi}\cdot\mathbf{B}}{B^2}\right) - 2\boldsymbol{\xi}\cdot\boldsymbol{\kappa}. \end{aligned} \tag{8.7}$$

Using the induction equation to express $\nabla\times(\boldsymbol{\xi}\times\mathbf{B})$ and noting that $\tau = \mu_0\boldsymbol{\xi}\cdot\nabla p + \mu_0\Gamma P\nabla\cdot\boldsymbol{\xi}$, this equation is further reduced to

$$-\frac{\eta}{\gamma}\frac{1}{B^2}\mathbf{B}\cdot\nabla^2\delta\mathbf{B} = -\left(1+\frac{\mu_0\Gamma P}{B^2}\right)\nabla\cdot\boldsymbol{\xi} + \mathbf{B}\cdot\nabla\left(\frac{\boldsymbol{\xi}\cdot\mathbf{B}}{B^2}\right) - 2\boldsymbol{\xi}\cdot\boldsymbol{\kappa}.$$

Using this equation for $\boldsymbol{\xi}_{\parallel}$, the parallel acoustic wave equation (8.6) is reduced to

$$\begin{aligned} &\Gamma P \mathbf{B}\cdot\nabla\left(\frac{1}{B^2}\mathbf{B}\cdot\nabla\nabla\cdot\boldsymbol{\xi}.\right) - \rho_m\gamma^2\left(1+\frac{\mu_0\Gamma P}{B^2}\right)\nabla\cdot\boldsymbol{\xi} \\ &+\rho_m\gamma^2\frac{\eta}{\gamma}\frac{1}{B^2}\mathbf{B}\cdot\nabla^2\delta\mathbf{B} - \frac{\eta}{\gamma}P'\mathbf{B}\cdot\nabla\left(\frac{1}{B^2}\nabla V\cdot\nabla^2\delta\mathbf{B}\right) \\ = &\ 2\rho_m\gamma^2\boldsymbol{\xi}\cdot\boldsymbol{\kappa}. \end{aligned} \tag{8.8}$$

These equations will be used for the resistive MHD mode analyses in the toroidal geometry later in this chapter.

8.1 RESISTIVE MHD THEORY BY GLASSER, GREENE, AND JOHNSON

In this section, we describe the resistive MHD singular layer equations by A. Glasser, J. Greene, and J. Johnson [1, 2]. The derivation is mainly based on the vorticity equation (8.5) as described in reference [5]. The theory is the extension to the cylinder one in Chapter 7 and the ideal MHD theory in Chapter 6. We will also describe another formulation of resistive MHD theory in the next section.

8.1.1 SINGULAR LAYER EQUATIONS

As in the ideal MHD case in Section 6.5, we consider the modes that localize around a particular rational (or singular) magnetic surface V_0 and then introduce the localized coordinates (x, u, θ), in which $x = V - V_0$ and $u = m\theta - n\zeta$. In this coordinate system, the parallel derivative becomes $\mathbf{B}\cdot\nabla = \psi'(\partial/\partial\zeta) + (\Lambda x/\Xi)(\partial/\partial u)$, where $\Lambda = \psi'\chi'' - \chi'\psi''$ and $\Xi_e = \psi'/m = \chi'/n$.

The perturbations is assumed to vary as $\exp\{ik_u u\}$ with $k_u = 2\pi n/\chi'$. Similarily, $\boldsymbol{\xi}$ and $\delta\mathbf{B}$ are decomposed in three directions as follows:

$$\boldsymbol{\xi} = \xi\frac{\nabla V}{|\nabla V|^2} + \mu\frac{\mathbf{B}\times\nabla V}{B^2} + \nu\frac{\mathbf{B}}{B^2}, \tag{8.9}$$

$$\delta\mathbf{B} = b\frac{\nabla V}{|\nabla V|^2} + v\frac{\mathbf{B}\times\nabla V}{B^2} + \tau\frac{\mathbf{B}}{B^2}. \tag{8.10}$$

We consider the singular layer modes whose wavelength across the magnetic surface $\lambda_\perp$ is much smaller than the parallel wavelength and the binormal one, $\lambda_\wedge$, that is on the surface and perpendicular to the magnetic field line. As reference [1], the ordering assumptions are given as follows:

$$\gamma\sim x\sim\varepsilon,\ \frac{\partial}{\partial V}\sim\varepsilon^{-1},\ \frac{\partial}{\partial u}\sim\frac{\partial}{\partial\theta}\sim 1,\ \widetilde{\nabla\cdot\boldsymbol{\xi}}\sim\varepsilon^2,$$
$$\langle\nabla\cdot\boldsymbol{\xi}\rangle\sim\varepsilon,\ \eta\sim\varepsilon^3,$$

and

$$\begin{aligned} \xi &= \varepsilon\xi^{(1)}+\cdots, & \mu &= \mu^{(0)}+\cdots, & \nu^{(0)} &= \nu^{(0)}+\cdots, \\ b &= \varepsilon^2 b^{(2)}+\cdots, & v &= \varepsilon v^{(1)}+\cdots, & \tau &= \varepsilon\tau^{(1)}+\cdots, \end{aligned}$$

where the small parameter $\varepsilon \ll 1$. Here, it is noted that the localization assumption in the binormal direction for $\lambda_\wedge$ is not required. This is a generalization of reference [1] from the earlier work in reference [2].

From the basic set of resistive MHD equations (8.1) - (8.4), one can see that the resistivity appears only in the induction equation (8.2). Therefore, the derivations in the ideal MHD case in Section 6.5 remain applicable as long as the induction equation is not involved. In view of this, we only briefly outline the derivation process. Readers are referred to Section 6.5 for details.

We first note that both perpendicular components of the momentum equation (8.1) in the lowest order yield

$$\tau^{(1)} + \delta p^{(1)} = \tau^{(1)} - P'\xi^{(1)} - \Gamma P(\nabla\cdot\boldsymbol{\xi})^{(1)} = 0. \tag{8.11}$$

Noting further equation (8.14), one can also find from equation (8.11) that $\tau^{(1)}$ and $\delta p^{(1)}$ are independent of θ. With these ordering assumptions described, we proceed to derive the resistive MHD singular layer equations. In reference [1], ξ, b, and τ are used as three unknowns to construct the singular layer equations.

As described in Section 6.5, one can reduce $\nabla \cdot \delta \mathbf{B} = 0$ to

$$\frac{\partial b^{(2)}}{\partial x} + \frac{1}{\Xi}\frac{\partial}{\partial u}v^{(1)} + \frac{J'}{P'}\frac{\partial v^{(1)}}{\partial \theta} - \frac{\chi'}{P'}\frac{\partial \sigma v^{(1)}}{\partial \theta} + \chi'\frac{\partial}{\partial \theta}\frac{\tau^{(1)}}{B^2} = 0. \tag{8.12}$$

After the surface average, one obtains

$$\frac{\partial \bar{b}^{(2)}}{\partial x} + \frac{1}{\Xi_e}\frac{\partial \bar{v}^{(1)}}{\partial u} = 0. \tag{8.13}$$

As described in Section 6.5 as well, the ∇V and ∇u projections of the induction equation (8.2) in the lowest order give respectively

$$\chi'\frac{\partial \xi^{(1)}}{\partial \theta} = 0 \quad \text{and} \quad \chi'\frac{\partial \mu^{(0)}}{\partial \theta} = 0. \tag{8.14}$$

Then, the parallel component of equation (8.2) to the magnetic field line in the lowest order yields

$$(\nabla \cdot \boldsymbol{\xi}_{\perp})^{(0)} + 2\boldsymbol{\kappa} \cdot \boldsymbol{\xi}^{(0)} = \frac{\partial \xi^{(1)}}{\partial x} + \frac{1}{\Xi_e}\frac{\partial \mu^{(0)}}{\partial u} = 0, \tag{8.15}$$

where equations (8.7), (8.11), and (8.14) have been used.

Now, let us reduce the induction equation (8.2), which will be summarized later into equation (8.30). With the ideal MHD results described in Section 6.5, we focus on the resistivity effects in this equation. Noting that

$$\nabla V \cdot \nabla \times \nabla \times \delta \mathbf{B} = \frac{\partial}{\partial \theta}(\cdots) + i\alpha \frac{|\nabla V|^2}{\Xi}\frac{\partial v^{(1)}}{\partial x}, \tag{8.16}$$

the ∇V projection of the induction equation (8.2) yields

$$\bar{b}^{(2)} - \frac{ik_u \Lambda x}{\Xi}\xi^{(1)} = -\frac{ik_u}{\Xi_e}\frac{\eta}{\gamma}\left\langle |\nabla V|^2 \frac{\partial v^{(1)}}{\partial x} \right\rangle. \tag{8.17}$$

To reduce equation (8.17) further, one needs to determine $v^{(1)}$. This can be obtained from the vorticity equation (8.5), which in the lowest order is reduced to

$$\chi'\frac{\partial}{\partial \theta}\left(\frac{|\nabla V|^2}{B^2}\frac{\partial v^{(1)}}{\partial x}\right) - \frac{1}{P'}\chi'\frac{\partial \sigma}{\partial \theta}\frac{\partial \delta p}{\partial x} = 0.$$

It can be further reduced using equation (8.11)

$$\frac{\partial}{\partial \theta}\left(\frac{|\nabla V|^2}{B^2}\frac{\partial v^{(1)}}{\partial x}\right) + \frac{1}{P'}\frac{\partial \sigma}{\partial \theta}\frac{\partial \tau^{(1)}}{\partial x} = 0. \tag{8.18}$$

After integrating over θ and using equation (8.13), one obtains

$$\begin{aligned}\frac{\partial v^{(1)}}{\partial x} &= \frac{B^2/|\nabla V|^2}{\langle B^2/|\nabla V|^2\rangle}\frac{\partial \bar{v}^{(1)}}{\partial x}+\frac{1}{P'}\left(\frac{B^2/|\nabla V|^2}{\langle B^2/|\nabla V|^2\rangle}\left\langle\frac{B^2\sigma}{|\nabla V|^2}\right\rangle-\frac{B^2\sigma}{|\nabla V|^2}\right)\frac{\partial \tau^{(1)}}{\partial x}\\ &= -\frac{\Xi_e}{ik_u}\frac{B^2/|\nabla V|^2}{\langle B^2/|\nabla V|^2\rangle}\frac{\partial^2 \bar{b}^{(2)}}{\partial x^2}\\ &\quad+\frac{1}{P'}\left(\frac{B^2/|\nabla V|^2}{\langle B^2/|\nabla V|^2\rangle}\left\langle\frac{B^2\sigma}{|\nabla V|^2}\right\rangle-\frac{B^2\sigma}{|\nabla V|^2}\right)\frac{\partial \tau^{(1)}}{\partial x}.\end{aligned} \tag{8.19}$$

Now, inserting equation (8.19) into equation (8.17) yields

$$\begin{aligned}&\bar{b}^{(2)}-\frac{ik_u\Lambda x}{\Xi}\xi^{(1)}\\ &= \frac{\eta}{\gamma}\left\{\frac{\langle B^2\rangle}{\langle B^2/|\nabla V|^2\rangle}\frac{\partial^2 \bar{b}^{(2)}}{\partial x^2}\right.\\ &\quad\left.-ik_u\frac{1}{\Xi_e P'}\left(\frac{\langle B^2\rangle}{\langle B^2/|\nabla V|^2\rangle}\left\langle\frac{B^2\sigma}{|\nabla V|^2}\right\rangle-\langle B^2\sigma\rangle\right)\frac{\partial \tau^{(1)}}{\partial x}\right\}.\end{aligned} \tag{8.20}$$

This is the reduced induction equation to be collected in equation (8.30) later.

Next, let us reduce the vorticity equation (8.5), which will be summarized later into equation (8.31). Note that the lowest-order equation has been given in equation (8.18). In the next order, using the decompositions in equations (8.9) and (8.10), one obtains

$$\begin{aligned}&\frac{\rho\gamma^2|\nabla V|^2}{B^2}\frac{\partial \mu^{(0)}}{\partial x}=\frac{\partial}{\partial\theta}(\cdots)\\ &+\frac{J'-\sigma\chi'}{P'}\frac{\partial\sigma}{\partial\theta}v^{(1)}+\frac{\chi'}{B^2}\frac{\partial\sigma}{\partial\theta}\tau^{(1)}+\frac{\Lambda x}{\Xi_e}\frac{\partial}{\partial u}\left(\frac{|\nabla V|^2}{B^2}\frac{\partial v^{(1)}}{\partial x}\right)-\frac{P'}{\Xi_e}\frac{\partial}{\partial u}\left(\frac{\tau^{(1)}}{B^2}\right)\\ &-\left(\frac{\nabla V\cdot\nabla(P+B^2)}{\Xi_e B^2|\nabla V|^2}-\frac{\chi'}{P'}\Theta\frac{\partial\sigma}{\partial\theta}\right)\frac{\partial\tau^{(1)}}{\partial u}-\frac{\chi'}{P'}\frac{\partial\sigma}{\partial\theta}\frac{\partial}{\partial x}\left(\delta p^{(2)}-\xi^{(2)}P'\right).\end{aligned} \tag{8.21}$$

We try to express the equation in terms of $\xi^{(1)}$. Therefore, the term on the left-hand side can be reduced using equation (8.15). Also, as shown in the derivation later on, the second and third terms on the right cancel each other (see the first term on the right-hand side of equation (8.25)). One just needs to determine the fourth term and the last term on the right.

Let us first work on the fourth term on the right hand side of equation (8.21).

Using equation (8.19), one obtains

$$
\begin{aligned}
&\left\langle \frac{\Lambda x}{\Xi_e}\frac{\partial}{\partial u}\left(\frac{|\nabla V|^2}{B^2}\frac{\partial v^{(1)}}{\partial x}\right)\right\rangle \\
= \;& -\frac{\Lambda x}{\Xi_e}\frac{1}{\langle B^2/|\nabla V|^2\rangle}\frac{\partial^2\bar{b}^{(2)}}{\partial x^2} \\
&+ik_u\frac{\Lambda x}{\Xi_e}\frac{1}{P'}\left(\frac{1}{\langle B^2/|\nabla V|^2\rangle}\left\langle\frac{B^2\sigma}{|\nabla V|^2}\right\rangle-\langle\sigma\rangle\right)\frac{\partial\tau^{(1)}}{\partial x}. \\
= \;& -\frac{\Lambda x}{\Xi_e}\frac{1}{\langle B^2/|\nabla V|^2\rangle}\left\{\frac{q}{\eta}\frac{\langle B^2/|\nabla V|^2\rangle}{\langle B^2\rangle}\left(\bar{b}^{(2)}-\frac{ik_u\Lambda x}{\Xi_e}\xi^{(1)}\right)\right. \\
&\left.+ik_u\frac{1}{\Xi_e P'}\left(\left\langle\frac{B^2\sigma}{|\nabla V|^2}\right\rangle-\frac{\langle B^2/|\nabla V|^2\rangle}{\langle B^2\rangle}\langle B^2\sigma\rangle\right)\frac{\partial\tau^{(1)}}{\partial x}\right\} \\
&+ik_u\frac{\Lambda x}{\Xi_e}\frac{1}{P'}\left(\frac{1}{\langle B^2/|\nabla V|^2\rangle}\left\langle\frac{B^2\sigma}{|\nabla V|^2}\right\rangle-\langle\sigma\rangle\right)\frac{\partial\tau^{(1)}}{\partial x} \\
= \;& -\frac{\gamma}{\eta}\frac{\Lambda x}{\Xi_e}\frac{1}{\langle B^2\rangle}\bar{b}^{(2)}+ik_u\frac{\gamma}{\eta}\frac{\Lambda^2x^2}{\Xi}\frac{1}{\langle B^2\rangle}\xi^{(1)} \\
&+ik_u\frac{\Lambda x}{\Xi_e}\frac{1}{P'}\left(\frac{\langle B^2\sigma\rangle}{\langle B^2\rangle}-\langle\sigma\rangle\right)\frac{\partial\tau^{(1)}}{\partial x}.
\end{aligned}
\tag{8.22}
$$

Next, let us work on the last term on the right of equation (8.21), which involves the coupling of parallel motion. The equation of parallel motion can be obtained from equation (8.1)

$$
\rho_m\gamma^2 v^{(0)} = -P'b^{(2)}+\chi'\frac{\partial}{\partial\theta}(P'\xi^{(2)}-\delta p^{(2)})+\frac{ik_u\Lambda}{\Xi_e}x\tau^{(1)}. \tag{8.23}
$$

Noting that $\nabla\cdot\boldsymbol{\xi}^{(0)}=0$, one obtains

$$
\begin{aligned}
v^{(0)} &= \left(\frac{B^2\sigma}{P'}-\frac{B^2}{P'}\frac{\langle B^2\sigma\rangle}{\langle B^2\rangle}\right)\mu^{(0)}+\frac{B^2}{\langle B^2\rangle}\bar{v}^{(0)} \\
&= -\frac{\Xi_e}{ik_u}\left(\frac{B^2\sigma}{P'}-\frac{B^2}{P'}\frac{\langle B^2\sigma\rangle}{\langle B^2\rangle}\right)\frac{\partial\xi^{(1)}}{\partial x} \\
&\quad+\frac{1}{\rho_m\gamma^2}\frac{B^2}{\langle B^2\rangle}\left(-P'b^{(2)}+\frac{ik_u\Lambda}{\Xi_e}x\tau^{(1)}\right).
\end{aligned}
$$

Here, $\bar{v}^{(0)}$ has been obtained from equation (8.23) and $\mu^{(0)}$ from equation (8.15). Inserting this equation into equation (8.23), multiplying it by σ, and taking the derivative with respect to x yield

$$\frac{\chi'}{P'}\frac{\partial}{\partial x}\left\langle\left(\delta p^{(2)}-P'\xi^{(2)}\right)\frac{\partial\sigma}{\partial\theta}\right\rangle$$
$$= -\left(\langle\sigma\rangle-\frac{\langle\sigma B^2\rangle}{\langle B^2\rangle}\right)\left[-\frac{\partial\bar{b}^{(2)}}{\partial x}+\frac{ik_u\Lambda}{P'\Xi_e}\left(x\frac{\partial\tau^{(1)}}{\partial x}+\tau^{(1)}\right)\right]$$
$$+\left\langle\sigma\frac{\partial\tilde{b}}{\partial x}\right\rangle-\frac{\rho\gamma^2\Xi_e}{ik_u}\frac{M_t}{\langle B^2/|\nabla V|^2\rangle}\frac{\partial^2\xi^{(1)}}{\partial x^2}. \tag{8.24}$$

The second term on the right-hand side of this equation can be reduced as follows, using equation (8.12),

$$\left\langle\sigma\frac{\partial\tilde{b}}{\partial x}\right\rangle = \left\langle\frac{J'-\sigma\chi'}{P'}\frac{\partial\sigma}{\partial\theta}v^{(1)}+\frac{\chi'}{B^2}\frac{\partial\sigma}{\partial\theta}\tau^{(1)}\right\rangle+i\frac{k_u}{\Xi_e}\left\langle\sigma\tilde{v}^{(1)}\right\rangle$$
$$= \left\langle\frac{J'-\sigma\chi'}{P'}\frac{\partial\sigma}{\partial\theta}v^{(1)}+\frac{\chi'}{B^2}\frac{\partial\sigma}{\partial\theta}\tau^{(1)}\right\rangle-\left(\frac{\langle B^2\sigma/|\nabla V|^2\rangle}{\langle B^2/|\nabla V|^2\rangle}-\langle\sigma\rangle\right)\frac{\partial\bar{b}^{(2)}}{\partial x}$$
$$+\frac{ik_u}{P'\Xi_e}\left(\frac{\langle B^2\sigma/|\nabla V|^2\rangle^2}{\langle B^2/|\nabla V|^2\rangle}-\left\langle\frac{B^2\sigma^2}{|\nabla V|^2}\right\rangle\right)\tau^{(1)}. \tag{8.25}$$

Here, in reducing the term $\left\langle\sigma\tilde{v}^{(1)}\right\rangle$, equation (8.13) and the following equation has been used

$$\frac{1}{P'}\left\langle\sigma v^{(1)}\right\rangle = -\frac{\Xi_e}{ik_uP'}\frac{\langle\sigma B^2/|\nabla V|^2\rangle}{\langle B^2/|\nabla V|^2\rangle}\frac{\partial\bar{b}^{(2)}}{\partial x}$$
$$+\frac{1}{P'^2}\left(\frac{\langle B^2\sigma/|\nabla V|^2\rangle^2}{\langle B^2/|\nabla V|^2\rangle}-\left\langle\frac{B^2\sigma^2}{|\nabla V|^2}\right\rangle\right)\tau^{(1)}, \tag{8.26}$$

where equation (8.19) has been used to express $v^{(1)}$.

The reduced vorticity equation is obtained by inserting equations (8.22) and (8.24) into equation (8.21) and averaging over θ. The results will be summarized in equation (8.31) later.

Next, we derive the third singular-layer equation of resistive MHD to be summarized later in equation (8.32). By the dot product with $\mathbf{B}\times\nabla V/|\nabla V|^2$, the poloidal projection of the induction equation (8.2) yields

$$v^{(1)}-\frac{\eta}{\gamma}|\nabla V|^2\frac{\partial^2 v^{(1)}}{\partial x^2}$$
$$= \mathbf{B}\cdot\nabla\left(\frac{\boldsymbol{\xi}\cdot\mathbf{B}\times\nabla V}{|\nabla V|^2}\right)-\frac{(\mathbf{B}\times\nabla V)\cdot\nabla\times(\mathbf{B}\times\nabla V)}{|\nabla V|^4}\boldsymbol{\xi}\cdot\nabla V$$
$$= \chi'\frac{\partial\mu^{(1)}}{\partial\theta}+i\frac{k_u}{\Xi_e}\Lambda x\mu^{(0)}-\frac{(\mathbf{B}\times\nabla V)\cdot\nabla\times(\mathbf{B}\times\nabla V)}{|\nabla V|^4}\xi^{(1)}, \tag{8.27}$$

The parallel projection of the induction equation (8.2) can be obtained by the dot product with **B**, yielding

$$\begin{aligned}
&\frac{1}{B^2}\tau^{(1)} - \frac{\eta}{\gamma}\frac{|\nabla V|^2}{B^2}\frac{\partial^2 \tau^{(1)}}{\partial x^2} \\
=\;& \frac{P'}{B^2}\xi^{(1)} - (\nabla\cdot\boldsymbol{\xi})^{(1)} + \mathbf{B}\cdot\nabla\frac{\nu^{(1)}}{B^2} - 2(\boldsymbol{\xi}\cdot\boldsymbol{\kappa})^{(1)} \\
=\;& \frac{P'}{B^2}\xi^{(1)} + \frac{P'}{\Gamma P}\xi^{(1)} - \frac{1}{\Gamma P}\tau^{(1)} + \chi'\frac{\partial}{\partial\theta}\frac{\nu^{(1)}}{B^2} + \frac{\Lambda x}{\Xi}\frac{\nu^{(0)}}{B^2} \\
&-\xi^{(1)}\frac{\nabla V\cdot\nabla(2P+B^2)}{B^2|\nabla V|^2} - \frac{\chi'}{P'}\mu^{(1)}\frac{\partial\sigma}{\partial\theta},
\end{aligned} \tag{8.28}$$

where equations (6.18) and (8.11) have been used.

Averaging over θ and using equation (8.27) to eliminate $\frac{\chi'}{P'}\mu^{(1)}\frac{\partial\sigma}{\partial\theta}$ term, equation (8.28) is reduced to

$$\begin{aligned}
&\left\langle\frac{1}{B^2}\right\rangle\tau^{(1)} - \frac{\eta}{\gamma}\left\langle\frac{|\nabla V|^2}{B^2}\right\rangle\frac{\partial^2\tau^{(1)}}{\partial x^2} - \frac{1}{P'}\left\langle\sigma\nu^{(1)}\right\rangle + \frac{\eta}{\gamma P'}\left\langle\sigma|\nabla V|^2\frac{\partial^2\nu^{(1)}}{\partial x^2}\right\rangle \\
=\;& P'\left\langle\frac{1}{B^2}\right\rangle\xi^{(1)} + \frac{P'}{\Gamma P}\xi^{(1)} - \frac{1}{\Gamma P}\tau^{(1)} + \frac{\Lambda x}{\Xi}\left\langle\frac{\nu^{(0)}}{B^2}\right\rangle - ik_u\frac{1}{P'}\langle\sigma\rangle\Lambda x\mu^{(0)} \\
&+\left\langle-\frac{\nabla V\cdot\nabla(2P+B^2)}{B^2|\nabla V|^2} + \frac{\sigma}{P'}\frac{(\mathbf{B}\times\nabla V)\cdot\nabla\times(\mathbf{B}\times\nabla V)}{|\nabla V|^4}\right\rangle\xi^{(1)}.
\end{aligned} \tag{8.29}$$

This equation can be further reduced. The last term on the right-hand side can be reduced by equation (6.111) in Section 6.5, the third term on the left is reduced in equation (8.26), and the fourth term on the left can be reduced as follows:

$$\begin{aligned}
&\frac{\eta}{\gamma P'}\left\langle|\nabla V|^2\sigma\frac{\partial^2\nu^{(1)}}{\partial x^2}\right\rangle \\
=\;& -\frac{\eta}{ik_uP'\gamma}\frac{\langle\sigma B^2\rangle}{\langle B^2/|\nabla V|^2\rangle}\frac{\partial^3\bar{b}^{(2)}}{\partial x^3} + \frac{\eta}{P'^2\gamma}\left(\frac{\langle B^2\sigma/|\nabla V|^2\rangle\langle\sigma B^2\rangle}{\langle B^2/|\nabla V|^2\rangle} - \left\langle B^2\sigma^2\right\rangle\right)\frac{\partial^2\tau^{(1)}}{\partial x^2}.
\end{aligned}$$

Here, $\frac{\partial^3\bar{b}^{(2)}}{\partial x^3}$ can be reduced by using equation (8.20) (or equation (8.30)). In this way, the third singular-layer equation of resistive MHD to be summarized later in equation (8.32) is obtained.

Summarizing the results above for the derivations or reductions of equations (8.20), (8.21), and (8.29), one obtains the set of resistive MHD singular layer equations as follows:

$$\Psi_{XX} - H\Upsilon_X = Q(\Psi - X\Xi), \tag{8.30}$$

$$Q^2\Xi_{XX} - QX^2\Xi + E\Upsilon + QX\Psi + \Gamma = 0, \tag{8.31}$$

$$Q\Upsilon_{XX} - X^2\Upsilon - GQ^2\Upsilon + (G-KE)Q^2\Xi + X\Psi - KQ^2\Gamma = 0, \tag{8.32}$$

$$\Gamma_X = H\Psi_{XX} + F\Upsilon_X, \tag{8.33}$$

where equation (8.33) may be viewed as the definition of Γ for the rest of equations (8.30)–(8.32), the parameters E, F, H, and M have been given in Section 6.5 as follows

$$
\begin{aligned}
E &\equiv \frac{\langle B^2/|\nabla V|^2\rangle}{\Lambda^2}\left(J'\psi'' - I'\chi'' + \Lambda\frac{\langle\sigma B^2\rangle}{\langle B^2\rangle}\right),\\
F &\equiv \frac{\langle B^2/|\nabla V|^2\rangle}{\Lambda^2}\left(\left\langle\frac{\sigma^2 B^2}{|\nabla V|^2}\right\rangle - \frac{\langle\sigma B^2/|\nabla V|^2\rangle^2}{\langle B^2/|\nabla V|^2\rangle} + P'^2\left\langle\frac{1}{B^2}\right\rangle\right),\\
H &\equiv \frac{\langle B^2/|\nabla V|^2\rangle}{\Lambda}\left(\frac{\langle\sigma B^2/|\nabla V|^2\rangle}{\langle B^2/|\nabla V|^2\rangle} - \frac{\langle\sigma B^2\rangle}{\langle B^2\rangle}\right),\\
M &= M_c + M_t,\\
M_c &\equiv \frac{\rho_m \Xi_e^2}{k_u^2\Lambda^2}\left\langle\frac{B^2}{|\nabla V|^2}\right\rangle\left\langle\frac{|\nabla V|^2}{B^2}\right\rangle, && (8.34)\\
M_t &\equiv \frac{\rho_m \Xi_e^2}{k_u^2\Lambda^2 P'^2}\left\langle\frac{B^2}{|\nabla V|^2}\right\rangle\left(\langle\sigma^2 B^2\rangle - \frac{\langle\sigma B^2\rangle^2}{\langle B^2\rangle}\right), && (8.35)
\end{aligned}
$$

and

$$
\begin{aligned}
G &= \frac{\langle B^2\rangle}{M\Gamma P}, \quad K = \frac{\Lambda^2}{MP'^2}\frac{\langle B^2\rangle}{\langle B^2/|\nabla V|^2\rangle},\\
Q_0 &= \left(\frac{\eta k_u^2\Lambda^2\langle B^2\rangle}{\rho M\Xi_e^2\langle B^2/|\nabla V|^2\rangle}\right)^{1/3}, \quad Q = \frac{\gamma}{Q_0}, && (8.36)\\
X_0 &= \left(\frac{\rho_m M\Xi_e^2\eta^2\langle B^2\rangle^2}{k_u^2\Lambda^2\langle B^2/|\nabla V|^2\rangle^2}\right)^{1/6}, \quad X = \frac{V - V_0}{X_0}. && (8.37)
\end{aligned}
$$

Here, the perturbed quantities are given as follows:

$$
\begin{aligned}
\Psi &= \langle\delta\mathbf{B}\cdot\nabla V\rangle / ik_u\Lambda X_0,\\
\Gamma &= \frac{\langle B^2/|\nabla V|^2\rangle}{\Lambda^2}\frac{p'}{\chi'}\left[\langle\delta\mathbf{B}\cdot\nabla\theta\rangle - \Lambda\frac{\langle\mathbf{B}\cdot\nabla\theta\times\nabla V\rangle}{\langle B^2\rangle}\Psi_X\right],\\
\Xi &= \langle\boldsymbol{\xi}\cdot\nabla V\rangle, \quad \Upsilon = \langle\mathbf{B}\cdot\delta\mathbf{B}\rangle / P'.
\end{aligned}
$$

It is also convenient to define

$$
D_I \equiv E + F + H - \frac{1}{4}, \tag{8.38}
$$

$$
D_R \equiv E + F - H^2, \tag{8.39}
$$

where the definition of D_I in equation (8.38) has been given in the ideal MHD case in Section 6.5.

Equation (8.33) can be solved. Inserting the solution Γ into equations (8.31) and (8.32). The set of singular layer equations (8.30)-(8.33) becomes

$$\Psi_{XX} - H\Upsilon_X = Q(\Psi - X\Xi), \tag{8.40}$$

$$Q^2\Xi_{XX} - QX^2\Xi + (E+F)\Upsilon + QX\Psi + H\Psi_X = 0, \tag{8.41}$$

$$Q\Upsilon_{XX} - X^2\Upsilon + X\Psi = Q^2\left[(G+KF)\Upsilon - (G-KE)\Xi - KH\Psi_X\right], \tag{8.42}$$

This set of resistive MHD singular layer equations will be used to study the resistive instabilities in the following subsections.

8.1.2 IDEAL MHD LIMIT

In this subsection, the ideal MHD limit of the set of resistive MHD equations (8.40)–(8.42) is derived. This is important to verify the matching to the outer ideal MHD regions.

From equations (8.36), and (8.37), one can find that Q, X, and ψ are proportional to $\eta^{-1/3}$. Therefore, equation (8.40) in the ideal MHD limit is reduced to

$$\Psi = X\Xi. \tag{8.43}$$

For reducing equation (8.41), we note that

$$\begin{aligned} -QX^2\Xi + QX\Psi &= XQ(\Psi - X\Xi) \\ &= X\Psi_{XX} - XH\Upsilon_X \\ &= X(X\Xi)_{XX} - XH\Upsilon_X. \end{aligned}$$

where equations (8.40) and (8.43) have been used. Using equation (8.43) to reduce the last term, equation (8.41) in the ideal MHD limit is reduced to

$$X(X\Xi)_{XX} + Q^2\Xi_{XX} + (E+F)\Upsilon + H\Xi - HX\Upsilon_X. + HX\Xi_X = 0. \tag{8.44}$$

Using equation (8.43), equation (8.42) in the ideal MHD limit is reduced to

$$-X^2\Upsilon + X^2\Xi - GQ^2\Upsilon + (G-KE)Q^2\Xi - KQ^2F\Upsilon - KQ^2H(X\Xi)_X = 0. \tag{8.45}$$

From equation (8.11) one can find that

$$\Upsilon = \Xi + \frac{1}{P'}\Gamma P\langle\nabla\cdot\xi\rangle. \tag{8.46}$$

Inserting it into equation (8.45) yields

$$\begin{aligned} &\frac{1}{P'}\Gamma P\langle\nabla\cdot\xi\rangle\left(X^2 + GQ^2 + KQ^2F\right) \\ &+Q^2\left(KE+KF\right)\Xi - KQ^2H(X\Xi)_X = 0. \end{aligned}$$

That is

$$\Gamma P\langle\nabla\cdot\xi\rangle = -\frac{K(E+F)\Xi + KH(X\Xi)_X}{X^2/Q^2 + G + KF}. \tag{8.47}$$

We now use equations (8.43), (8.44), and (8.47) to discuss the ideal MHD limit of the resistive singular layer equations in reference [1]. If $\langle \nabla \cdot \xi \rangle$ is $\mathscr{O}(\varepsilon^2)$, equation (8.46) yields $\Upsilon = \Xi$. Inserting it to equation (8.45), one readily obtains the ideal MHD singular layer equation, equation (6.121), in Section 6.5. However, equation (8.47) indicates that $\langle \nabla \cdot \xi \rangle$ is $\mathscr{O}(\varepsilon)$ and therefore one has to keep the plasma compressibility effect $(\Gamma P/P') \langle \nabla \cdot \xi \rangle$ in equation (8.46). The results under the ideal MHD limit are in perfect agreement with those obtained using the acoustic equation in Section 6.5.1.

8.1.3 EXTERNAL LIMITS AND ORDERING SCHEMES

In this subsection, we discuss the external limits of the resistive singular layer equations (8.40)–(8.42) and the ordering schemes to solve them. In particular, a subsidiary small parameter ordering is introduced so that the set of singular layer equations can be solved perturbatively in the following subsections. The solution is then matched to the solutions in the outer ideal MHD regions to obtain the dispersion relation.

The basic set of resistive singular layer equations (8.40)–(8.42) constitutes a sixth-order differential equation and therefore has six independent solutions. Structurally, it is similar to that in the cylinder model described in Chapter 7. With similar analyses, one can find that four of these six independent solutions are exponentially increasing or decreasing as

$$\exp\{\pm X^2/2Q^{1/2}\} \quad \text{as} \quad |X| \to \infty. \tag{8.48}$$

The remaining two behave as powers of X as $|X| \to \infty$ as follows:

$$\Psi \approx X^s, \quad \Xi = \Upsilon = \Psi/X, \quad s = \frac{1}{2} \pm (-D_I)^{1/2}. \tag{8.49}$$

The two exponentially increasing solutions must be discarded since they grow exponentially and cannot be matched to the outer ideal MHD solutions. The two exponentially decreasing solutions can be neglected in the matching to the outer ideal MHD solutions. One, therefore, only needs to consider the two asymptotic limits in equation (8.49). For studying the slower resistive instabilities, one also focuses on considering the Mercier stable case with

$$D_I < 0. \tag{8.50}$$

In this case, s is real. Compared with the ideal MHD case in Section 6.5, equation (8.49) agrees with the inner limits toward the singular layer of the outer ideal MHD solutions.

To solve the resistive singular layer equations, a subsidiary small parameter is introduced according to the proximity to the marginal stability. As discussed in the cylinder model given in Chapter 7, reference [1] has shown that when the curvature driving force for interchange modes is small, i.e., s in equation (8.49) is small, one has that QX^2 becomes small while X^4/Q remains finite. One can therefore introduce the small parameter δ such that

$$Q \sim \delta^{2/3}, \quad X \sim \delta^{1/6}. \tag{8.51}$$

With equation (8.49), this yields

$$\Psi \sim 1, \quad \Xi \sim \Upsilon \sim \delta^{-1/6}. \tag{8.52}$$

This yields $X \sim Q^{1/4}$, which is the same as that considered in the cylindrical equilibrium case in equation (7.58). All equilibrium quantities are assumed to be of order unity, except the so-called constant-Ψ case to be discussed later on. With these orderings, equations (8.40) and (8.41) in the leading order become

$$\Psi_{XX} - H\Upsilon_X = 0, \tag{8.53}$$

$$(E+F)\Upsilon^{(0)} + H\Psi_X^{(0)} = 0. \tag{8.54}$$

In the constant-Ψ case, it is assumed that

$$E + F \sim H \sim \delta. \tag{8.55}$$

Consequently, equation (8.53) yields $\Psi_{XX} = 0$. In this case, equations (8.40)–(8.42) can be solved by the procedure based on the so-called constant-Ψ approximation as given in subsection 8.1.4. One can find in subsection 8.1.4 that H does not enter the dispersion relation under the constant-Ψ approximation.

To get the H contribution, one must let

$$E + F \sim H \sim 1. \tag{8.56}$$

Using equation (8.54) to eliminate Υ_X from equation (8.53), one obtains

$$D_R \Psi_{XX}^{(0)} = 0. \tag{8.57}$$

If D_R is of order one, one again gets $\Psi_{XX}^{(0)} = 0$, leading to the constant-Ψ approximation. We therefore further assume that

$$D_R \sim \delta \tag{8.58}$$

with E, F, and H separately allowed to be finite. With the ordering in equation (8.58), the leading-order equations (8.53) and (8.54) are no longer independent. To obtain another leading-order equation, we differentiate equation (8.41) with respect to X and use equation (8.40) to eliminate Ψ_{XX}, yielding

$$\left(Q\Xi_{XX} - X^2\Xi\right)_X - HX\Xi + (X\Psi)_X + H\Psi + (D_R/Q)\Upsilon_X = 0. \tag{8.59}$$

All terms in this equation are of the same order in δ. In equations (8.40) and (8.41), the zeroth-order terms are grouped on the left, while the first-order terms on the right. Equations (8.40), (8.41), and (8.59) will be solved iteratively in Section 8.1.5 to derive the dispersion relation in the non-constant-Ψ procedure.

By far, the ordering consideration of constant-Ψ approximation and that with the H contribution included in the dispersion relation have been reviewed for the analyses in the following subsections. There are some more subtle points about the asymptotic matching. Interested readers are referred to reference [1]. The asymptotic matching is different from the matching for solving differential equations region by region. The asymptotic matching requires the solutions to overlap in an intermediate region. Instead, the matching in solving differential equations is only required on the interface between two regions. In some cases, the intermediate regions are required to build smooth matchings. This is related to the concerns about whether the internal solutions under the ordering in equation (8.51) can well match the external solutions as described in equation (8.49). Reference [1] points out that no particular concern needs to be addressed in the current situation.

8.1.4 SOLUTION WITH THE CONSTANT-ψ ASSUMPTION

In this subsection, we describe the procedure to solve the resistive singular layer equations (8.40)–(8.42) to derive the dispersion relation under the so-called constant ψ approximation. As discussed in subsection 8.1.3, the constant Ψ approximation leads to the omission of the effects from the parameter H. We describe it nevertheless for completeness. The treatment is similar to the case in the cylinder geometry in Section 7.5.

As the ordering schemes discussed in subsection 8.1.3, we consider the following orderings as given in equations (8.51) and (8.52)

$$Q \sim \delta^{2/3}, \quad X \sim \delta^{1/6} \tag{8.60}$$

and

$$\psi \sim 1, \quad \Xi \sim \Upsilon \sim \delta^{-1/6}. \tag{8.61}$$

In the constant-ψ treatment in particular, it is further assumed, as given in equation (8.55), that

$$(E+F) \sim H \sim \delta. \tag{8.62}$$

With the ordering assumptions in equations (8.60)–(8.62), terms in equation (8.40) are ordered as follows:

$$\underbrace{\Psi_{XX}}_{\mathscr{O}(1)} - \underbrace{H\Upsilon_X}_{\delta^{7/6}} = \underbrace{Q(\Psi - X\Xi)}_{\delta}.$$

Therefore, one has the lowest order in δ

$$\Psi_{xx}^{(0)} = 0.$$

Its solution to the first order in $X \sim \delta^{1/6}$ is as follows:

$$\Psi^{(0)} = \Psi_0 + \Psi_1 X, \tag{8.63}$$

where Ψ_0 and Ψ_1 are constants. To consider the even modes, it is assumed that $\Psi_1 = 0$.

Because equation (8.40) is solved, one can use its solution $\Psi^{(0)} = \Psi_0 + \mathscr{O}(\delta^{1/3})$ to solve equation (8.41) and equation (8.42). Similar to the cylinder case in Section 6.5, the expansions in the Hermite polynomials are employed to solve these two equations

$$\Psi^{(0)}(z) = e^{z^2/2} \sum_{n=0}^{+\infty} \psi_n H_n(z), \tag{8.64}$$

$$\Xi^{(0)}(z) = e^{z^2/2} \sum_{n=0}^{+\infty} \xi_n H_n(z), \tag{8.65}$$

$$\Upsilon^{(0)}(z) = e^{z^2/2} \sum_{n=0}^{+\infty} \upsilon_n H_n(z), \tag{8.66}$$

where $z = X/Q^{1/4}$ and

$$\begin{aligned}
\psi_{2m} &= \psi_0/(2^{2m-1/2}m!), \\
\xi_{2m+1} &= \frac{\psi_{2m} Q^{-1/4}\left[(4m+3) + D_R/Q^{3/2} + CQ^{3/2}\right]}{(4m+3)\left[(4m+3) + CQ^{3/2}\right] - (C - KD_R)D_R}, \\
\upsilon_{2m+1} &= \frac{\psi_{2m} Q^{-1/4}\left[(4m+3) + (C - KD_R)Q^{3/2}\right]}{(4m+3)\left[(4m+3) + CQ^{3/2}\right] - (C - KD_R)D_R}, \\
\psi_{2m+1} &= \xi_{2m} = \upsilon_{2m} = 0.
\end{aligned}$$

Here, $D_R \equiv E + F$, as in equation (8.39) with $H = 0$, and $C \equiv G + KF$. Note here that the z-dependence part of $\Psi^{(0)}$ in equation (8.64) is of order $X^2 \sim \delta^{1/3}$, which is consistent with the solution in equation (8.63).

The dispersion relation is just $\Delta' = \Delta$ as shown in the cylinder case in section 6.5. To calculate Δ, we note that equation (8.40) in the first order becomes

$$\Psi^{(1)}_{XX} = Q\left(\Psi^{(0)} - X\Xi^{(0)}\right) + H\Upsilon^{(0)}_X.$$

Integrating it yields

$$\begin{aligned}
\Delta(Q) &= (V_s/X_0\Psi_0)\,\Psi^{(1)}_X\Big|_{-\infty}^{+\infty} \\
&= (V_s/X_0\Psi_0)\left[\int_{-\infty}^{+\infty} dX Q\left(\Psi^{(0)} - X\Xi^{(0)}\right) + H\Upsilon^{(0)}\Big|_{-\infty}^{+\infty}\right].
\end{aligned}$$

After some calculations, one obtains

$$\begin{aligned}
\Delta(Q) = {}& \frac{2V_s Q^{5/4}}{X_0} \sum_{m=0}^{+\infty} \frac{\Gamma(m+1}{\Gamma(m+1)} \\
\times {}& \frac{(4m+3)(1 - D_R/Q^{3/2} + D_R/Q^{3/2} + CQ3/2 - (C - KD_R)D_r}{(4m+1)\left[(4m+3) + CQ^{3/2}\right] - (C - KD_R)D_r},
\end{aligned} \tag{8.67}$$

where

$$
\begin{aligned}
a &\equiv 1 - D_r/Q^{3/2}, \\
b &\equiv D_R/Q^{3/2} + CQ^{3/2} - (C - KD_R)D_R, \\
\tau &= \left[C^2Q^3 + 4(C - KD_r)D_R\right]^{1/2}, \\
R_\pm &\equiv \frac{1}{2}\left(-CQ^{3/2} \pm \tau\right).
\end{aligned}
$$

With Δ' given in the outer region and $\Delta(Q)$ in equation (8.67), one can use the dispersion relation $\Delta' = \Delta(Q)$ to determine the normalized eigenfrequency Q in the constant-Ψ approximation.

8.1.5 GENERAL SOLUTION WITHOUT CONSTANT-Ψ APPROXIMATION

In this subsection, we describe the procedure to solve the resistive singular layer equations to derive the dispersion relation without invoking the so-called constant ψ approximation.

As the ordering schemes discussed in subsection 8.1.3, we consider the following orderings as given in equations (8.51) and (8.52)

$$Q \sim \delta^{2/3}, \; X \sim \delta^{1/6}, \tag{8.68}$$

and

$$\psi \sim 1, \;\; \Xi \sim \Upsilon \sim \delta^{-1/6}. \tag{8.69}$$

Different from the constant-ψ treatment, it is further assumed, as given in equations (8.56) and (8.58), that

$$E + F \sim H \sim 1, \tag{8.70}$$

$$D_R \sim \delta. \tag{8.71}$$

As discussed in subsection 8.1.3, in the ordering scheme in equations (8.68)–(8.71) we can use equations (8.40), (8.42), and (8.74) as recited as follows to derive the dispersion relation

$$\Psi_{XX} - H\Upsilon_X = Q(\Psi - X\Xi), \tag{8.72}$$

$$Q\Upsilon_{XX} - X^2\Upsilon + X\Psi = Q^2\left[(G+KF)\Upsilon - (G-KE)\Xi + X\Psi - KH\Psi_X\right], \tag{8.73}$$

$$\left(Q\Xi_{XX} - X^2\Xi\right)_X - HX\Xi + (X\Psi)_X + H\Psi + (D_R/Q)\Upsilon_X = 0, \tag{8.74}$$

In the following subsections, we will first solve for the lowest-order solutions for $\Psi^{(0)}$, $\Upsilon^{(0)}$, and $\Xi^{(0)}$. Then, the lowest-order solutions are used to solve for the first-order solutions $\Psi^{(1)}$ and $\Upsilon^{(1)}$. The solutions are then used to calculate $\Delta(Q)$ to construct the dispersion relation. The analyses of the dispersion relation will be given in the last subsection.

8.1.5.1 Lowest-order solutions

In this subsection, we first solve for the lowest-order solutions for $\Psi^{(0)}$ and $\Upsilon^{(0)}$ from equations (8.72) and (8.73).

The lowest-order terms are grouped on the left-hand side in these two equations, which read

$$\Psi_{XX}^{(0)} - H\Upsilon_X^{(0)} = 0, \tag{8.75}$$

$$Q\Upsilon_{XX}^{(0)} - X^2\Upsilon^{(0)} + X\Psi^{(0)} = 0. \tag{8.76}$$

Integration of equation (8.75) yields

$$\Upsilon^{(0)} = \Psi_X^{(0)}/H. \tag{8.77}$$

Here, the integration constant is ignored as it does not yield a substantial effect. The solution in equation (8.77) can be used to eliminate $\Upsilon^{(0)}$ from equation (8.76), yielding

$$Q\Psi_{XXX}^{(0)} - X^2\Psi_X^{(0)} + XH\Psi^{(0)} = 0. \tag{8.78}$$

We now introduce the coordinate transformation

$$z = X/Q^{1/4} \tag{8.79}$$

with a branch cut at $\arg(Q) = \pi$ in the complex Q plane for making z single-valued. Since Q can vary from $-\pi$ to π, the new coordinate z has a range $z \leq \pi/4$. Thus, equation (8.78) becomes

$$\Psi_{zzz}^{(0)} - z^2\Psi_z^{(0)} + Hz\Psi^{(0)} = 0. \tag{8.80}$$

Because the boundary conditions need to be taken into account, (8.80) is solved by the integral transform method, or the Laplace transform method in reference [6] or in Section 11.7 in reference [7], instead of the Fourier transform method. Let

$$\Psi^{(0)}(z) = \int_C dk \exp\{ikz\}\psi(k), \tag{8.81}$$

where "C" is a contour in the k plane to be determined to suit the problem. Inserting this ansatz into equation (8.80), one obtains

$$\begin{aligned}&\int_C dk \exp\{ikz\}\left[k\psi_{kk} + (2+H)\psi_k - k^3\psi\right] \\ &- \{\exp\{ikz\}\left[(1+H)\psi + k\psi_k - izk\psi\right]\}|_C = 0.\end{aligned} \tag{8.82}$$

The second term in this equation results from the integration by parts, which specifies the difference of the quantities in the brackets between the two ends of the contour C. In this method, the third-order differential equation (8.80) is transformed to the second-order differential equation

$$k\psi_{kk} + (2+H)\psi_k - k^3\psi = 0 \tag{8.83}$$

with the boundary conditions

$$\{\exp\{ikz\}[(1+H)\psi + k\psi_k - izk\psi]\}|_C = 0. \tag{8.84}$$

For solving a differential equation, one usually can first resort to the so-called look-up method. From the list of various types of differential equations in reference [8], one can find in item 67 on p. 526 that the solution of equation (8.83) can be expressed in terms of the Bessel functions: $I_\nu(s)$ and $K_\nu(s)$, where

$$s = k^2/2, \quad \nu = (1+H)/4.$$

Besides, the solutions based on $I_\nu(s)$ and $K_\nu(s)$ need to satisfy the boundary condition in equation (8.84). For further analyses, we need to know the asymptotic behavior of the Bessel function K_ν. From Item 9.7.1 and 9.7.2 in reference [9], one has

$$I_\nu(z) \sim \sqrt{\frac{1}{2\pi z}} e^z \left(1 + \frac{\mu - 1}{8z} + \cdots\right), \quad |\arg(z)| < \pi/2, \tag{8.85}$$

$$K_\nu(z) \sim \sqrt{\frac{\pi}{2z}} e^{-z} \left(1 + \frac{\mu - 1}{8z} + \cdots\right), \quad |\arg(z)| < 3\pi/2, \tag{8.86}$$

where $\mu = 4\nu^2$. From Items 9.6.2, 9.6.10, 9.6.30, and 9.6.31 in reference [9], one also has

$$K_\nu(z) = \frac{\pi}{2} \frac{I_{-\nu}(z) - I_\nu(z)}{\sin(\nu\pi)}, \tag{8.87}$$

$$I_\nu(z) = \left(\frac{z}{2}\right)^\nu \sum_{n=0}^{+\infty} \frac{(z^2/2)^n}{n!\Gamma(\nu+n+1)}, \tag{8.88}$$

$$I_\nu(ze^{m\pi i}) = e^{m\nu\pi i} I_\nu(z), \quad m\text{: an integer}, \tag{8.89}$$

$$K_\nu(ze^{m\pi i}) = e^{-m\nu\pi i} K_\nu(z) - \pi \sin(m\nu\pi)\cos(\nu\pi) I_\nu(z). \tag{8.90}$$

Note that equation (8.80) is a differential equation of third order. There are three independent solutions. They are constructed by choosing suitable integration contours and integrands in equation (8.81). Reference [1] considers the following three independent solutions for $X > 0$, $|\arg z| < \pi/4$:

$$\Psi_1(z) = \exp\{2\nu\pi i\} \int_{C_1} dk \exp\{ikz\} s^{-\nu} K_\nu(s), \tag{8.91}$$

$$\Psi_2(z) = \exp\{-2\nu\pi i\} \int_{C_2} dk \exp\{ikz\} s^{-\nu} K_\nu(s), \tag{8.92}$$

$$\Psi_3(z) = \int_{C_3} dk \exp\{ikz\} s^{-\nu} K_\nu(s \exp\{i\pi\}), \tag{8.93}$$

where Contours C_1, C_2, and C_3 are shown in figure 8.1. Both Contours C_1 and C_2 have the endpoints at $|k| \to \infty$ in the upper k plane. But C_2 surrounds the origin

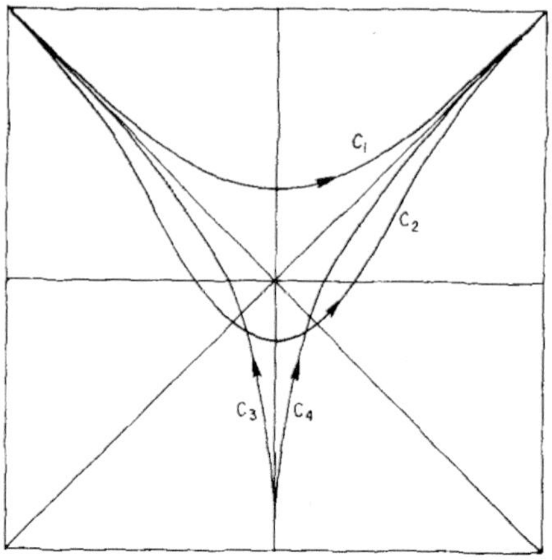

Figure 8.1 Integration contours for three independent solutions. Reproduced from reference [1] with the permission of AIP Publishing.

$k = 0$, while C_1 not. This makes $\arg(k)$ for two contours in the second quadrant different. In the first quadrant, both contours end with the same angle $\arg(k) = \pi/4$. To reach the endpoint in the second quadrant, Contour C_t swipes counter-clockwise so that its endpoint in the second quadrant is $3\pi/4$. Contour C_2 is different. It has to surround the origin $k = 0$ and go clockwise. Therefore, its endpoint in the second quadrant becomes $-5/4\pi$. Since the integrals are multivalued as one can see from the asymptotic behavior of the Bessel function in equations (8.85) and (8.86), $\Psi_1(z)$ and $\Psi_2(z)$ in equations (8.91) and (8.92) are independent.

Using equation (8.87), it can be proved by direct substitution that the solutions Ψ_1 and Ψ_2 in equations (8.91) and (8.92) satisfy the boundary condition in equation (8.84) with $\arg(z) < \pi/4$. For example at the end point $\arg(k) = \pi/4$, one just needs to note that $e^{-s} = e^{-|s|\exp\{i\pi/2\}}$ is purely oscillating and $e^{ikz} = e^{|kz|\exp\{\pi/2+\pi/4+\arg(z)\}}$ goes to zero exponentially as $|k| \to +\infty$. Contour C_3 has one endpoint at $\arg(k) = -5\pi/4$ and the other anywhere in the quadrant $-3\pi/4 < \arg(k) < -\pi/4$. As discussed in reference [1], one can alternatively use the reflection of C_3 about the axis $Real\{k\} = 0$, i.e., Contour C_4 in figure 8.1 for the solution Ψ_3.

Next, let us determine the asymptotic behavior of the independent solutions in equations (8.91)-(8.93) for large $|z|$. The asymptotic behavior of Ψ_1 and Ψ_3 can be obtained by the so-called saddle point method, or the method of steepest descent [7]. For this, one needs to determine the saddle point of the integrand. For Ψ_1, one requires

$$\frac{d}{dk}\left(e^{ikz}s^{-\nu}\frac{1}{\sqrt{z}}e^{-s}\right) = 0.$$

This yields

$$k^2 - ikz + (2\nu+1) = 0. \tag{8.94}$$

Its solution at large $|z|$ is the saddle point for Ψ_1:

$$k_1 \approx z\exp\{i\pi/2\}\left[1+(1+2\nu)/z^2\right]. \tag{8.95}$$

Similarly, for Ψ_3, one has the saddle point

$$k_3 \approx z\exp\{-i\pi/2\}\left[1-(1+2\nu)/z^2\right]. \tag{8.96}$$

Using the saddle point method, one generally has [7]

$$\int_c g(k)e^{zh(k)}dk \approx i\sqrt{\frac{2\pi}{az}}g(k_0)e^{zh(k_0)-i\theta_0/2}, \tag{8.97}$$

where k_0 is the saddle point, $h''(t_0) = ae^{i\theta_0}$, and $\theta_0 = arg[h''(t_0)]$. Using the saddle points in equations (8.91)–(8.92), and the formula in equation (8.97), one obtains

$$\Psi_1(z) \approx -i\pi\exp\{i\pi\nu\}(z^2/2)^{-(\nu+1/2)}\exp\{-z^2/2\}, \tag{8.98}$$

$$\Psi_3(z) \approx i\pi\exp\{i\pi\nu\}(z^2/2)^{-(\nu+1/2)}\exp\{z^2/2\}. \tag{8.99}$$

For computing the asymptotic behavior of Ψ_2 in equation (8.91), we first use equation (8.87) to express K_ν in terms of I_ν and then introduce the expansion of I_ν in equation (8.88). Using items 6.1.4 and 6.1.29 in reference [9]

$$\frac{1}{\Gamma(z)} = \frac{i}{2\pi}\int_{C_\Gamma}(-t)^{-z}e^{-t}dt, \tag{8.100}$$

$$\Gamma(x)\Gamma(1-x) = \frac{\pi}{\sin\pi x}, \tag{8.101}$$

one for example can prove

$$\begin{aligned}
&\exp\{-2\nu\pi i\}\int dke^{ikz}s^{-\nu}K_\nu(s)\\
=\ &\exp\{-2\nu\pi i\}\int_C dke^{ikz}s^{-\nu}\frac{\pi}{2}\frac{I_{-\nu}(s)-I_\nu(s)}{\sin(\nu\pi)}\\
\approx\ &\exp\{-2\nu\pi i\}\int_C dke^{ikz}s^{-\nu}\frac{\pi}{2}\frac{I_{-\nu}(s)}{\sin(\nu\pi)}\\
=\ &\int_C dke^{ikz}s^{-\nu}\frac{\pi}{2}\frac{I_{-\nu}(se^{-2\pi i})}{\sin(\nu\pi)}\\
=\ &\int_{C_\Gamma} dt\frac{i}{z}\frac{\pi}{2}\frac{e^{-t}\left(\frac{2z^2}{-t^2}\right)^\nu\frac{1}{2}}{\sin(\nu\pi)}\left(\frac{2z^2}{-t^2}\right)^\nu\left(\frac{1}{\Gamma(1-\nu)}+\frac{z^2/2}{\Gamma(2-\nu)}+\cdots\right),
\end{aligned} \tag{8.102}$$

where equations (8.87) - (8.89) have been used and it has been noted that the I_ν contribution is negligible. Using equations (8.100) and (8.101), equation (8.102) yields

$$\begin{aligned}\Psi_2(z) &\approx 2^{3\nu}\pi[\Gamma(\nu)/\Gamma(4\nu)]z^{4\nu-1} \\ &\quad \times\left[1-(4\nu-3)(4\nu-2)(4\nu-1)/4z^4+\cdots\right]. \end{aligned} \tag{8.103}$$

Note here that this asymptotic behavior has the form given in equation (8.49).

We have obtained three independent solutions in equations (8.91)-(8.93) together with their asymptotic limits, respectively, in equations (8.98), (8.103), and (8.99). The three asymptotic limits are those we expect to have, as we see in the cylinder analyses in Chapter 7. However, the independent solutions in the toroidal analyses rely on the Laplace transform. This makes the treatment complicated.

Note that z is complex. According to equations (8.98) and (8.99), both of the solutions Ψ_1 and Ψ_3 can diverge as $|z| \to \infty$. We discard the independent solution Ψ_3 in considering that it grows exponentially with z^2 for $\arg(z) < \pi/4$. For the independent solution Ψ_2, instead, we try to make it convergent as $|z| \to \infty$ with $3\pi/4 < \arg(z) < 5\pi/4$ by a suitable summation of two remaining independent solutions Ψ_1 and Ψ_2 and properly detort the integration contours in equations (8.91)-(8.92).

First, we note that if the contours C_1 and C_2 are detorted such that their right-hand ends in figure 8.1 lie in the quadrant $\arg(k) < \pi/4$, instead of on the ray $\arg(k) = \pi/4$, the integrals converge for all values of z due to the factor $K_\nu(s)$. Satisfaction of the boundary condition in equation (8.84) is obvious using these contours. The exponential convergence for z can also be verified. However, similar contour detortions to $3\pi/4 < \arg(z) < 5\pi/4$ do not work on the left since $K_\nu(s)$ becomes exponentially large in this case. Nevertheless, the summation of two remaining independent solutions, Ψ_1 and Ψ_2, can avoid this difficulty

$$\psi^{(0)}(z) = \Psi_1(z)+\Psi_2(z). \tag{8.104}$$

Certainly, to make the desired cancellation of their integrands possible, the two contours C_1 and C_2 are required to join together at some point on the left.

The summation of two independent solutions in equation (8.104) works since the suitable coefficients $\exp\{\pm 2\nu\pi i\}$ have been introduced in the definitions of Ψ_1 and Ψ_2 in equations (8.91) and (8.92). Using equation (8.90), one can prove that

$$\begin{aligned} & e^{ikz}\left[\psi_1(k)+\psi_2(ke^{-2\pi i})\right] \\ = \; & e^{2\nu\pi i}e^{ikz}s^{-\nu}\left[K_\nu(s)+K_\nu(se^{-4\pi i})\right] \\ = \; & 2\cos(2\nu\pi)e^{2\nu\pi i}e^{ikz}s^{-\nu}K_\nu(se^{-2\pi i}). \end{aligned}$$

Note that $K_\nu(se^{-2\pi i})$ makes the convergences in k and z for the contour detortions of the left ends to $3\pi/4 < \arg(z) < 5\pi/4$. Similar analyses can be made for the $-z$ case. Therefore, we obtain the desired solution

$$\begin{aligned}\psi^{(0)}(z) &= \int_{C_1} dk \exp\{ikz\}s^{-\nu}K_\nu(s\exp\{-2\pi i\}) \\ &\quad + \int_{C_2} dk \exp\{ikz\}s^{-\nu}K_\nu(s\exp\{2\pi i\}). \end{aligned} \tag{8.105}$$

Here, it has been noted that adding to ψ_1 and subtracting from ψ_2 the single-valued function $2\pi\cos(\nu\pi)s^{-\nu}I_\nu(s)$ does not change $\Psi^{(0)}$.

Note that changing the sign of z can be realized by changing the integration variable k and the interchange of Contours C_1 and C_2. Using equation (8.105), one can prove that $\psi^{(0)}(z)$ is an even function, i.e.,

$$\psi^{(0)}(-z) = \psi^{(0)}(z). \tag{8.106}$$

We also need to solve for the lowest-order solution $\Xi^{(0)}$ of equation (8.59). Using equations (8.77) and (8.79), the equation is reduced to

$$\begin{aligned}&\left(\Xi^{(0)}_{zz} - z^2\Xi^{(0)}\right)_z - Hz\Xi^{(0)}\\ =\ & -Q^{-1/4}\left[\left(z\Psi^{(0)}\right)_z + H\Psi^{(0)} + \left(D_R/HQ^{3/2}\right)\Psi^{(0)}_{zz}\right].\end{aligned} \tag{8.107}$$

The solution of equation (8.107) contains the homogeneous and inhomogeneous parts.

We first solve for the homogeneous solution. The homogeneous part of equation (8.107) is structurally similar to equation (8.80). Likewise, the integral transform method can be used to solve it. Let

$$\Xi^{(0)}(z) = \int_{C_1} dk \exp\{ikz\}\xi_1(s) + \int_{C_2} dk \exp\{ikz\}\xi_2(s). \tag{8.108}$$

One obtains the solutions for $\Psi^{(0)}$ and $\Upsilon^{(0)}$ in subsubsetion 8.1.5.1

$$\begin{aligned}&\int_c dk \exp\{ikz\}\left\{k\left(\xi_{kk} - k^2\xi\right) - H\xi_k\right.\\ &\left.+iQ^{-1/4}\left[k\psi_k - H\psi + \left(D_R/HQ^{3/2}\right)k^2\psi\right]\right\}\\ &+\left\{\exp\{ikz\}\left[(1H)\xi - k\xi_k - iQ^{-1/4}k\psi + izk\xi\right]\right\}\Big|_C = 0..\end{aligned} \tag{8.109}$$

Here, without ambiguity, contour C is used to represent the contours C_1 and C_2. For $\arg(z) < \pi/4$, the solution for equation (8.109) is

$$\begin{aligned}\Xi^{(0)}(z) =\ & \exp\{-2\nu\pi i\}\int_{C_1} dk \exp\{ikz\}s^\nu K_\nu(s)\\ &+\exp\{2\nu\pi i\}\int_{C_2} dk \exp\{ikz\}s^\nu K_\nu(s).\end{aligned} \tag{8.110}$$

Similar to the derivation of equation (8.9), one can obtain the asymptotic limit as $\arg(z) < \pi/4$

$$\begin{aligned}\Xi(z) \approx\ & 2^{-3\nu}\pi[\Gamma(3nu)/\Gamma(34\nu)]z^{-4\nu-1}\\ &\times\left[1 + (4\nu+3)(4\nu+2)(4\nu+1)/4z^4 + \cdots\right].\end{aligned} \tag{8.111}$$

We next derive the inhomogeneous solution. One can construct a Green's function by the method of parameter variation and obtains

$$\begin{aligned}
\Xi^{(0)}(z) &= \int_{C_1} dk\, \exp\{ikz\} s^{\nu} \Bigg\{ K_\nu(s) \left[\int_{\infty \exp\{2\pi i\}}^{a \exp\{2\pi i\}} dt\, t^{-\nu} K_\nu\left[t \exp\{-2\pi i\}\right] R_1^-(t) \right. \\
&\quad \left. + \int_{a \exp\{2\pi i\}}^{s} dt\, t^{-\nu} K_\nu\left[t \exp\{-2\pi i\}\right] \bar{R}_1(t) \right] + K_\nu(s \exp\{-2\pi i\}) \\
&\quad \times \left(\int_s^a dt\, t^{-\nu} K_\nu(t) \bar{R}_1(t) + \int_a^{\infty} dt\, t^{-\nu} K_\nu(t) R_1^*(t) \right) \Bigg\} \\
&\quad + \int_{C_2} dk\, \exp\{ikz\} s^{\nu} \Bigg\{ K_\nu(s) \left[\int_{\infty \exp\{-2\pi i\}}^{a \exp\{-2\pi i\}} dt\, t^{-\nu} K_\nu\left[t \exp\{2\pi i\}\right] R_2^-(t) \right. \\
&\quad \left. + \int_{a \exp\{-2\pi i\}}^{s} dt\, t^{-\nu} K_\nu\left[t \exp\{2\pi i\}\right] \bar{R}_2(t) \right] + K_\nu(s \exp\{2\pi i\}) \\
&\quad \times \left(\int_s^a dt\, t^{-\nu} K_\nu(t) \bar{R}_2(t) + \int_a^{\infty} dt\, t^{-\nu} K_\nu(t) R_2^*(t) \right) \Bigg\}, \qquad (8.112)
\end{aligned}$$

where

$$\begin{aligned}
R_1^* &\equiv \left[2\sqrt{2}\pi Q^{1/4} \cos(\nu\pi) \right]^{-1} \exp\{2\nu\pi i\} t^{-\nu-1/2} \\
&\quad \times \left[K_\nu'(t) + (1/2 - 3\nu) K_\nu(t)/t + (D_R/HQ^{3/2}) K_\nu(t) \right], \\
R_2^* &\equiv -\exp\{-4\nu\pi i\} R_1^*(t), \quad R_1^- \equiv R_2^*[t \exp(2\pi i\}], \quad R_2^- \equiv R_1^*[t \exp(2\pi i\}], \\
\bar{R}_1(t) &\equiv \frac{1}{2}\left[R_1^*(t) + R_1^-(t) \right], \quad \bar{R}_2(t) \equiv \frac{1}{2}\left[R_2^*(t) + R_2^-(t) \right],
\end{aligned}$$

and a is an arbitrary real positive constant. One can also verify this solution by the direct substitution of the solution back into the original equation. Differentiation of equation (8.112) with respect to a gives rise to the difference between two integrals with the same single-valued integrand. Thus, one obtains

$$\partial \Xi^{(0)} / \partial a = 0.$$

Similar to the proof of the parity in equation (8.106), one can show that

$$\Xi^{(0)}(-z) \quad = \quad -\Xi^{(0)}(z).$$

This can also be verified by the governing equation for $\Xi^{(0)}$, equation (8.107). An even solution $\Psi^{(0)}$ gives rise to an odd solution $\Xi^{(0)}$.

To obtain the dispersion relation, the asymptotic behavior of $\Xi^{(0)}$ in equation (8.112) as $z \to \infty$ with $\arg(z) < \pi/4$ needs to be determined. First, it is noted that the contribution of the integral along the contour C_1 is negligible since it derives from the inhomogeneity, which is exponentially small. To determine the asymptotic behavior of the integral along the contour C_2 next, the value of the constant a can be chosen to simplify the calculation. One can choose $a = 0$ and calculate the integrals for $\nu < 1/4$, Note that the results are meromorphic functions of ν and therefore are valid for $\nu > 1/4$ as well. Similar to the derivation of equation (8.103) for $\Psi^{(0)}$, the infinite integrals are evaluated by expressing K_ν in terms of I_ν as given in equation (8.87) and using the series expansion for I_ν as given in equation (8.88). The contributions from the limit $t = a = 0$ vanish, while the contributions from the limit $t = s$ are evaluated using the contour integral representation of the Gamma function as given in equation (8.100). The definite integrals are evaluated directly using the table of integrals. The final result is as follows:

$$\Xi^{(0)}(z) \approx \Psi^{(0)}(z)/zQ^{1/4} + C_\nu z^{-4\nu-1}, \tag{8.113}$$

where

$$\begin{aligned} C_\nu \equiv & \left[4\cos^2(2\nu\pi)\Gamma(4\nu)/2^{s\nu}Q^{1/4}(\nu-1/4)\Gamma(\nu)\Gamma(1/2-2\nu)\right] \\ & \times\left[\Gamma(1/4)\Gamma^2(5/4-\nu)\Gamma(5/4-2\nu)\right. \\ & \left.-\Gamma(3/4)\Gamma^2(3/4-\nu)\Gamma(3/4-2\nu)(D_R/4Q^{3/2})\right]. \end{aligned} \tag{8.114}$$

Note here that equation (8.113) has the form given in equation (8.49).

8.1.5.2 First-order Solutions

As shown in the cylindrical theory in Chapter 7, to derive the dispersion relation one also needs to obtain the solutions for $\Psi^{(1)}$ and $\Upsilon^{(1)}$. For this, one needs to consider equations (8.40) and (8.42) in the next order.

Noting the ordering assumption in equation (8.58), one has $F \approx -(E+H^2)$. Using equation (8.77), equations (8.40) and (8.42) in the first-order are reduced to

$$\Psi^{(1)}_{XX} - H\Upsilon^{(1)}_{X} = Q\left(\Psi^{(0)} - X\Xi^{(0)}\right), \tag{8.115}$$

$$Q\Upsilon^{(1)}_{XX} - X^2\Upsilon^{(1)} + X\Psi^{(1)} = Q^2(G-KE)\left(\Psi^{(0)}/H - \Xi^{(0)}\right). \tag{8.116}$$

We only need the asymptotic solutions for these two equations in the limit z (or X) $\to \infty$ with $|\arg(z)| < \pi/4$. Using the asymptotic limits for $\Psi^{(0)}$ and $\Xi^{(0)}$ in equations (8.103) and (8.113), equations (8.98) and (8.103) are further reduced to

$$\Psi^{(1)}_{z} - H\Upsilon^{(1)}_{z} \approx -Q^{7/4}C_\nu z^{-H-1}, \tag{8.117}$$

$$Q\Upsilon^{(1)}_{zz} - z^2\Upsilon^{(1)} + z\Psi^{(1)} = Q^{3/2}(G-KE)C_\nu z^{-H-2}. \tag{8.118}$$

Their leading-order solutions are

$$\Psi^{(1)} \approx X\Upsilon^{(1)} = -\left[Q^{7/4}C_\nu/H(2H-1)\right]z^{1-H}. \tag{8.119}$$

Here, it has been noted that the term containing $G-KE$ does not contribute.

8.1.5.3 Dispersion relation

Having derived the solutions of the singular equations (8.40)–(8.42) to sufficient order in the previous subsections, we are going to match the solutions between the inner and outer regions and derive the dispersion relation in this subsection. The procedure is similar to the cylindrical case described in Chapter 7.

The outer solutions are described by the ideal MHD. According to equations (8.49) and (8.58), the outer solutions to the left and right of the singular layer are given, respectively,

$$\Psi_L = A_L|Y|^H + B_L|Y|^{1-H}, \tag{8.120}$$

$$\Psi_R = A_R|Y|^H + B_R|Y|^{1-H}, \tag{8.121}$$

where

$$Y \equiv (V - V_0)/V_s = \left(X_0 Q^{1/4}/V_s\right) z \tag{8.122}$$

with V_s being an arbitrary macroscopic scale factor. The inner solution on either side has the asymptotic form

$$\Psi_I \approx A_I|Y|^H + (B_I + \bar{B}_I \mathrm{sgn}(Y))\,|Y|^{1-H}, \tag{8.123}$$

where A_I and B_I are given according to equations (8.103) and (8.113), and $\tilde{B}_I$ is an arbitrary coefficient resulting from the homogeneous even solution for Ξ in equation (8.110).

Matching the outer solutions in equations (8.120) and (8.121) with the inner solution in (8.123) yield

$$A_L = A_R = A_I, \quad B_L = B_I - \bar{B}_I, \quad B_R = B_I + \bar{B}_I. \tag{8.124}$$

Introducing the definitions

$$\Delta' \equiv B_R/A_R + B_L/A_L,$$

$$\Delta(Q) \equiv 2(B_I/A_I),$$

one obtains the dispersion relation

$$\Delta' = \Delta(Q). \tag{8.125}$$

Substituting the definition $\nu = (1+H)/4$ in the expressions for A_I and B_I, the dispersion relation in equation (8.125) becomes

$$\Delta' = \Delta(Q). \equiv \frac{(2V_s/X_0)^{1-2H}\pi\Gamma\left(\frac{1}{4}\right)\Gamma^2\left(1-\frac{H}{4}\right)\Gamma\left(\frac{3}{4}-\frac{H}{2}\right)}{(1-2H)\Gamma\left(1-\frac{H}{2}\right)\left[\cos(H\pi/2)\Gamma\left(\frac{1+H}{4}\right)\Gamma(1-H)\right]^2} \times Q^{(2H+5)/4}\left[1 - \frac{\Gamma\left(\frac{3}{4}\right)\Gamma^2\left(\frac{1}{2}-\frac{H}{4}\right)\Gamma\left(\frac{1}{4}-\frac{H}{2}\right)}{\Gamma\left(\frac{1}{4}\right)\Gamma^2\left(1-\frac{H}{4}\right)\Gamma\left(\frac{3}{4}-\frac{H}{2}\right)}\frac{D_R}{4Q^{3/2}}\right]. \tag{8.126}$$

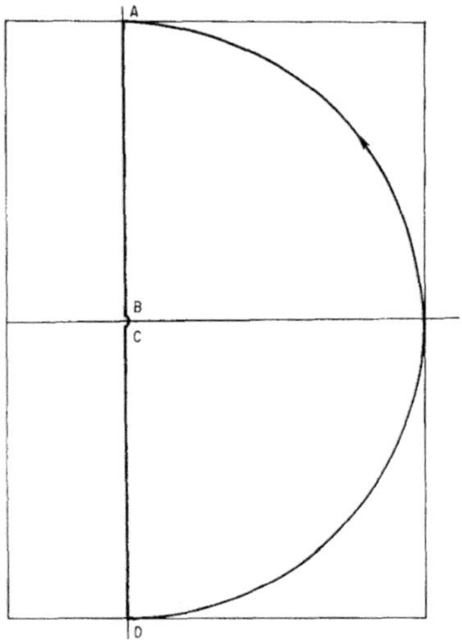

Figure 8.2 The contour encircling the unstable half of Q plane for Nyquist diagram analyses. Reproduced from reference [1] with the permission of AIP Publishing.

In the case with $H = 0$, one recovers the result in the constant Ψ approximation described in Section 8.1.4:

$$\Delta' = 2\pi \frac{V_s}{X_0} \frac{\Gamma(3/4)}{\Gamma(1/4)} Q^{5/4} \left(1 - \frac{\pi D_R}{4Q^{3/2}}\right). \tag{8.127}$$

In the dispersion relation in equation (8.126), Δ' must be determined by the solutions of the ideal MHD equations in the outer regions. It is real and independent of Q and η. The system is unstable if, for a given Δ', there is a solution of equation (8.126) with $\Re e\{Q\} > 0$.

8.1.5.4 Analyses of the dispersion relation

In this subsection, we are going to analyze the dispersion relation derived in the last subsection 8.1.5.3. Only the case $H = 0$ is presented. Interested readers may refer to reference [1] for the case with $H \neq 0$. We will discuss three cases: $D_R > 0$, $D_R = 0$, and $D_R < 0$.

The Nyquist diagram method is used for the stability analyses. For this, we first draw the contour encircling the unstable half of Q plane in figure 8.2. The mapping of Δ' in equation (8.127) for the case with $H = 0$ will then be drawn for various cases. The contour is divided into four segments with letters $A - D$ in order to help determine the mappings. In figure 8.2, points A and D go to infinity, and points B and C go to zero.

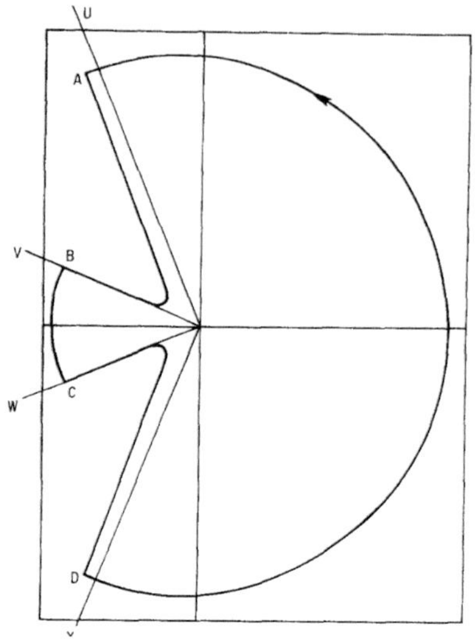

Figure 8.3 Mapping of Δ with the contour plot for Q given in figure 8.2 for the case $D_R > 0$. It is used for the Nyquist diagram analyses. Reproduced from reference [1] with the permission of AIP Publishing.

We first discuss the case with $D_R > 0$. For the case with $H = 0$, the dispersion relation in equation (8.127) can be reduced to

$$\Delta' = aQ^{5/4} - bQ^{-1/4}, \tag{8.128}$$

where a and b are positive since $D_R > 0$. In the case with $a = b = 1$, Δ mapping is drawn in figure 8.3. Other cases with a and b not being 1 are similar. For the segment AB, the phase angle in the Q contour in figure 8.2 is $\pi/2$. When $|Q|$ is large, the first term $aQ^{5/4}$ in equation (8.128) is much larger than the second term. The mapping in the Δ plane has the phase angle $(\pi/2)(5/4) = 5\pi/8$ as shown in figure 8.2. When $|Q|$ is small, the second term $-bQ^{-1/4}$ in equation (8.128) is much larger than the first term. The mapping in the Δ plane has the phase angle $\pi + (\pi/2)(-1/4) = 7\pi/8$, where the first term, π, results from the negative sign in the case with $D_R > 0$, as shown in figure 8.2. The analyses for the rest of the segments are similar.

The Nyquist diagram usually uses the mapping of the dispersion relation $\Delta - \Delta'$. Since Δ' is real and independent of Q, one can use the mapping of Δ in figure 8.3 for the analyses. In the Nyquist diagram analyses, if the contour in figure 8.3 shifted by $-\Delta'$ encircles the zero, the system is unstable, noting that the dispersion relation does not contain poles. The number of unstable roots equals the number of times the origin is encircled. Because the segments BC and CD go to infinity, there is an unstable root for any given Δ'.

Let us consider the limit $\eta \to 0$. Note that V_s/X_0 in equation (8.127) varies as $\eta^{-1/3}$ and becomes large as $\eta \to 0$. Therefore, for a finite Δ', equation (8.127) yields

$1 - \pi D_R / 4Q^{3/2} = 0$. i.e.,

$$Q = (\pi D_R / 4)^{2/3}. \tag{8.129}$$

As a result, the actual growth rate sacles as

$$\gamma = Q_0 Q \sim \eta^{1/3}.$$

This is the usual scale for the resistive interchange modes as discussed in the cylinder case in Section 7.1.

We next discuss the case with $D_R = 0$. In this case, the segment *BC*' in figure 8.4 merges to zero. The unstable root only occurs for the case

$$\Delta' > 0.$$

The dispersion relation in equation (8.127) yields in this case

$$Q = \left(\frac{1}{2\pi} \frac{\Gamma(1/4)}{\Gamma(3/4)} \frac{X_0}{V_s} \Delta' \right)^{4/5}.$$

Thus, one has

$$q \sim Q_0 X^{4/5} \sim \eta^{3/5}.$$

This is the tearing mode scale as discussed in the cylinder case in Section 7.1. Its growth rate is slower than the interchange mode as $\eta \to 0$.

Further on, let us discuss the case with $D_R < 0$. In this case, the dispersion relation in equation (8.127) can be written briefly as follows

$$\Delta' = aQ^{5/4} + bQ^{-1/4}, \tag{8.130}$$

where a and b are positive since $D_R < 0$. Similar to the analyses of figure 8.3, one can find that, in the case $a = b = 1$, the Δ mapping to the contour in figure 8.2 is given in figure 8.4. Other cases with a and b not being 1 are similar. Because of the sign change of D_R, the segment *BC* is flipped left to right and upside down.

From figure 8.4 one can see that the unstable roots occur only for Δ to the right of the point Δ_C, i.e.,

$$\Delta > \Delta_C,$$

and that there are two unstable roots in this case, in which $\Delta_C = \Delta(Q_C)$. To determine Q_C so that Δ_C, we first note that, since $\pm Q_C$ lie in the segments *AB* and *CD* in figure 8.2, Q_C must be purely imaginary, i.e.,

$$Q_C = \pm i|Q_C|.$$

By requiring $\Im m\{\Delta(Q_C)\} = 0$, one obtains

$$|Q_C| \equiv [(\pi/4)\tan(\pi/8)|D_R|]^{2/3} \approx 0.743|D_R|^{2/3}.$$

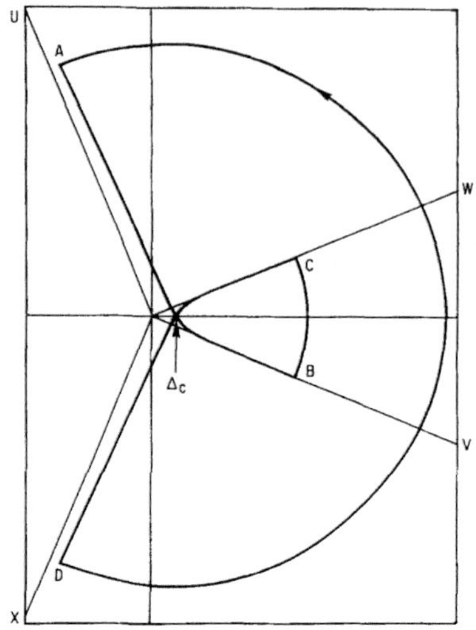

Figure 8.4 Mapping of Δ with the Q contour plot given in figure 8.2 for the case $D_R < 0$. It is used for the Nyquist diagram analyses. Reproduced from reference [1] with the permission of AIP Publishing.

Inserting it back to equation (8.127) yields

$$\begin{aligned}\Delta_C &\equiv \sqrt{2}\pi \csc(\pi/8)\,[\Gamma(3/4)\Gamma(1/4)]\,(V_s/X_0)Q_C^{3/4}.\\ &\approx 1/54(V_s/X_0)|D_R|^{5/6}.\end{aligned}$$

Note that Q_C scales as the interchange mode given in equation (8.129) and Δ_C scales as

$$\Delta_C \sim X_0^{-1} \sim \eta^{-1/3}.$$

One has $\Delta_C \to \infty$ as $\eta \to 0$. This indicates that the negative value of D_R stabilizes larger values of Δ'. This case is similar to the tearing mode in that it is driven by the unfavorable Δ', i.e., $\Delta' > 0$. It is similar to the interchange mode in that the scaling of Q and the role played by D_R. Nevertheless, it is different from both since Q can be complex. This mode is referred to as the modified tearing mode in view that the growth rate depends sensitively on Δ'.

In summary, the theory in reference [1] has been reviewed in this section. The resistive stability criteria for the case with $H = 0$ have been described. Reference [1] especially extended the analyses to the case with $H \neq 0$. Reference [1] classifies the mode types according to the growth rate scales with the resistivity as discussed in the cylinder case in Section 7.1. The conclusions are as follows. The interchange mode occurs if $D_R > 0$, the tearing mode occurs when $D_R = 0$ and $\Delta' > 0$, and the modified tearing mode occurs if $D_R < 0$ and $\Delta' > \Delta_C$. Actually, the mode classification should consider the parity of the modes. The even modes always cause reconnection as shown in Section 9.1 and, therefore, should be classified as the tearing modes.

8.2 RESISTIVE MHD BALLOONING MODES

In this section, we review the resistive ballooning mode theory. Historically, considerable efforts have been made on this topic. In this section, we focus on reviewing two systematic works, references [3] and [4]. They constitute a systematic theory of resistive MHD ballooning modes. We first derive the resistive ballooning mode equations using the formulation in references [3] and [4] and then derive the singular layer equations from the asymptotic analyses of the ballooning equations using the formulation in references [4]. The stability criteria of resistive MHD ballooning modes are given in the end.

8.2.1 RESISTIVE MHD BALLOONING MODE EQUATIONS

Let us now describe the derivation of the resistive MHD ballooning mode equations. The equations are presented in reference [3], consisting of the vorticity and acoustic or sound wave equations. The acoustic wave equation was further reduced in reference [4].

As in the ideal MHD case described in Section 6.6, Chapter 6, high n modes are considered, and the ballooning mode representation is employed. The flux coordinates (χ, θ, ζ) as described in Chapter 4 are employed. One thus has $\nabla_\perp \gg \nabla_\parallel$, $\nabla_\perp = in\nabla\beta$, with $\beta = n\zeta - q\theta$, and $\mathbf{B}\cdot\nabla = \chi'\partial/\partial\theta$.

As in the ideal MHD case in Section 6.6, since the compressional Alfvén modes are suppressed, one can introduce the so-called stream function $\delta\varphi$, such that

$$\boldsymbol{\xi}_\perp = \mathbf{B}\times\nabla\delta\varphi/B^2. \tag{8.131}$$

The two coupled resistive ballooning mode equations can be obtained from the vorticity and the parallel acoustic wave equations (8.5) and (8.8) with the reductions using the induction equation (8.2),

$$\delta\mathbf{B} \;=\; \frac{\nabla\times\boldsymbol{\xi}\times\mathbf{B}}{1+(\eta^*/\gamma)|\nabla\beta|^2} = -\frac{\nabla\times(\mathbf{BB}\cdot\nabla\delta\varphi/B^2)}{1+(\eta^*/\gamma)|\nabla\beta|^2}, \tag{8.132}$$

where $\eta^* = n^2\eta$. Note that for high n ballooning ordering, the second and third terms on the right-hand side of the vorticity equation (8.5) can be neglected. Using equation (8.132), one can obtain

$$\mathbf{B}\cdot\delta\mathbf{J} = \frac{\nabla^2(\mathbf{B}\cdot\nabla\delta\varphi)}{1+(\eta^*/\gamma)|\nabla\beta|^2}.$$

Using this result, the field line bending term, the first term on the right in equation (8.5), can be reduced. The reductions of test terms in the vorticity equation (8.5) are trivial. The reduction of the parallel acoustic wave equation (8.8) needs some explanations. The third term on the left can be reduced by noting that $\mathbf{B} \cdot \nabla^2 \delta \mathbf{B} = \nabla^2(\boldsymbol{\xi} \cdot \nabla P + \Gamma P \nabla \cdot \boldsymbol{\xi})$. To reduce the fourth term on the left-hand side of equation (8.8), it is noted that

$$\left(1 + \frac{\eta}{\gamma}\nabla^2\right) \nabla V \cdot \delta \mathbf{B} = \mathbf{B} \cdot \nabla(\boldsymbol{\xi} \cdot \nabla V) \rightarrow \mathbf{B} \cdot \nabla \delta \varphi.$$

This term can then be further reduced by substituting the bending term in the vorticity equation. Finally, by introducing the ballooning representation, equations (8.5) and (8.8) are reduced to

$$\begin{aligned} & \frac{\partial}{\partial \theta}\left(\frac{|\nabla \beta|^2}{1+(\eta^*/\gamma)|\nabla \beta|^2} \frac{1}{B^2} \frac{\partial}{\partial \theta} \delta \varphi\right) + \frac{2P'}{\chi'^4}(\kappa_n + q\theta \kappa_g)\delta \varphi - \frac{\rho_m \gamma^2}{\chi'^2 B^2}|\nabla \beta|^2 \delta \varphi \\ = & -\frac{2P'}{\chi'^4}(\kappa_n + q\theta\kappa_g)D, \qquad (8.133) \\ & \frac{\partial}{\partial \theta}\left(\frac{1}{B^2}\frac{\partial D}{\partial \theta}\right) - \frac{\rho_m \gamma^2}{\chi'^2} \frac{B^2+\Gamma P}{B^2 \Gamma P} D - \frac{\rho_m \eta^* \gamma}{\chi'^2 B^2}|\nabla \beta|^2 D - \frac{2P'}{\chi'^4}\frac{\eta^*}{\gamma}(\kappa_n + q\theta\kappa_g)D \\ = & \frac{2\rho_m}{P' \chi'^2}\left(\frac{P'^2 \eta^*}{\chi'^2 \gamma \rho_m} + \gamma^2\right)(\kappa_n + q\theta\kappa_g)\delta\varphi, \qquad (8.134) \end{aligned}$$

where $D = \Gamma P \nabla \cdot \boldsymbol{\xi}$, $\kappa_n = \boldsymbol{\kappa} \cdot \nabla \theta \times \nabla \zeta$, and $\kappa_g = \boldsymbol{\kappa} \cdot \nabla v \times \nabla \theta = -(\chi'/2P')\mathbf{B} \cdot \nabla \sigma$. Note also $|\nabla \beta|^2 = |\nabla \zeta|^2 - 2q'\theta \nabla V \cdot \nabla \zeta + q'^2 \theta^2 |\nabla V|^2$.

The basic set of equations (8.133) and (8.134) can be solved with the shooting method to determine the eigenvalue γ. This gives the global solution in the high n limit through the ballooning mode representation. In passing, the numerical scheme based on the decomposition of independent solutions is outlined as follows. Equations (8.133) and (8.134) can be transformed into a set of four first-order differential equations. The general solution can be expressed as the sum of four independent solutions:

$$\mathbf{u} = \sum_{k=1}^{4} c_k \mathbf{u}^k,$$

where $\mathbf{u}$ and $\mathbf{u}^k$ are vectors with four components, describing $(\varphi, d\varphi/d\theta; D, dD/d\theta)$, $\mathbf{u}^k$ $(k = 1, \cdots 4)$ are four independent solutions, and c_k are four constants to be determined by the boundary conditions. The solubility condition for c_k gives rise to the dispersion relation. Usually, multiple region matchings are required to achieve good numerical convergence.

8.2.2 ASYMPTOTIC ANALYSES AND SINGULAR LAYER EQUATIONS

In this subsection, the asymptotic analyses of the resistive MHD ballooning mode equations (8.133) and (8.134) are described. The analyses are similar to the ideal MHD case in Section 6.6. It is noted that the resistivity effects are only important in the singular layer. In the ballooning mode representation space, this corresponds to

$\theta \sim 1/\varepsilon_\theta \to \infty$ and is therefore referred to as the outer region. In the inner region with $\theta \sim 1$, the ideal MHD theory applies. Here, it is noted that the outer region in the ballooning formalism corresponds to the inner layer in the configuration description as treated in Section 8.1. We first lay out the ordering scheme. Next, the inner ideal MHD region is analyzed. The singular layer equations will be derived in the end of this subsection.

8.2.2.1 Inner ideal MHD region

In this subsection, we first analyze the inner ideal MHD region with $\theta \sim 1$. Since the resistivity and the inertia effects can be neglected in this region, the asymptotic analyses for the Mercier criterion in Section 6.5 can be used.

The basic set of equations (8.133) and (8.134) in the inner ideal MHD region is reduced to

$$\begin{aligned} &\frac{\partial}{\partial\theta}\left(|\nabla\beta|^2\frac{\partial}{\partial\theta}\delta\varphi\right)+\frac{2P'}{\chi'^4}(\kappa_n+q\theta\kappa_g)\delta\varphi \\ = &-\frac{2P'}{\chi'^4}(\kappa_n+q\theta\kappa_g)D, \qquad (8.135) \\ &\frac{\partial}{\partial\theta}\left(\frac{1}{B^2}\frac{\partial D}{\partial\theta}\right)=0. \qquad (8.136) \end{aligned}$$

Equation (8.136) yields $D = 0$ as discussed in Section 6.6.3. In this case, equation (8.135) becomes a second-order equation for $\delta\varphi$ with its right-hand side vanishes. The reduced equation has two independent solutions. From the asymptotic analyses for Mercier's criterion in Section 6.5, it is known that in the limit $\theta \to \infty$ the two independent solutions behave as the small and large solutions, so that the general solution can be expressed as

$$\delta\varphi = a_1|\theta|^s + a_2|\theta|^{-1-s}, \qquad (8.137)$$

where a_1 and a_2 are constants and $s = -1/2+\sqrt{-D_I}$ with $-D_I = 1/4+H^2-H-D_R$ and the index for resistive interchange mode being defined as follows:

$$D_R \quad = \quad F+E+H^2. \qquad (8.138)$$

Here, the definitions of parameters D_I, D_R, F, E, and H are the same as those in the Mercier criterion in Section 6.5 and in the resistive MHD singular layer equations in Section 8.1.

8.2.2.2 Outer resistive MHD region

To proceed with asymptotic analyses in the outer resistive MHD region, let us first lay out the ordering scheme. We consider the optimal ordering scheme: $(\eta^*/\gamma)|\nabla\beta|^2 \sim 1$. In this case, two terms in the denominator of the first term in equation (8.133) are of the same order. Besides, the low-frequency regime is considered. The balance between the field line bending term and the inertia term determines the width of singular layer, i.e., the small parameter ε_θ, as can be seen in the ordering analyses later on in equation (8.143). These lead to the following ordering assumptions:

$$a^2\eta^*/\varepsilon_\theta^2 \sim \omega, \quad \varepsilon_\theta \sim \beta_t^{1/2}(\omega/\omega_{si}), \; \omega \ll \omega_{si}. \tag{8.139}$$

Here, the minor radius a is introduced for dimensionless processing, β_t is the plasma beta, and $\omega_{si} \sim v_{thi}/Rq$ is the ion acoustic frequency, with q assumed to be of order unity and v_{thi} being the ion thermal velocity. Noting that the plasma beta is usually small, it is assumed in addition that

$$\beta_t \lesssim a/R. \tag{8.140}$$

With the basic ordering assumptions given, one can analyze the basic set of equations (8.133) and (8.134). Let us first examine the sound wave equation (8.134). Using the orderings in equations (8.139) and (8.140), one can find the relative orderings of four terms on the left-hand side of equation (8.134) as follows:

$$\omega_{si}^2, \; \omega^2, \; \omega^2\beta_t, \; \omega_{si}\omega\beta_t^{1/2}, \tag{8.141}$$

One can also find the ordering for the ratio of two terms in the bracket on the right-hand side of (8.134) as follows

$$\frac{\omega^2}{p'^2\eta^*/\omega\rho_m} \sim \frac{a/R}{\beta_t}. \tag{8.142}$$

Next, let us consider equation (8.133). Using the orderings in equations (8.139) and (8.140), one can also estimate the orders of the first, second, and last terms in equation (8.133), respectively, as follows

$$\frac{1}{\varepsilon_\theta^2}, \; \beta_t\frac{R}{a}\frac{1}{\varepsilon_\theta}, \; \beta_t\frac{\omega^2}{\omega_{si}^2}\frac{1}{\varepsilon_\theta^2}. \tag{8.143}$$

The term on the right of equation (8.133), which gives the so-called apparent mass effects, can be proved *a posteriori* to be of the same order as the last term on the left, i.e., the perpendicular inertia term.

Here, it should be pointed out that in these ordering considerations, the orderings for beta and aspect ratio are not essential. In the ordering sequence in equation (8.141) for the sound wave, the only necessary ordering is the low frequency regime, i.e., $\omega_{si} \gg \omega$. In the ordering sequence in equation (8.143) for the vorticity equation or shear Alfvén wave, the only necessary ordering is the mode localization, i.e., $\varepsilon_\theta \gg 1$.

With the ordering scheme given, we can proceed to analyze the ballooning mode equations. As in the ideal MHD case, two length scale analyses are performed with $(\theta, \varepsilon_\theta z)$, where z corresponds to the slowly varying scale. The two unknowns are expanded as follows:

$$\begin{aligned} \delta\varphi &= \delta\varphi_0 + \varepsilon_\theta\delta\varphi_1 + \cdots, \\ D &= D_0 + \varepsilon_\theta D_1 + \cdots. \end{aligned}$$

We first treat the acoustic wave equation (8.134). In the lowest order, the equation becomes:

$$\frac{\partial}{\partial\theta}\left(\frac{1}{B^2}\frac{\partial D_0}{\partial\theta}\right)=0.$$

Its solution is simply

$$D_0 = D_0(z). \tag{8.144}$$

In the next order, the sound wave equation (8.134) becomes

$$\begin{aligned}&\frac{\partial}{\partial\theta}\left[\frac{1}{B^2}\left(\frac{\partial D_1}{\partial\theta}+\frac{\partial D_0}{\partial z}\right)\right]+\rho\omega^2\frac{1}{\gamma P}D_0-\frac{2P'}{\chi'^4}\frac{\eta^*}{\gamma}q\theta\kappa_g D_0\\ = \;&\frac{2\rho_m}{P'\chi'^2}\left(\frac{P'^2\eta^*}{\chi'^2\gamma\rho_m}+\gamma^2\right)q\theta\kappa_g\delta\varphi_0.\end{aligned} \tag{8.145}$$

Note here that the second term in this equation is of order $(\omega_{si}/\omega)\beta_t^{1/2}$, which is much larger than ε_θ, and is therefore kept. To solve equation (8.145), we first point out that it can be proved *a posteriori* that $\delta\varphi_0$ in the lowest order is independent of θ. For simplicity, we only discuss the case with the assumption $D_0 = 0$ in reference [4]. Note that $\kappa_g \propto \partial\sigma/\partial\theta$. As shown in the ideal MHD case in Section 6.6.3, equation (8.145) can be solved, yielding

$$\frac{\partial D_1}{\partial\theta} = \frac{\rho_m}{P'^2}\left(\frac{P'^2\eta^*}{\chi'^2\gamma\rho_m}+\gamma^2\right)\left(B^2\sigma-\frac{B^2\langle B^2\sigma\rangle}{\langle B^2\rangle}\right)qz\delta\varphi_0. \tag{8.146}$$

Let us now deal with the vorticity equation (8.133). In the lowest order, one has

$$\frac{\partial\delta\varphi_0}{\partial\theta}=0.$$

This yields

$$\delta\varphi_0 = \delta\varphi_0(z). \tag{8.147}$$

Using the solution in equation (8.147), equation (8.133) in the first order can be solved

$$\begin{aligned}&\delta\varphi_1(\theta,z)\\ =\;& f_1(z)+\frac{d\delta\varphi_0/dz}{\langle B^2/|\nabla V|^2\rangle+(\eta^*/\gamma)\langle B^2\rangle q'^2z^2}\int\left[\widetilde{\left(\frac{B^2}{|\nabla V|^2}\right)}+\frac{\eta^*}{\gamma}q'^2z^2\widetilde{B^2}\right]d\theta\\ &+\frac{\delta\varphi_0}{\chi'^2q'z}\int\left[\widetilde{\left(\frac{B^2}{|\nabla V|^2}\right)}+\widetilde{\left(\frac{B^2\sigma}{|\nabla V|^2}\right)}+\frac{\eta^*}{\gamma}\left(\widetilde{B^2}+\widetilde{B^2\sigma z^2}\right)\right]d\theta,\end{aligned} \tag{8.148}$$

where $\tilde{a} = a - \langle a \rangle$. In the second order, the average of equation (8.133) over the fast variable θ gives the solubility condition for $\delta\varphi_2$:

$$\frac{d}{dz}\frac{x^2}{1+x^2}\frac{d\delta\varphi_0}{dx} + \frac{H(1-H)}{(1+x^2)^2}\delta\varphi_0 - \frac{H(1+H)x^2}{(1+x^2)^2}\delta\varphi_0$$
$$+D_R\delta\varphi_0 - Q^3x^2\delta\varphi_0 = 0. \tag{8.149}$$

Here, equations (8.146) and (8.147) have been used; the definition of Q is given in equation (8.36) in Section 8.1, except with the replacement

$$M_t \to \left(1 + \frac{p'^2\eta^*}{\gamma^3\chi'^2\rho_m}\right)M_t \tag{8.150}$$

for the parallel inertia parameter M_t inside the definition of M, and $x = [\eta^* q'^2 \langle B^2\rangle/\gamma\langle B^2/|\nabla V|^2\rangle]^{1/2}z$.

Now, let us proceed to solve equation (8.149). Introducing the following transformation

$$\begin{aligned}\delta\varphi_0 &= |x|^s \exp\left\{(1+s-H)\frac{x^2}{2}\right\}\\ &\times\frac{d}{d(x^2)}\left[\exp\left\{(H-1-s-Q^{3/2})\frac{x^2}{2}\right\}P(r),\right],\end{aligned} \tag{8.151}$$

where $r = Q^{3/2}x^2$, equation (8.149) is transformed to the Kummer equation [9]:

$$r\frac{dP}{dr} + \left(\frac{1}{2}+s-r\right)\frac{dP}{dr} - \frac{1}{4}\left(Q^{3/2}+1+2s-\frac{D_R}{Q^{3/2}}\right)P = 0. \tag{8.152}$$

The solution, which remains finite and thus is adaptable to the boundary conditions at ∞, is

$$P(r) \propto \frac{\pi}{\sin\pi\nu}\left[\frac{{}_1F_1(a^*,\nu,r)}{\Gamma(1+a^*-\nu)\Gamma(\nu)} - r^{1-\nu}\frac{{}_1F_1(1+a^*-\nu,2-\nu,r)}{\Gamma(a^*)\Gamma(2-\nu)}\right], \tag{8.153}$$

where $\nu = 1/2+s$, $a^* = (Q^{3/2}+2\nu - D_R/Q^{3/2})/4$, Γ is the gamma function, and ${}_1F_1$ is the Kummer function

$${}_1F_1(a^*,\nu,r) = 1 + \frac{a^*}{\nu}r + \frac{a^*(a^*+1)}{\nu(\nu+1)}\frac{r^2}{2!} + \cdots.$$

Therefore, the solution of equation (8.151) becomes

$$\begin{aligned}\delta\varphi_0 &= \frac{\pi|x|^s\exp\{-Q^{3/2}z^2/2\}}{\sin\pi\nu\Gamma(1+a^*-\nu)\Gamma(\nu)} \\ &\times\left\{\frac{Q^{3/2}-1-s+H}{2}{}_1F_1(a^*,\nu,r)+\frac{a^*-\nu}{\nu}{}_1F_1(a^*,\nu+1,r)\right. \\ &-\frac{\Gamma(\nu)\Gamma(1+a^*-\nu)}{\Gamma(1-\nu)\Gamma(a^*)}Q^{\frac{3}{2}\left(\frac{1}{2}-s\right)}|x|^{-1-2s}\left[{}_1F_1(a^*,1-nu,1-\nu,r)\right. \\ &\left.\left.+\frac{Q^{3/2}-1-s+H}{2(1-\nu)}x^2{}_1F_1(1+a^*-\nu,2-\nu,r)\right]\right\}. \end{aligned} \tag{8.154}$$

Note here that in the inner limit $|x| \to 0$, the last term, which is proportional x^2, can be neglected, so that the solution in equation (8.154) exhibits the familiar behavior $|x|^s$ and $|x|^{-1-s}$ as the outer limit of inner solution in equation (8.137), which will be used to study the stability condition in the next subsection.

8.2.3 DISPERSION RELATION

In this subsection, the inner and outer solutions obtained in the last two subsections are matched to derive the dispersion relation. The dispersion relation is then used to derive the stability conditions in various scenarios [4].

To construct a global solution, the ideal MHD solution in the inner region in equation (8.137) for $\theta \to \infty$ and the resistive MHD solution in the outer region in equation (8.154) for $x \to 0$ are matched to derive the dispersion relation. For the matchings, it is noted that the coordinates in the two regions are related as follows: $x^2 = (1/\theta_0^2 Q)\theta$, where $\theta_0^2 = \left\langle B^2/|\nabla V|^2\right\rangle Q_0/q'^2\left\langle B^2\right\rangle \eta^*$ and Q_0 is given in equation (8.36). Also, it is noted that the matching between the nearby regions is carried out with the ratio of the big to the small solutions.

In the ideal MHD region, from equation (8.137) in the large θ limit, one obtains ratio

$$\Delta' = a_2/a_1. \tag{8.155}$$

In the resistive region, the ratio can be found from equation (8.154) in the small x limit:

$$\Delta(Q) = \frac{4y_0^{1+2s}Q^{(5-2s)/4}}{Q^3-(1+s-H)^2}\frac{\Gamma(1/2+s)}{\Gamma(-1/2-s)}\frac{\Gamma[(Q^{3/2}+3-2s-D_R/Q^{3/2})/4]}{\Gamma[(Q^{3/2}+1+2s-D_R/Q^{3/2})/4]}. \tag{8.156}$$

Therefore, matching (8.155) and (8.156), one obtains the solubility condition or dispersion relation

$$\Delta' = \Delta(Q). \tag{8.157}$$

In fact, the matching between equations (8.155) and (8.156) needs to be made on two sides, i.e., $\theta \to \pm\infty$. This is doable. Note that the differential equation (8.135) (with $D = 0$) in the inner ideal MHD region is of the second order. Therefore, there are two independent solutions, which we choose to be even and odd solutions: f_e and f_o. Their outer limits can be expressed as

$$\lim_{|y|\to\pm\infty} f_{e,o} \quad = \quad A^{\pm}_{e,o}|y|^s + B^{\pm}_{e,o}|y|^{-1-s}.$$

One then determines a constant factor c from the following equation

$$\frac{B_e^+ + cB_o^+}{A_e^+ + cA_e^+} = \frac{B_e^- + cB_o^-}{A_e^- + cA_e^-}.$$

It can be proved that this equation always has a real solution c using the property of Wronskian of equation (8.135) (with $D = 0$). With the contact c determined, one can construct the inner solution as

$$F_0 = f_e(y) + cf_o(y). \tag{8.158}$$

This makes the matching conditions on the left and right identical. Besides, one can determine the parity of the eigenmode.

The dispersion relation, equation (8.157), can be used to study the resistive MHD stability condition. We first consider the case with $D_R > 0$. In this case, the Gamma functions in equation (8.156) have an infinite sequence of poles due to the term $-D_R/4Q^{3/2}$. The parameter Δ in the resistive MHD region vanishes and diverges alternately, passing through all values many times. There are infinitely many roots Q for the dispersion relation, equation (8.157), for any given Δ' in the ideal MHD region. This indicates that there are resistive instabilities. Since y_0 scales as $\eta^{*-1/3}$, the factor y_0^{1+2s} in equation (8.157) is large. Consequently, the roots of equation (8.157) are in general near the poles of $\Gamma[(Q^{3/2} + 3 - 2s - D_R/Q^{3/2})/4]$, that is

$$Q_k \quad = \quad -(1/2 + s - 2k) + [(1/2 + s + 2k)^2 + D_R]^{1/2},$$

where k labels the root sequence. Noting that $\gamma \sim \eta^{*1/3}Q$, the actual growth rate γ scales as $\eta^{*1/3}$. This kind of scale corresponds to the resistive interchange modes mentioned in reference [1]. However, its mode parity is actually determined by the parity of eigenfunction expressed in equation (8.158).

Next, we consider the case with $D_R = 0$. In this case, the term $(D_R/Q^{3/2})/4$ no longer appears in the Gamma function in equation (8.156). One cannot use the analyses for the case with $D_R > 0$ discussed earlier. To study this case, we first need to determine the range s. From the definition $s = -1/2 + \sqrt{-D_I}$, one can conclude $s > -1/2$. The upper bound of s can be determined by the matching conditions. There are three asymptotic solutions in equation (8.154): x^s, x^{-1-s}, and x^{1-s}. Only the first two match the inner ideal MHD asymptotic solution in equation (8.137). To eliminate the last solution as compared to the first two solutions as $|x| \to 0$, it is required $s < 1/2$. Therefore, in summary, one has $-1/2 < s < 1/2$. This avoids the case with the arguments of the Gamma function equation (8.156) being negative integers or zero — the undefined cases for Gamma function. Since $\Gamma(1/2+s)/\Gamma(-1/2-s) < 0$, one can find that, when $0 \leq Q^3 \leq (1+s-H)^2$, Δ in equation (8.156) takes all positive values between 0 and $+\infty$. Therefore, one can always find the unstable modes as soon as $\Delta' > 0$ in this case.

Lastly, let us discuss the case with $D_R > 0$. In this case, the scenario for the case $D_R < 9$ about the vanishing and diverging of the Gamma function does not apply. The case with small Q discussed for the case $D_R =$ leads to an interesting phenomenon. When Q is small, the term $|D_R|/4Q^{3/2}$ in the arguments of the Gamma functions in equation (8.156) becomes very large. Using the method of steepest descent [7], one obtains the asymptotic limit for the Gamma function ratio in equation (8.156)

$$\frac{\Gamma[(Q^{3/2}+3-2s-D_R/Q^{3/2})/4]}{\Gamma[(Q^{3/2}+1+2s-D_R/Q^{3/2})/4]} \sim \left(\frac{|D_R|}{4Q^{3/2}}\right)^{1/2-s}. \tag{8.159}$$

Using this result, one can obtain from equation (8.156)

$$\Delta \propto \left(y_0^2 Q\right)^{1/2+s}.$$

With this result, it seems that the unstable modes could result with $\Delta' > 0$. However, one should note that there is a condition for these analyses to apply: $D_R/4Q^{3/2} \gg 1$, not to mention that the ordering scheme described in subsection 8.2.2.2 can also impose some limitations on Q. Therefore, one usually has the instability condition: $\Delta' > \Delta(Q_c)$ with a critical value Q_c. Note that when $D_R/4Q^{3/2}$ is moderately larger than 1, equation (8.159) represents a good asymptotic estimate. One can use $D_R/4Q_c^{3/2} \approx 1$ to determine Q_c.

8.3 CONCLUSIONS AND DISCUSSION

In this chapter, the resistive MHD theories are reviewed. Mainly, two theories are described: the resistive MHD theory by Glasser, Greene, and Johnson [1], in which the vorticity equation approach is used [5], and the ballooning mode theory by Correa-Restrepo [4]. There are more efforts on this topic. Nevertheless, these two works represent systematic theories in this field. These theories extend the cylinder theory in Chapter 7 to toroidal geometry.

Here, it should be pointed out that using the growth rate scaling $\eta^1/3$ or $\eta^{3/5}$ cannot be used to classify the interchange or tearing modes. The even modes can always cause field line reconnection. The ballooning mode theory in reference [4] indicates that both interchange and tearing modes, or hybrid, can have either type of scaling, *i.e.*, $\eta^1/3$ or $\eta^{3/5}$. This seems to extend the resistive MHD theory in reference [1] from the ballooning mode formalism. Reference [1] only analyzes the even modes.

Another difference between the singular layer theory in reference [1] and the resistive ballooning mode theory in reference [4] lies in the treatment of the parallel momentum equation. Reference [4] uses the acoustic wave equation (8.8) and instead reference [1] uses the parallel momentum equation directly. In the ideal MHD case, it is shown in Section 6.5.1 that both approaches give identical results. In the resistive MHD case, they should do the same, although the analyses remain to be detailed. Using the acoustic wave equation (8.8), one can solve the oscillating part $\widetilde{\nabla \cdot \boldsymbol{\xi}}$. This can modify the apparent mass and cause the changes of the scaling laws $\eta^1/3$ or $\eta^{3/5}$.

From the vorticity and acoustic wave equations, (8.5) and (8.8), one can see that there are two frequency scalings: ω^2/ω_A^2 and ω^2/ω_s^2. Note that $\omega_s^2/\omega_A^2 \sim \beta$. Since plasma beta β usually is small, one may use it as auxiliary ordering for analyzing the resistive MHD modes.

Let us now discuss the kinetic effects, especially to the so-called small-parallel-ion-velocity effects [10, 11, 12]. In discussing the optimal resitivity ordering scheme in equation (8.139) in Section 8.2.2.2, one can see that two main factors are considered. The first is related to the induction equation, and the order is associated with the vorticity equation. The balance between the field line bending term and the inertia term determines the width of the singular layer. Note that there is a sound wave coupling in the vorticity equation, which gives rise to the so-called apparent mass effect. Due to the so-called small-parallel-ion-velocity effects, the apparent mass effect is enhanced in the kinetic description. Therefore, in the ordering scheme given in equation (8.139) in Section 8.2.2.2, ε_θ should be rescaled. This affects the optimal resistive MHD orderings. Certainly, there are also the finite Larmor radius and the parallel electric effects in the kinetic description. They have been discussed in reference [13].

References

1. Glasser A H, Greene J M, Johnson J L 1975 Resistive instabilities in general toroidal plasma configurations, *Phys. Fluids 18*, 875–88. https://doi.org/10.1063/1.861224
2. Johnson J L, Greene J M 1967 Resistive interchanges and the negative v'' criterion, *Plasma Phys. 9*, 611. DOI 10.1088/0032-1028/9/5/311
3. Chance M S *et al.* 1979 MHD stability limits on high-beta tokamaks, Proc. 7th Int. Conf. Plasma Physics and Controlled Fusion Research (Innsbruck, Austria, 23–30 August, 1978) vol 1 (Vienna: International Atomic Energy Agency) p 677.
4. Correa-Restrepo D 1982 Resistive ballooning modes in three-dimensional configurations, *Z. Naturforsch. A 37*, 848.
5. Zheng L 2015 Advanced Tokamak Stability Theory (London: IOP Publishing & San Rafael, CA: Morgan & Claypool). https://dx.doi.org/10.1088/978-1-6270-5423-2
6. Watson G N, 1944 A treatise on the theory of Bessel Functions (London: Cambridge University Press) 2nd ed. p. 160.
7. Wang Z X, Guo D R 1989 Special Functions (Singapore: World Scientific Publishing).
8. Polyanin A D, Zaitsev V F 2018 Handbook of ordinary differential equations, exact solutions, method, and problems, (Boca Raton, FL: CRC Press, Taylor & Francis Group).
9. Abramowitz M, Stegun I A (ed) 1972 Handbook of Mathematical Functions (Washington, D.C.: National Bureau of Standards).
10. Mikhailovsky A 1974 Theory of Plasma Instabilities (New York: Consultants Bureau).
11. Zheng L-J, Tessarotto M 1994 Collisionless kinetic ballooning mode equation in the low-frequency regime *Phys. Plasmas 1*, 3928–35. https://doi.org/10.1063/1.870865.

12. Zheng L-J, Tessarotto M 1996 Collisional effect on the magnetohydrodynamic modes of low frequency *Phys. Plasmas 3*, 1029–37. https://doi.org/10.1063/1.871758.
13. Zheng L 2022 Magnetically Confined Fusion Plasma Physics, Volume 3, Kinetic theory (London: IOP Publishing). https://doi.org/10.1088/978-0-7503-3779-3.

9 Nonlinear stability theories

In this chapter, major nonlinear stability theories are reviewed. We first give a general description of the formation of a magnetic island in the presence of a radial magnetic perturbation. This depicts the basic picture of magnetic reconnection in a shear magnetic field. The nonlinear tearing mode theory by P. H. Rutherford will be described next. This brings about the well-known Rutherford equation. The neoclassical tearing mode theory is described afterward. The Kadomtsev theory for the interpretation of giant sawteeth will be described further next, followed by the theory of current interchange tearing modes. After reviewing these major nonlinear tearing mode theories, the couplings to the external wall/divertor structure and the inductive electric field are discussed. These provide interpretations of the so-called edge localized modes and the ear-like reversed pressure profiles during the off-axis heating. The blob transport and edge harmonic oscillations will be described further on. The forced reconnection will also be discussed. The conclusions and discussion are given in the end.

9.1 MAGNETIC ISLAND

In this section, the magnetic island formation and geometry are discussed [1]. These help understand the nonlinear stability theories to be described in this chapter, especially for the nonlinear tearing modes.

Let us first explain the so-called helical magnetic field in a toroidal system with magnetic shear. In figure 9.1, the composition of magnetic field is displayed on a magnetic surface, in which B_ϕ and B_θ are respectively toroidal and poloidal magnetic fields, R and r are respectively major and minor radii, and $q = m/n$ with m and n are respectively the numbers of poloidal and toroidal transits of a magnetic field line. The helical magnetic field, $\mathbf{B}^*$, is the projection of magnetic field onto the direction perpendicular to the magnetic field on the rational surface. Noting that the magnetic field contains toroidal and poloidal components, the projection should contain the respective two parts:

$$B^* = B_\phi \sin\theta_\phi - B_\theta \sin\theta_\theta. \tag{9.1}$$

From figure 9.1, one can see that

$$\sin\theta_\theta = \frac{2\pi r/m}{[(2\pi R/n)^2 + (2\pi r/m)^2]^{1/2}},$$

$$\sin\theta_\phi = \frac{2\pi R/n}{[(2\pi R/n)^2 + (2\pi r/m)^2]^{1/2}}.$$

DOI: 10.1201/9781003522133-9

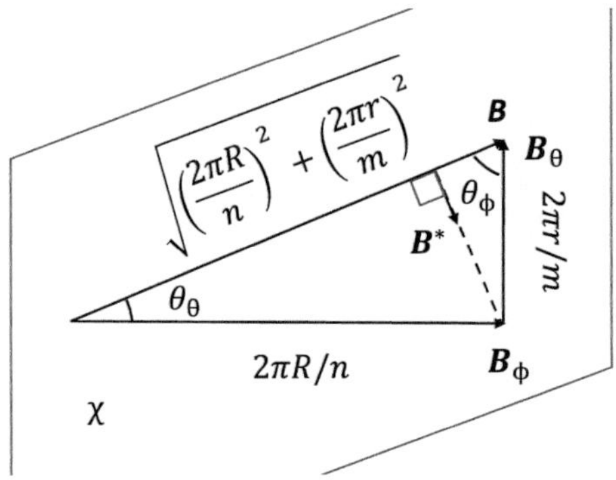

Figure 9.1 Helical magnetic field $\mathbf{B}^*$.

Using these formulas, the helical magnetic field in equation (9.1) is reduced to

$$\begin{aligned} B^* &= B_\phi \frac{n}{m}\frac{r}{R} - B_\theta \\ &= -B_\theta\left(1-\frac{n}{m}q\right) = B_\theta \frac{q'}{q}x, \end{aligned} \tag{9.2}$$

where $x = r - r_s$ with r_s being the minor radius of the rational surface.

Next, let us discuss how the magnetic island is formed when a perturbed radial magnetic field is present in the helical magnetic field

$$B_x = -B_x(r)\sin ky. \tag{9.3}$$

The change of magnetic topology in the presence of the radial magnetic perturbation is illustrated in figure 9.2. To gain a more intuitive understanding of island structures,

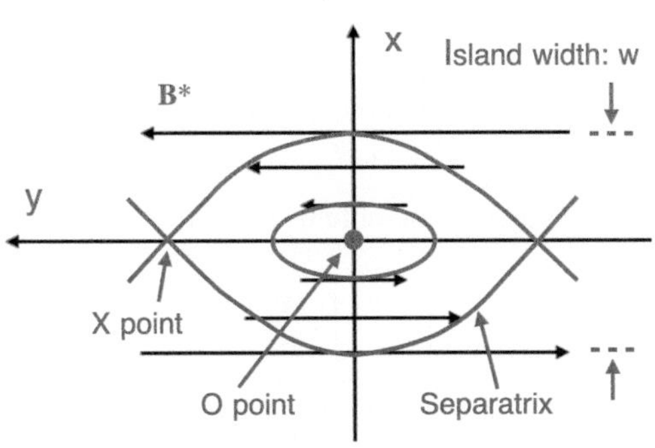

Figure 9.2 Magnetic island.

try adding $\mathbf{B}^*$ and $\mathbf{B}_x$ together to see how the sland structure is formed in figure 9.2. Mathematically, the trajectory of a magnetic field line is determined by the equation

$$\frac{1}{r_s}\frac{dx}{dy} = \frac{B_x}{B^*}.$$

Inserting equations (9.2) and (9.3) yields

$$B_\theta \frac{q'}{q} x dx = -r_s B_x \sin ky dy.$$

This can be integrated to obtain

$$x^2 = \frac{w^2}{8}\left(\cos ky - \cos ky_0\right), \tag{9.4}$$

where the island width as shown in figure 9.2 is given by

$$w = 4\left(\frac{r_s q B_r}{k q' B_\theta}\right)^{1/2}.$$

In equation (9.4), the integration constant labels the magnetic surfaces of the island. $y_0 = 0$ specifies the "O" point and $ky_0 = \pm\pi$ dtermine the "X" points .

The magnetic island can also be described by its local helical flux ψ

$$\mathbf{B} = \mathbf{z} \times \nabla\psi + I\mathbf{z},$$

where I represents the guiding field along the magnetic island. It can be found that the helical flux

$$\psi = \psi_0(r) + \hat{\psi}\cos(ky),$$

where

$$\psi_0 = \frac{1}{2}\left(B_\theta \frac{q'}{q}\right)x^2 \quad \text{and} \quad \hat{\psi} = -\frac{rB_r}{k}.$$

Therefore, one has

$$x^2(\psi) = \frac{2q}{B_\theta q'}\left(\psi - \hat{\psi}\cos my\right).$$

Since $\psi = -\psi$ on the separatrix , one obtains the island width

$$w = 4\left|\frac{q\hat{\psi}}{q' B_\theta}\right|^{1/2}.$$

We show the formation of a magnetic island in a sheared magnetic field by assuming there is a radial magnetic perturbation. The radial magnetic perturbation can be an error field or the perturbed magnetic field due to the excitation of the tearing modes, as discussed in the following subsections.

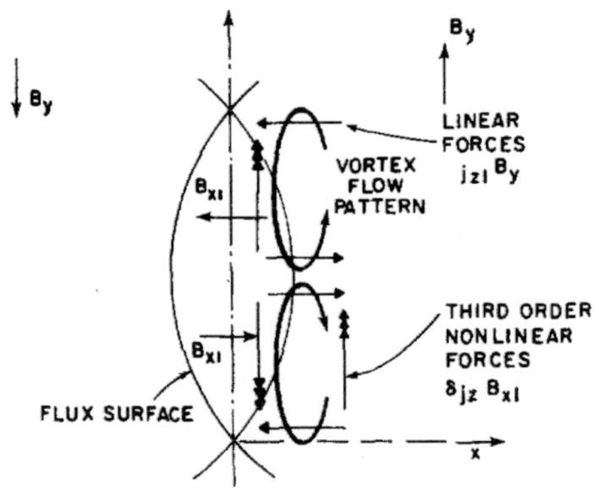

Figure 9.3 Tearing mode structure with ψ being even and v_x odd in x. Reproduced from reference [2] with the permission of AIP Publishing.

9.2 NONLINEAR TEARING MODES: RUTHERFORD'S EQUATION

In chapters 7 and 8, we have discussed the importance of the resistivity effects in the singular layer. In this section, we describe the nonlinear tearing theory by P. H. Rutherford [2]. The nonlinear phase of tearing mode is of interest in tokamak physics [1], especially it is related to the major tokamak MHD instabilities. We first derive the so-called Rutherford equation. The nonlinear effects will be discussed afterward.

The equilibrium model in reference [2] is similar to that described in Section 9.1. Figure 9.3 shows the local coordinate system (x,y,z), in which x labels the unperturbed magnetic surface, and y is the binormal direction to the guiding magnetic field, which is in the z direction. Without losing generality, one can use the relative field to study the tearing modes by subtracting the guiding field. In this case, the equilibrium magnetic field can be expressed as $\mathbf{B}^* = (0, B_y' x, 0)$. Here, prime denotes the derivative with respect to x. It is assumed that there is a finite magnetic shear with $B_y' > 0$.

In view of the divergence-free constraint of magnetic field $\nabla \cdot \mathbf{B} = 0$, the poloidal flux ψ can be introduced to represent the magnetic field: $\mathbf{B}^* = \nabla z \times \nabla \psi$, *i.e.*, $B_x = -\partial\psi/\partial y$, $B_y = \partial\psi/\partial x$. Also, due to the suppression of compressional Alfvén wave $\delta\mathbf{B} + \delta P = 0$, one has $\nabla \cdot \mathbf{v}_\perp = 0$ in the leading order (using for instance equation (7.6)). Therefore, the velocity stream function φ can be introduced to represent the perpendicular velocity: $\mathbf{v}_\perp = \nabla z \times \nabla \phi$, *i.e.*, $v_x = -\partial\varphi/\partial y$, $v_y = \partial\varphi/\partial x$.

We decompose the total magnetic flux as follows

$$\psi(x,y,t) \quad = \quad \psi_0(x) + \delta\psi(y,t), \tag{9.5}$$

where $\psi_0(x)$ is the equilibrium value with subscript 0 and $\delta\psi(y,t)$ is the perturbed value tagged by "δ". In the case of no ambiguity to the total value, the subscript 0 may be dropped. Note that the equilibrium magnetic field with the guiding field along ∇z subtracted contains only the $B_y(x)$ component and reverses its direction across the singular layer $x = 0$ as shown in figure 9.3, *i.e.*, $B_y(x) = B'_y x$. It corresponds to the helical field described in Section 9.1. One can find that the equilibrium magnetic flux can be expressed as

$$\psi_0(x) = B'_y x^2/2. \tag{9.6}$$

We introduce the Fourier decomposition for perturbed quantities in the helical direction, such that

$$\delta\psi(y,t) = \delta\psi_1(t)\cos ky \tag{9.7}$$

with k the poloidal wave number. For the current interest, it is sufficient to keep only the first harmonic.

To understand the island structure in the nonlinear analyses in reference [2], one needs to relate the total flux to the Cartesian coordinates. Using equations (9.6) and (9.7), one can transform the total flux coordinate ψ to the radial Cartesian coordinate

$$x = \pm\sqrt{\frac{2}{B'_y}}\sqrt{\psi - \delta\psi}. \tag{9.8}$$

Introducing further the definitions of the normalized flux coordinate $w = \psi/\delta\psi_1$ and $x_T = 2\sqrt{\delta\psi_1/B'_y}$ and using equation (9.7), equation (9.8) becomes

$$x = \pm\frac{x_T}{2}\sqrt{w - \cos ky}. \tag{9.9}$$

The picture described here is similar to what is laid out in Section 9.1.

Equation (9.9) can be used to discuss the island structure in the Cartesian coordinates. As shown in figure 9.3, inside the magnetic island, magnetic field lines are confined within the island's boundary, the separatrix. These field lines are closed and form a loop, rotating around a central point (the O-point). The separatrix separates the surrounding open field lines from the reconnected magnetic field lines within the island. In equation (9.9), the "O" point of the island is located at $w = -1$, and $ky = \pi$, and the two "X" points are at $w = 1$ and $ky = 0, 2\pi$, respectively. Inside the magnetic island, one has $-1 \leq w \leq 1$ and ky is constrained by $w - \cos ky \geq 0$. Outside the magnetic island, w ranges from 1 to $+\infty$ and $0 \leq ky \leq 2\pi$. One can see that the half island width is x_T, which can be obtained by choosing $w = 1$ and $ky = \pi$.

Let us now derive the so-called Ruthford equation, *i.e.*, the nonlinear tearing mode equation. Ampere's law, Ohm's law, and Faraday's law are used to construct the basic set of equations

$$\mu_0 \mathbf{J} = \nabla \times \mathbf{B}, \tag{9.10}$$

$$\eta \mathbf{J} = \mathbf{E} + \mathbf{v} \times \mathbf{B}, \tag{9.11}$$

$$\partial \mathbf{B}/\partial t = \nabla \times \mathbf{E}. \tag{9.12}$$

Linearizing equation (9.10) and projecting it on ∇z yields

$$\frac{d^2\delta\psi}{dx^2} = \mu_0 \delta j_z, \tag{9.13}$$

Here, one has noted that $\nabla z \cdot \nabla \times \delta\mathbf{B} = \nabla x \cdot \nabla \times (\nabla z \times \nabla\delta\psi) = \nabla^2\delta\psi$, and the derivative on x is dominant. Applying the curl operation on the linearized Ohm's law in equation (9.11), using Faraday's law in equation (9.12) to express $\nabla \times \delta\mathbf{E}$, and the projection of the resulting equation on ∇x, one obtains

$$\frac{\partial\delta\psi}{\partial t} - \left.\frac{\partial\delta\varphi}{\partial y}\right|_{\psi} B_y' x = \eta\delta j_z. \tag{9.14}$$

Here, it has been noted that $\nabla x \cdot \nabla \times \mathbf{v} \times \mathbf{B} = \mathbf{B} \cdot \nabla v_x$. Thus, the derivative of $\delta\varphi$ in the second term is on the surface with constant ψ. Also, the current or resistivity gradient effect has not been included. Furthermore, using the vorticity equation (7.5) in the lowest order, one can obtain $\mathbf{B} \cdot \nabla j_z = 0$, i.e.,

$$j_z = j_z(\psi). \tag{9.15}$$

In deriving the Rutherford equation, we tried to eliminate the second term in equation (9.14). Dividing by x and averaging over y at constant ψ and using equation (9.15), equation (9.14) becomes

$$j_z(\psi) = \frac{\langle \partial\delta\psi/\partial t/(\psi-\delta\psi)^{1/2}\rangle}{\eta\langle 1/(\psi-\delta\psi)^{1/2}\rangle}. \tag{9.16}$$

where $\langle\cdots\rangle = (k/2\pi)\int_0^{2\pi/k}\{\cdots\}dy$ with ψ or w holding constant. Here, we have used equation (9.9) to express x, and noted the periodicity of φ in y. Inserting equation (9.13), equation (9.16) becomes

$$\frac{1}{\mu_0}\frac{\partial^2\delta\psi}{\partial x^2} = \frac{\langle \partial\delta\psi/\partial t/(\psi-\delta\psi)^{1/2}\rangle}{\eta\langle 1/(\psi-\delta\psi)^{1/2}\rangle}. \tag{9.17}$$

Further on, integrating over x from $-\infty \to +\infty$, equation (9.17) becomes

$$\left.\frac{\partial\delta\psi}{\partial x}\right|_{-\infty}^{\infty} = -\frac{\mu_0}{\eta\sqrt{2B_y'}}\int_{-\infty}^{\infty}\frac{d\psi}{(\psi-\delta\psi)^{1/2}}\frac{\langle \partial\delta\psi/\partial t/(\psi-\delta\psi)^{1/2}\rangle}{\langle 1/(\psi-\delta\psi)^{1/2}\rangle}.$$

Multiplying $\cos ky$ and averaging over y, this equation is reduced to

$$\frac{\left.\frac{\partial\delta\psi_1}{\partial x}\right|_{-\infty}^{\infty}\delta\psi_1}{\delta\psi_1} = -\frac{2\mu_0}{\sqrt{2B_y'}}\frac{1}{\eta}\frac{\partial\delta\psi_1}{\partial t}\int_{-\infty}^{\infty}d\psi\frac{\langle\cos ky/(\psi-\delta\psi)^{1/2}\rangle^2}{\langle 1/(\psi-\delta\psi)^{1/2}\rangle}.$$

Introducing the dimensionless parameter $\Delta' = \left.\frac{\partial\delta\psi_1}{\partial x}\right|_{-\infty}^{+\infty}/\delta\psi_1$, which is determined by the outer solutions, one obtains the nonlinear tearing mode equation:

$$\frac{\partial x_T}{\partial t} = \frac{\eta}{2\sqrt{2}\mu_0 A_0}\Delta', \tag{9.18}$$

where

$$A_0 = \int_{-1}^{+\infty} dw \frac{\langle \cos ky/(w-\cos ky)^{1/2}\rangle^2}{\langle 1/(w-\cos ky)^{1/2}\rangle} = 0.55.$$

Equation (9.18) is the so-called Rutherford equation. It indicates that the tearing modes are nonlinearly unstable as soon as $\Delta' > 0$.

Let us now discuss the underlying physics for layer width in figure 9.3 by the torque balance. We first consider the linear case. Suppose there is a radial field perturbation $B_x = B_{x1}\sin(ky)$, which grows with the rate γ. The Faraday law in equation (9.12) yields $E_z = (\gamma B_{x1}/k)\cos ky$. Here, the time dependence is assumed to follow the argument $ky - i\gamma$. With Ohm's law in equation (9.11), one has $j_z = (\gamma B_{x1}/k\eta)\cos ky$. This provides a linear force in the x direction, $-j_z B'_y x$. Applying the momentum equation for the force yields

$$v_x \sim \frac{1}{\gamma\rho} j_z B'_y x \sim (B_{x1}B'_y x/\rho k\eta)\cos ky. \tag{9.19}$$

Note also that the induced electric field produces a flow in the x direction,

$$v_x = -E_z/B_y = -(\gamma B_{x1}/kB'_y x)\cos ky. \tag{9.20}$$

As shown in figure 9.3, these drive a flow pattern of narrow vortices due to the torque generated. Note that $v_y \sim v_x/(kx_T)$. By balancing v_y from equations (9.19) and (9.20), one obtains the estimate of the layer width in the linear description

$$x_T \sim (\gamma\rho\eta)^{1/4}/(kB'_y)^{1/2}.$$

Next, we consider the nonlinear case. Due to the phase difference in y, one can conclude that the second-order force $j_z B_x$ does not contribute a torque that drives the flow vortex. Going to the third order, one can see that the flow v_y and B_x with the same phase can generate the second-order eddy current

$$\delta j_z = -v_y B_x/\eta \sim \gamma B_{x1}^2/k^2 B'_y \eta x_T^2,$$

which is y-independent. The third-order nonlinear force in the y direction, $\delta j_z B_x$, as shown in figure 9.3 with triple arrows, does provide a torque opposing the vortex flow. The balance of the torque produced by the linear force and the opposing torque of the third-order force yields

$$k j_{z1} B'_y x_T \sim \delta j_z B_{x1}/x_T \sim \gamma B_{x1}^2/k^2 B'_y \eta x_T^2.$$

Noting that $j_{z1} \sim \gamma B_{x1}/k\eta$, one obtains the island layer width in the nonlinear case: $X_T \sim (B_{x1}/kB'y)^{1/2}$. This shows that the singular layer width is comparable to the width of the separatrix or "magnetic island" in the nonlinear description. In the end, it is noted that, as shown in figure 9.3, ψ is even and v_x is odd in x in the Rutherford theory.

9.3 NEOCLASSICAL TEARING MODES

In this section, we discuss the neoclassical tearing modes (NTMs) [3, 4]. In the conventional tearing mode theory, one only considers the inductive current. For NTMs, the bootstrap current is included. As magnetic islands flatten the pressure profile, they eliminate the bootstrap current. The localized, helical deficit in the bootstrap current drives the modes. We mainly review the systematic NTM theory developed in reference [3].

They describe the magnetic islands that grow due to a localized, helical deficit in the bootstrap current within the plasma. As magnetic islands flatten the pressure profile, they eliminate the bootstrap current—a key source of plasma current. This missing current (a "current hole") is the primary driver of the mode.

Unlike the Rutherfod theory in Section 9.2, tokamak configurations with large aspect ratio and circular cross section are considered here. This leads slightly different notations be used. The magnetic field is represented as

$$\mathbf{B} = \mathbf{B}_0 + \mathbf{B}_1, \tag{9.21}$$

where

$$\mathbf{B}_0 = \nabla\zeta \times \nabla\chi + q_s \nabla\chi \times \nabla\theta, \tag{9.22}$$

$$\mathbf{B}_1 = \nabla\zeta \times \nabla\psi + q\nabla\psi \times \nabla\theta + (q - q_s)\nabla\chi \times \nabla\theta, \tag{9.23}$$

and q_s is the safety factor on the equilibrium rational surface.

To reduce the drift kinetic equation, one needs to evaluate $\mathbf{B} \cdot \nabla$. We use the coordinate system (χ, θ, α), where $\alpha = \zeta - q_s\theta$. It is assumed that the modes are radially localized. Therefore, one has

$$\begin{aligned}
\mathbf{B}_0 \cdot \nabla &= \frac{1}{Rq_s}\frac{\partial}{\partial\theta}, \\
(q - q_s)\nabla\chi \times \nabla\theta \cdot \nabla &= \frac{1}{Rq_s}(q - q_s)\frac{\partial}{\partial\alpha}, \\
\nabla\zeta \times \nabla\psi \cdot \nabla &= \frac{1}{Rq}\frac{\partial\psi}{\partial\theta}\frac{\partial}{\partial\chi}, \\
q\nabla\psi \times \nabla\theta \cdot \nabla &= -\frac{q_s}{Rq}\frac{\partial\psi}{\partial\alpha}\frac{\partial}{\partial\chi}.
\end{aligned}$$

Using these formulas, the drift kinetic equation is reduced to

$$\frac{\partial f}{\partial t} + v_\parallel \nabla_\parallel f + v_D \cdot \nabla f - \frac{r}{m} v_\parallel E_\parallel \frac{\partial f}{\partial \omega} = C(f, f), \tag{9.24}$$

where

$$\begin{aligned}
\nabla_{\|} f &= \frac{1}{qR}\left(\frac{\partial f}{\partial \theta} + W_{\|} f\right), \\
W_{\|} &= (q-q_s)\frac{\partial}{\partial \alpha} - \left(\frac{\partial \psi}{\partial \theta} - q_s\frac{\partial \psi}{\partial \alpha}\right)\frac{\partial}{\partial \chi}, \\
E_{\|} &= \frac{1}{qR}\left[-\frac{\partial \bar{\Phi}}{\partial \theta} - (q-q_s)\frac{\partial}{\partial \alpha} + \left(\frac{\partial \psi}{\partial \theta} - q_s\frac{\partial \psi}{\partial \alpha}\right)\frac{\partial \Phi}{\partial \chi} + q\frac{\partial \psi}{\partial t}\right],
\end{aligned}$$

$\tilde{\Phi} = \Phi - \bar{\Phi}$ with the overhar representing the average around the island magnetic surface, ω is the particle kinetic energy, and $\mathbf{v}_D$ is the magnetic drift velocity. The Lorentz collisional operator is used

$$C^1(f) = \nu\frac{\iota}{B}\frac{\partial}{\partial \lambda}\left(\lambda \iota \frac{\partial}{\partial \lambda}\right)f, \tag{9.25}$$

where

$$\nu = 2\nu_e(v_t/v)^3,\ \ \iota = \sigma(1-\lambda B)^{1/2},\ \ \sigma = \mathrm{sgn}(v_{\|}),\ \ \lambda = \mu/\varepsilon, \tag{9.26}$$

ν_e is the collisional frequency, v_t denotes the thermal velocity, and v represents the particle velocity. In equation (9.24), $\partial \ln\{f,\Phi\}/\partial \alpha \ll \partial \ln \psi/\partial \alpha$ has been assumed.

We intend to derive Ohm's law with the bootstrap current included. For this, the drift kinetic equation, equation (9.24), for electrons is used. Two ordering parameters are introduced: $\delta \sim \rho_e/w$ and $\Delta \sim \nu_e/\omega_T$, where ρ_e is the electron gyroradius and $\omega_T \sim v_t/Rq$. Ohm's law with the Spiter resistivity is described in the lowest order $\delta^{(0)}$, and the bootstrap current due to $\mathbf{v}_D$ is recovered in the first order $\delta^{(1)}$. To solve equation (9.24), auxiliary ordering assumptions are introduced:

$$W_{\|} \sim \Delta^2,\ \ \omega_T^{-1}\frac{\partial}{\partial t} \sim \Delta^2,\ \ \frac{e\bar{\Phi}}{T} \sim \Delta^2,\ \ \frac{\tilde{\Phi}}{\bar{\Phi}} \sim \Delta. \tag{9.27}$$

The distribution function is then expanded as follows

$$f = \sum_n \delta^n \left(\sum_m \Delta^m f_n^{(m)}\right).$$

With the ordering scheme described above, equation (9.24) in order $\delta^{(0)}$ becomes

$$v_{\|}\nabla_{\|} f_0 - \frac{e}{m} v_{\|} E_{\|} \frac{\partial f_0}{\partial \omega} = C(f_0, f_0). \tag{9.28}$$

The lowest order solution in Δ is just a Maxwellian distribution, and the first and second orders are the perturbed Maxwellians $f_M^{(1)} = f_M^{(1)}(\psi)$ and $f_0^2 = f_M^{(2)}$, in which the temperature gradient is ignored. Neither of them gives rise to the collisional operator contribution as the lowest order Maxwellian. The third-order solution, $\mathscr{O}(\delta^{(0)}\Delta^{(3)})$, is as follows

$$f_0^{(3)} = (e\tilde{\Phi}/T) f_M + g_0^{(3)} \tag{9.29}$$

with $g_0^{(3)}$, the integration constant, being determined by the next order equation, $\mathscr{O}(\delta^{(0)}\Delta^{(4)})$, by bounce average. The equation in this order is

$$v_\| \nabla_\| f_0^{(4)} - \frac{e}{m} v_\| E_\| \frac{\partial f_0^{(2)}}{\partial \omega} = C(f_0^{(3)}, f_0) + C(f_0, f_0^{(3)}). \tag{9.30}$$

The bounce average eliminates the first term, so that one obtains

$$g_0^{(3)} = \sigma \frac{\pi v (2\varepsilon)^{1/2}}{v q R_0} \left[\frac{e}{T} \left(\bar{W}_\| \bar{\Phi} - q \frac{\partial \psi}{\partial t} \right) f_M - W_\| f_M^{(2)} \right] I(\lambda) \theta(\lambda_c - \lambda), \tag{9.31}$$

where R_0 denotes the major radius at the axis, $I(\lambda)$ is an integral function of λ, θ is the step function, and $\lambda_c = [B_0(1+\varepsilon)]^{-1}$. The solution can be used to calculate the current moment, yielding Ohm's law

$$j_s = 2.3(1 - 2.1\sqrt{\varepsilon}) \sigma \bar{E}_\|. \tag{9.32}$$

The drift kinetic equation (9.24) in order $\delta^{(1)}$ becomes

$$\frac{\partial f_0}{\partial t} + v_\| \nabla_\| f_1 + \mathbf{v}_D \cdot \nabla f_0 - \frac{e}{m} v_\| E_\| \frac{\partial f_1}{\partial \omega} = C(f_1, f_0) + C(f_0, f_1). \tag{9.33}$$

Note that [5]

$$\mathbf{v}_D \cdot \nabla F_M(\chi) = -\mathbf{v}_\| \cdot \nabla \left(v_\| B_\alpha \frac{1}{\Omega_e} \frac{\partial F_M}{\partial \chi} \right),$$

where B_α is the α covariant component of $\mathbf{B}$. To the lowest order in Δ, one obtains

$$f_1^{(1)} = v_\| B_\alpha \frac{1}{\Omega_e} \frac{\partial F_M}{\partial \chi} + g_1^{(0)}. \tag{9.34}$$

Here, $g_1^{(0)}$ is an integration constant to be determined by the next order equation, $\mathscr{O}(\delta^{(1)}\Delta^{(1)})$, by bounce average, which yields $\oint (d\theta/|v_\|Rq) \left(C(f_1^{(0)}, f_0) + C(f_0, f_1^{(0)}) \right)$. The solution is

$$f_1^{(0)} = \sigma \frac{B_\alpha}{\Omega_e} v \frac{\partial f_M}{\partial \chi} \left(|\iota| - \theta(\lambda_c - \lambda) \int_\lambda^{\lambda_c} \frac{B_0 d\lambda}{2|\iota|} \right). \tag{9.35}$$

This solution can be used to calculate the current moment, yielding the bootstrap current contribution

$$J_b = -1.46\sqrt{\varepsilon} \frac{IT}{B} \frac{q_s' w}{q_s I_2} \frac{\partial n}{\partial \psi}, \tag{9.36}$$

where $I_2(\Omega) = \oint_\Omega d\alpha \, [2/(\Omega + \cos\alpha)]^{1/2} / 2\pi$.

Combining the Spitzer and bootstrap current contributions in equations (9.32) and (9.36), one obtains the total current. Using the total current, one can extend the

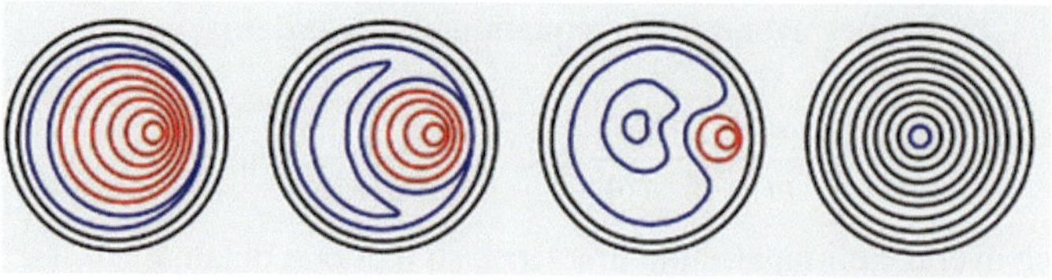

Figure 9.4 Kadomtsev theoretical picture for sawtooth oscillations. Reproduced from reference [7] with the permission of IAEA.

Rutherford theory in Section 9.2 to obtain the nonlinear tearing mode equation with bootstrap current included.

$$\frac{dx_T}{dt} = \frac{0.27}{1-2.1\sqrt{\varepsilon}}\frac{\Delta'\eta_s}{\mu_0}|\nabla\chi|_s^2 - 1.23\sqrt{\varepsilon}\frac{\eta_s l_s}{B_p}\frac{r_s}{q_s}\frac{p'}{x_T}|\nabla\chi|_s^2. \tag{9.37}$$

Comparing the Rutherford theory reviewed in Section 9.2, one can see that an additional driving term, the last term on the right, appears.

9.4 KADOMTSEV'S THEORY FOR SAWTOOTH OSCILLATIONS

Sawtooth oscillations and disruptions are of concern for tokamak confinement. It is deeply rooted in the presence of toroidal current in a tokamak. This is particularly concerning when the safety factor is below one in the plasma. It is related to the internal kink modes as reviewed in subSection 5.4.2. Conventionally, the Kadomtsev theory is used for the interpretation [6]. In this section, we review the theory.

The basic picture in Kadomtsev's theory is shown in figure 9.4. As depicted in frame a in this figure, plasma starts with a dashed circular column surrounded by the $q = 1$ surface. When instability occurs, the plasma inside the $q = 1$ surface moves toward the $q = 1$ surface as shown in frame b. Field line reconnection occurs when the plasmas inside and outside the $q = 1$ surface meet. This leads the island to be formed as shown in frame c. The island grows as the original plasma column moves further outward and shrinks until the original state is recovered, as shown in frame d. This explains the sawtooth oscillation circle.

The theoretical analyses are as follows. As described in Section 9.1, the helical magnetic field is expressed as

$$B^* = B_\theta - \frac{r}{R}B_\phi = B_\theta(1-q), \tag{9.38}$$

which is the magnetic field subtracted by that on the resonance surface $q = 1$.

Suppose that the plasma column inside the $q = 1$ surface moves outward radially with the velocity v_1. It induces an electric field $v_1 B^*$. Ohm's law for the induced field yields

$$j \sim v_1 B^* / \eta. \tag{9.39}$$

Also, using Ampere's law, one obtains

$$j \sim B^* / \mu_0 \delta, \tag{9.40}$$

where δ is the thickness of the resistive layer, in which the reconnection occurs. Combining equations (9.39) and (9.40) yields

$$v_1 \sim \eta / \mu_0 \delta. \tag{9.41}$$

From the momentum equation in the poloidal direction or energy balance for poloidal motion in the reconnection sheet, one obtains

$$\rho v_2^2 \sim B^{*2} / 2\mu_0, \tag{9.42}$$

where v_2 is the poloidal velocity. Also, noting the continuity equation, one obtains

$$v_1 r_1 \sim v_2 \delta. \tag{9.43}$$

Inserting v_1 and v_2 in equations (9.41) and (9.42), equation (9.43) yields

$$\delta^2 \sim \frac{\eta}{\mu_0} \frac{\sqrt{\mu_0 \rho}}{B^*}.$$

Defining the resistive and Alfvén times, respectively

$$\tau_R = \frac{\mu_0}{\eta} r_1^2 \quad \text{and} \quad \tau_A = \frac{r_1}{B^* / \sqrt{\mu_0 \rho}}.$$

One finally obtains the thickness of the reconnection layer

$$\delta \sim \left(\frac{\tau_A}{\tau_R} \right)^{1/2} r_1. \tag{9.44}$$

With δ determined in equation (9.44), one can determine v_1 in equation (9.41)

$$v_1 \sim \frac{\eta}{\mu_0 \left(\tau_A / \tau_R \right)^{1/2} r_1} = \frac{r_1}{\left(\tau_R \tau_A \right)^{1/2}}. \tag{9.45}$$

According to the physics picture in figure 9.4, the sawtooth oscillation time is

$$\tau_K \sim \frac{r_1}{v_1} \sim \left(\tau_R \tau_A \right)^{1/2}.$$

An important feature of Kadomtsev's theory lies in its fast time scale of oscillation. It is not proportional to the resistive time, but involves the fast Alfvén time scale.

However, the theory has relied on the self-rebuilding circle. Tokamak plasma is actually an open system, through Ohmic heating, wall interaction, etc. These need to be taken into account. There are two more factors. First, comparing with the Rutherford theory described in Section 9.2, one can see that equation (9.39) does not coincide with complete Ohm's law description, which reads $\eta \mathbf{j} = \mathbf{E} + \mathbf{v} \times \mathbf{B}$. Only the $\mathbf{v} \times \mathbf{B}$ term is included here in equation (9.39), which is suppressed by the helical average in Rutherford's theory. Second, the current interchange effects to be described in Section 9.5 have not been taken into account. With coupling between even (tearing type) and odd (interchange type) modes taken into account, one can explain the beta limit phenomenon in the disruption.

9.5 CURRENT INTERCHANGE TEARING MODES

The Kadomtsev theory reviewed in Section 9.4 has some similarity to the conventional forced reconnection picture in the Sweet-Parker model [8, 9], as reviewed in Section 9.10. The plasma in the $q < 1$ region is pushed out and meets with the plasma exterior to the $q < 0$ region. Because plasma carries the magnetic field, it leads to the reconnection at the $q = 1$ surface. Besides the forced reconnection picture, we discuss the current interchange tearing mode picture in this section [10]. Note that the kink or interchange type of modes exchange not only thermal and magnetic energies between flux tubes, but also current, as shown in figure 9.5. As soon as there is a current gradient, the interchange or kink modes and the tearing modes are coupled. This leads to the perturbed current sheet to appear on the $q = 1$ resonance surface during the development of the $n = 1$ modes. The current sheet subsequently induces the tearing modes. This gives a natural explanation of the development of the tearing modes that cause the tokamak disruption. The current interchange tearing modes can also explain why the electron transport in tokamaks is stronger than the ion one. Because of the coupling between the tearing and interchange type of modes, the stochastic field transport is always present. Electrons escape faster than ions along the stochastic field lines.

Theoretically, we demonstrate the current interchange tearing modes by modifying the Rutherford theory reviewed in Section 9.2. The same coordinates, equilibrium, and notations as those given in Section 9.2 are employed. We start with Ampere's law, Faraday's law, and Ohm's law

$$\begin{aligned} \mu_0 \mathbf{J} &= \nabla \times \mathbf{B}, && (9.46) \\ \partial \mathbf{B}/\partial t &= \nabla \times \mathbf{E}, && (9.47) \\ \eta \mathbf{J} &= \mathbf{E} + \mathbf{v} \times \mathbf{B}. && (9.48) \end{aligned}$$

Noting the Faraday's law in equation (9.47), the application of curl operation on Ohm's law in equation (9.48) and the projection of the resulting equation on the x direction yield

$$\frac{\partial \delta\psi}{\partial t} - \left.\frac{\partial \delta\varphi}{\partial y}\right|_{\psi} B'_y x = \eta \delta j_z + \delta\eta\, j_z. \qquad (9.49)$$

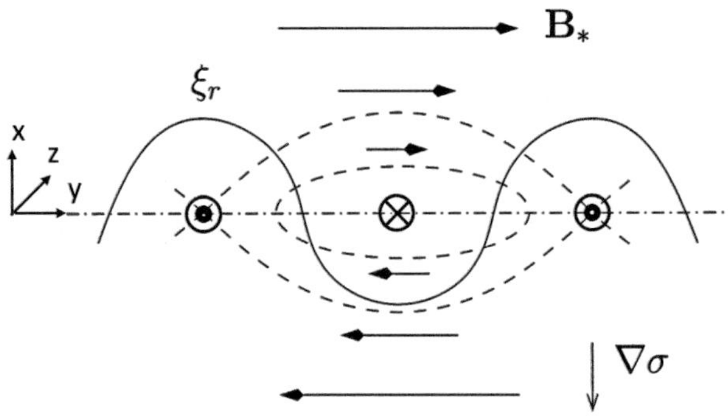

Figure 9.5 Current interchange tearing modes. Reproduced from reference [10] with the permission of AIP Publishing.

Here, the last term on the right-hand side represents the convective current interchange effect, with $\delta\eta = -\xi_x(d\eta/dx)$ and ξ_x representing the perturbation of interchange-type modes. It is assumed that the interchange-type modes vary poloidally as $\xi_x = \xi_1 \cos ky$.

Comparing equation (9.49) with equation (9.14), one can see that an additional current density, the last term, is included to describe the current interchange modes. One can proceed with a similar derivation of the Rutherford equation in Section 9.2 to obtain the tearing mode equation

$$\frac{\partial x_T}{\partial t} = \frac{\eta}{2\sqrt{2}\mu_0 A_0}\Delta' + \frac{2\eta j_z}{B'_y}\frac{d\ln\eta}{dx}\frac{\xi_1}{x_T}. \tag{9.50}$$

The last term on the right-hand side of equation (9.50) represents the current interchange effect [10]. The resulting modes are referred to as the current interchange tearing modes. Without the resistivity gradient effect, equation (9.50) reduces to the Rutherford equation (9.18), which indicates that the tearing modes are nonlinearly unstable as soon as $\Delta' > 0$. When there is a current gradient, the interchange-type modes can carry over the current and, consequently, a current sheet results in the singular layer, leading to the presence of the last term in equation (9.50). Note further that this last term is inversely proportional to the island width x_T. Therefore, it is particularly significant for the onset (small x_T case) of the tearing modes. In the derivation, only the Ohmic current is taken into account. The current interchange effects can also result from other types of currents, such as the bootstrap current.

9.6 EDGE LOCALIZED MODES: EXTERNAL COUPLINGS

We have reviewed various internal mode theories for explaining the disruption, etc. Actually, the extreme MHD instabilities involve the interaction with external components, such as the wall, divertors, etc. Edge localized modes look like the small-scale disruptions occurring at the plasma edge. We will describe the theory with the external coupling of the scrape-off-layer (SOL) current [11].

The underlying physics picture is as follows. Due to the divertor sheath effects, the SOL current is usually saturated. When the instabilities develop in the pedestal as plasma parameters reach the limit, they drive particles and heat to SOL. This breaks the original saturated status of the SOL current and causes the SOL current to rise. The increase in the SOL current feeds back to the instabilities. This constitutes a positive feedback process that leads to the pedestal's collapse. Afterwards, plasma heating starts to rebuild the pedestal. This loop explains the bursting of the edge localized modes and the recovery process after the bursting. This was seen in several experiments, for example, references [12] and [13].

A complete numerical simulation of the positive feedback process requires a global code which couples the nonlinear MHD, transport, SOL current, divertor sheath physics, etc. Since we focus only on revealing the amplification phenomenon, not on the details of the process, we instead use a simplified model to explain the physics. The magnetic diffusion type of equation is used as the starting equation.

$$\tau_w \frac{\partial \delta B_r}{\partial t} - \gamma_s \delta B_r = i\mu_0 w k_\theta \delta J_\parallel, \tag{9.51}$$

where $\delta J_\parallel$ represents the SOL current and γ_s denotes the growth rate in the absence of SOL current effects $\delta J_\parallel = 0$. The system is unstable when $\gamma_s > 0$ and stable when $\gamma_s \leq 0$. We can relate γ_s to the external kink mode growth rate in equation (6.87), *i.e.*, $\gamma_s \propto \delta W_\infty / \delta W_b$. We further specify various energy integrals as follows: $\Re e\{\delta W_\infty\} = C_w(-\beta + \beta_c)$ and $\delta W_b = \delta W_\infty + \delta W_v$, where C_w is a constant, $\beta = 2\mu_0 n^p T^p / B^2$, β_c denotes the critical beta for no-wall stability, and δW_v represents the vacuum energy integral. In the numerical simulation, δW_v and $\Im m\{\delta W_{\infty,b}\}$ are employed as parameters.

Compared with the usual induction equation for tearing modes, equation (9.51) contains a new term on the right-hand side. It takes into account the fact that the radial transport from the pedestal plasma can change the thermal properties in the SOL and consequently the SOL current changes. It is due to this new term that equation (9.51) contains a positive feedback process between the resistive SOL modes δB_r and the SOL current $\delta J_\parallel$. The positive feedback process leads to a mode bursting and provides a physical explanation for ELMs.

Using the theory developed in reference [14], we can express the SOL current in equation (9.51). Noting that electrons and ions have vastly different mobilities, we extend the SOL current expression in reference [14] to a two-fluid description for temperature. In this way, the SOL current can be expressed as $J_{\parallel} = J_{SAT}\hat{J}_{\parallel}$, where

$$\hat{J}_{\parallel} = -\gamma_0 \left\{ \frac{eV_0}{kT_{eh}} + (\ln 2 + 0.71)\left(\frac{T_{ec}}{T_{eh}} - 1\right) + \frac{T_{ec}}{T_{eh}} \ln \alpha_c - \ln \alpha_h + \ln \left[\frac{1+\hat{J}_{\parallel}}{\left(1 - \frac{j_{SAT,h}}{j_{SAT,c}}\hat{J}_{\parallel}\right)^{\frac{T_{ec}}{T_{eh}}}} \right] + \frac{1}{kT_{eh}} \int_h^c \frac{1}{n}\frac{dp_{ec}}{ds} ds \right\}. \tag{9.52}$$

Here,

$$J_{SAT} = en\left[k(T_i+T_e)/m_i\right]^{1/2},$$

where k is the Boltzmann constant, V_0 represents the applied voltage between the hot and cold divertors, $\alpha = [m_i T_e / 2\pi m_e (T_e + T_i)]^{1/2}$, $\gamma_0 = \sigma k T_{eh}/(eLJ_{SAT})$, s represents the field line arc length, and the subscripts h and c denote the sheaths respectively at the hot and cold divertors.

To describe the link between the thermal parameters of the pedestal and SOL, the pedestal temperature and density are modeled as follows:

$$T^p(t) = T^{p0}\left(1 + Q^h t - Q^{tr}(t)\right), \tag{9.53}$$

$$n^p(t) = n^{p0}\left(1 + D^h t - D^{tr}(t)\right), \tag{9.54}$$

where the constants Q^h and D^h describe the temperature and density increasing rates due to the heating, and $Q^{tr}(t)$ and $D^{tr}(t)$ describe the temperature and density losses due to the radial transport are described as follows

$$Q^{tr}(t) = Q_0^{tr} \int_{t_c}^{t} \frac{dt}{\Delta t} \left|\frac{\delta B_r}{\Delta B}\right|^{\alpha_q} \left(\frac{T^p n^p}{T^{p0} n^{p0}}\right)^{\alpha_{pt}},$$

$$D^{tr}(t) = D_0^{tr} \int_{t_c}^{t} \frac{dt}{\Delta t} \left|\frac{\delta B_r}{\Delta B}\right|^{\alpha_d} \left(\frac{T^p n^p}{T^{p0} n^{p0}}\right)^{\alpha_{pn}}.$$

Here, t_c is the critical time when the linear resistive SOL modes become unstable, T^{p0} and n^{p0} are the initial pedestal temperature and density, Q_0^{tr} and D_0^{tr} are constants specifying the strength of the radial transport, $\alpha_{...}$ describe the dependences of radial transport on the magnetic field and pressure, respectively, Δt is the normalization constant to reflect the ELM pulse duration, and ΔB is the normalization factor for δB_r to be specified by the typical experimentally observed value. The ion and electron species in the pedestal are assumed to have the same temperature and density.

We also need to specify the increments of the SOL temperature and density due to the transport from the pedestal to SOL. The temperature and density for electron and ion species at the divertor sheaths for $t \geq t_c + \delta t_e$ can be expressed respectively as follows:

$$T_{e,i}^{d}(t) = T_{e,i}^{d0}\left[1 + R_{e,i}^{t} Q^{tr}(t - \delta t_{e,i}) - C_{e,i}^{t}\int_0^t \frac{dt}{\Delta t}\left(\frac{T_{e,i}^{d}}{T_{e,i}^{d0}}\right)^{\kappa_{e,i}^{t}} H(T_{e,i}^{d} - T_{e,i}^{d0})\right], \quad (9.55)$$

$$n^{d}(t) = S_n n^{d0}\left[1 + R^{n} D^{tr}(t - \delta t_e) - C^{n}\int_0^t \frac{dt}{\Delta t}\left(\frac{n^{d}}{n^{d0}}\right)^{\kappa^{n}} H(n^{d} - n^{d0})\right], \quad (9.56)$$

Here, $T_{e,i}^{d0}$ and n^{d0} are the initial divertor sheath temperature and density with $S_n(t;\delta t_{e,i}) = S_n^0(\delta t_{e,i})\{2 - \exp[-\lambda_s(t - t_c - \delta t_e)]\}$ to simulate the density increment due to the dislodgement of neutrals from the saturated divertor plates, where λ_s is a constant and $S_n^0(\delta t_{e,i})$ specifies the density amplification factor. The constants $R_{e,i}^{t}$ and R^{n} describe respectively the coupling factors of the radial temperature and density transport. The constants $C_{e,i}^{t}$ and C^{d} describe the temperature and density cooling rates. The Heaviside function $H(x) = 1$ for $x > 0$, otherwise 0. Note that the nonaxisymmetric component coupled to the MHD modes in equation (9.57) is weak. One can expect $S_n^0(\delta t_{e,i})$ to be a decreasing function of $\delta t_{e,i}$. The ratio of the temperatures in hot and cold divertor sheaths is employed as a parameter.

The analytical solution of equation (9.51) can be expressed as

$$\delta B_r = e^{\int^t \gamma_s dt/\tau_w}\left[i\mu_0 dk_\theta \int^t \delta J_{\|} e^{-\int^t \gamma_s dt/\tau_w} dt/\tau_w + \delta B_{r0}\right], \quad (9.57)$$

where δB_{r0} is a constant. This solution can be used to explain the positive feedback process. The solution in equation (9.57) contains two parts: the solo resistive SOL mode solution and the $\delta J_{\|}$ driven part, respectively, by the second and first terms in the square parentheses on the right-hand side.

The set of equations (9.52)–(9.57) can be used to simulate the edge localized modes. Typical numerical results are given in figure 9.6. In this figure, the ELM bursting and quenching are reproduced. They are a consequence of changing pedestal beta and varying SOL current. In the simulation, it is shown that the δB_r growth exhibits two phases. As plasma beta exceeds the critical beta value at $t = t_c = 3.46$ ms, the resistive SOL modes become unstable. Due to the transport time delay, the divertor sheath temperature, density, and the SOL current remain at the equilibrium values at the beginning $t < t_c + \delta t_e = 3.47$ ms. The magnetic perturbation is small and reduces gradually due to the reduction of the pedestal pressure at this initial phase. Only when the pedestal heat and particles arrive at the divertor sheaths at $t = t_c + \delta t_e$, the divertor sheath temperature and especially density increase suddenly, and consequently the SOL current surges. The rapid increase of the SOL current leads to a rapid increase of the magnetic perturbations at the second phase $t > t_c + \delta t_e$. This positive feedback process defines the bursting feature of ELMs in the numerical results as observed experimentally. The simulation results also show that as the pedestal density and temperature drop to a certain level, the particle and heat sources for the SOL are outbalanced by the cooling, and the SOL current reduces. A smaller SOL current leads to a smaller magnetic perturbation. Consequently, an even smaller SOL current

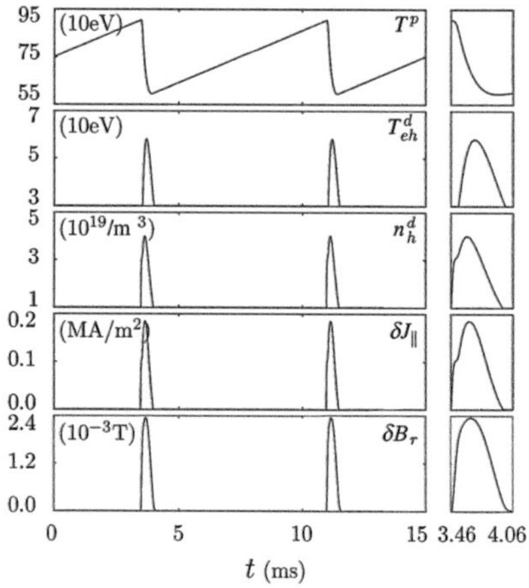

Figure 9.6 Edge localized modes. Reproduced from reference [11] with the permission of APS.

results, due to the reduction of the particle and heat sources. This inverted type of positive feedback process leads ELMs to quench sharply. The further heating shown in the first frame of figure 9.6 leads to the next ELM cycle. This process is observed in the experiments.

9.7 COUPLING TO THE INDUCTIVE ELECTRIC FIELD: OFF-AXIS HEATING

In the previous section, we discussed the coupling to the divertor or wall. Note that tokamak is a transformer. The plasma torus is coupled to the central solenoid. Therefore, tokamak plasma is immersed in an induced electric field generated by the central solenoid [15]. In this section, let us discuss this coupling, especially in the off-axis heating case.

Without Ohmic heating, there is no surprise if the electron cyclotron heating (ECH) is applied in an off-axis region, the electron temperature profile becomes peaked around the radius where the ECH deposits. However, in the early tokamak experiments, it was observed that the electron and ion temperature profiles always have roughly the same shape, regardless of the method and localization of the heating. This is referred to as the profile resilience [16]. It is the experiment in the Rijnhuizen Tokamak Project (RTP) that showed that a steady-state hollow electron temperature profile can be sustained with strong off-axis ECH, accompanied by a reversed magnetic shear in the region where ICH deposits [17]. The temperature profiles and heating locations (the vertical arrows) are shown in figure 9.7.

As pointed out in reference [16], the key difference between the RTP experiment and earlier ones, for example DIII-D, lies in that the electron collisionality in high-density discharges in RTP is high $\nu_* \sim 1$, whereas the collisionality in DIII-D is

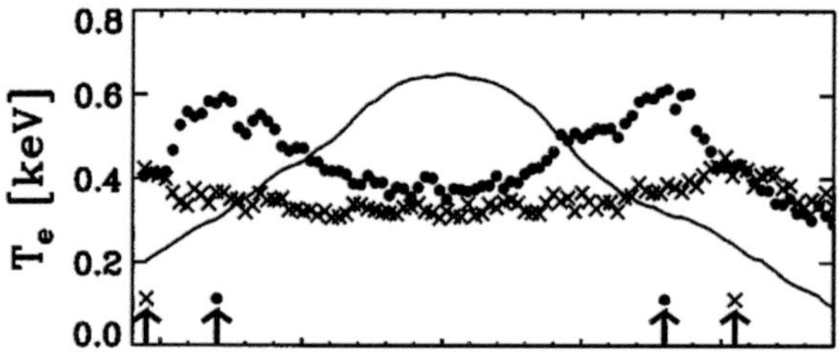

Figure 9.7 Temperature profiles during the off-axis heatings by the electron cyclotron wave observed in the Rijnhuizen Tokamak. Reproduced from reference [17] with the permission of APS.

much lower $\nu_* \ll 1$. One can imagine that when resistivity is high, the loop voltage is high as well. An off-axis ECH causes the local conductivity to drop, so that the Ohmic current redistributes readily, so does the Ohmic heating power. That is why it is accompanied by the magnetic shear reversal. This process is more pronounced in the high resistivity case since the loop voltage is high. Note that there can be a self-amplification process here. The off-axis heating increases the conductivity locally at the heating location. Consequently, the loop voltage decreases, which leads to a decrease in current and an increase in resistivity elsewhere in the heating location. This further enhances the current redistribution. The current redistribution can also lead to an inward pinch process. Nowadays, the hollow electron temperature profiles during an off-axis heating have been observed in many tokamaks, especially when ECH is applied during the Ohmic current ramping up process, when the loop voltage is high.

The inward pinch phenomenon in the off-axis heating can be explained by Ohm's law

$$J_\phi = \sigma(E_\phi + v_r B_\theta + v_\theta B_\phi), \tag{9.58}$$

where symbols assume the usual meanings. From equation (9.58) one can see that increasing J_ϕ and decreasing E_ϕ must be compensated by the $\mathbf{v} \times \mathbf{B}$ effects. Consequently, both inward pinch and poloidal rotation result. The inward shift of the temperature peak can be the consequence of the pinch effect from $-v_r$. From equation (9.58), one can see that the off-axis heating can also result in the poloidal rotation. This may also explain the so-called spontaneous rotation observed experimentally [18]. These analyses are applicable to other types of off-axis heating.

9.8 BLOB TRANSPORT

In this section, we discuss the so-called blob transport. As shown in figure 9.8, the blob transport is characterized by the convective movement toward the wall of the localized, high-density plasma structures (blobs) across magnetic field lines in the

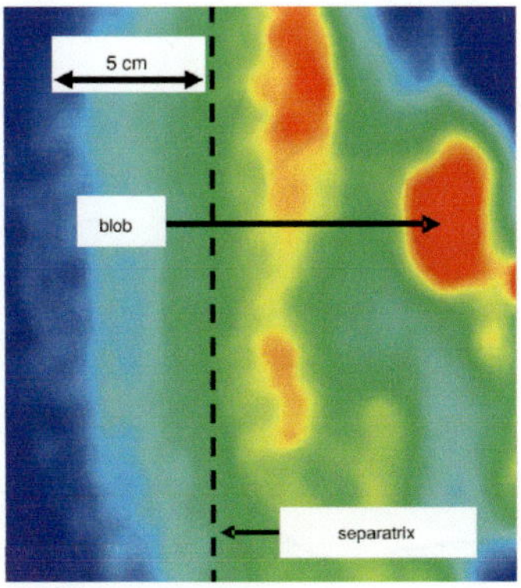

Figure 9.8 Localized, high-density plasma structures (blobs) across magnetic field lines in the scrape-off layer and vacuum region. Reproduced from reference [19] with the permission of AIP Publishing.

scrape-off layer and vacuum region [19]. The structure tends to align with the magnetic field lines and has different thermal properties, like density and temperature, from its ambient plasma. It is a major contributor to the so-called intermittent. The non-diffusive blob transport significantly impacts the edge particle and energy transport.

The blob transport is different from the conventional anonymous transport and the stochastic field transport. Traditionally, the blob transport is interpreted by the extended resistive interchange mode theories [20]. As shown in figure 9.9, it is noted that the magnetic field lines terminate at the divertors in the scrape-off layer. Due to the divertor sheath effects, the virtual resistivity in this case is big. Consequently, the charge separation can build up across the blob due to the curvature drifts. It induces the electric field, so that the $\mathbf{E} \times \mathbf{B}$ drift motion drives the blob toward the wall.

The conventional picture works well for the toroidally axisymmetric blob without carrying current. Note that there is a loop voltage in tokamak, and the blob has a different temperature and density from its ambient plasma as shown by the brightness of the blob in figure 9.8. The blob can carry current. As discussed in Section 9.6, the difference in divertor sheet temperatures can also contribute to a current. In this case, the blob can have its own magnetic-field-line rotational transform, so that the charge separation can be neutralized by the field pitch. The interpretation does not work in this case. Even if the blob does not carry current, the axis of blob filament itself has a rotation transform like a stellarator. The particle magnetic drift direction varies as the blob filament turns around the tokamak plasma torus. This makes the conventional interpretation not work either.

Drawing a comparison with the tokamak equilibrium, one can find a natural explanation of blob transport. Without introducing the field-line rotational transform, there are particle losses due to the charge separation that causes the outward $\mathbf{E} \times \mathbf{B}$ drift. But, even with the field line rotational transform, the tokamak plasma torus can

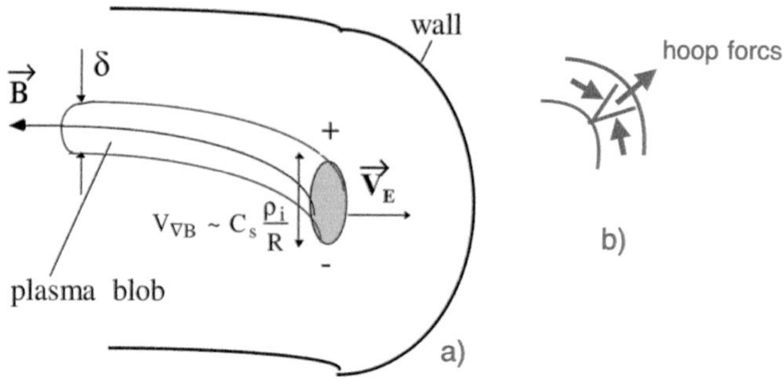

Figure 9.9 Interpretation of the blob transport by the extended resistive interchange mode theories (part a) and the hoop force (part b). Part a is reproduced from reference [20] with the permission of AIP Publishing.

expand outward. There is a so-called hoop force. Tokamak equilibrium needs a vertical magnetic field, which interacts with the toroidal current to generate an inward force. Since the blobs are aligned with the magnetic field lines, they are curved spatially. Consequently, there is a hoop force on the filament, and there is no mechanism to counterbalance the hoop force. Therefore, it expands toward the wall. This is a natural explanation of the blob transport. Note that considerable blob simulations are made in the straight system. If a curved system were used, one should recover the hoop force effects and reproduce the blob transport phenomenon readily.

Combining the current interchange mode theory in Section 9.5 and the edge localized mode theory in Section 9.6, one can construct an overall picture. The peel-ballooning modes can turn into the tearing modes due to the current interchange effect, noting that the current gradient at the edge is big due to the change of field line topology at the edge. The development of tearing modes is amplified due to coupling to the scrape-off-layer current. The tearing modes can become a blob filament, due to the blob not being in the force balance state. The hoop force pushes them toward the wall.

As shown in figure 9.8, the blobs appear brighter than the surrounding area. The current interchange tearing mode picture can also offer an explanation for this. The current interchange tearing modes bring cold and hot plasmas together in the magnetic island. This leads the charge-recombination process in the islands. This leads to the brightness of the blobs.

9.9 EDGE HARMONIC OSCILLATIONS

In this section, the edge harmonic oscillations (EHOs) in the quiescent-H (QH) modes are discussed. This is a typical nonlinear process that can be described. It is a nonlinear process but grows slowly.

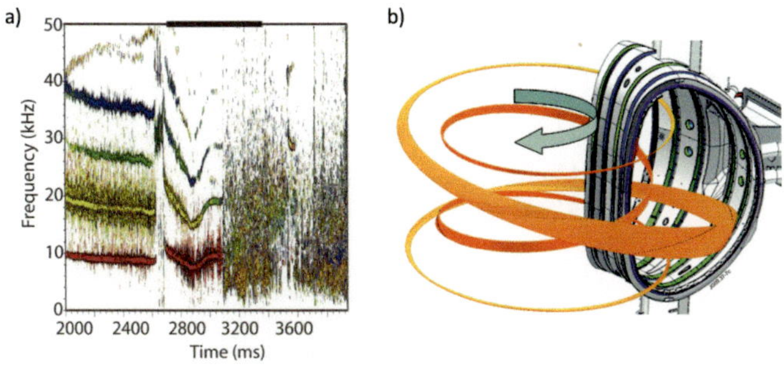

Figure 9.10 Frequency spectrum observed in DIII-D tokamak in the quiescent mode discharge (part a, reproduced from [21] with the permission of AIP Publishing) and the snaky fluctuation observed in JET (part b, reproduced from [22] with the permission of APS).

Figure 9.10 shows the frequency spectrum observed in DIII-D tokamak in the quiescent mode discharge [21] and the snaky fluctuation observed in JET [22]. In DIII-D, it is referred to as EHO, while in JET, it is referred to as outer modes (OMs). There are some important features of this type of modes. First, from figure 9.10, one can see that the modes are localized. Second, the $n = 1$ mode frequency is close to the rotation frequency Ω_p at the pedestal top; Third, the mode frequencies ω_n with the toroidal mode number n are $-n\Omega_p$, and fourth, the EHOs (or OMs) tend to develop when the local safety-factor value is equal to or slightly larger than an integer number (for example $q = 4$). Besides, the QH mode occurs in the case of low density and high temperature.

From the current ribbon observed in JET, one can think about the localized infernal modes. It also occurs around the q minimum in the core plasma. We note that the QH mode occurs at a steep pedestal with low collisionality. One can expect a high bootstrap current is present at the plasma edge. Including the bootstrap current effects, one can find that indeed a q stationary point occurs at the pedestal as shown in figure 9.11.

One can anticipate that when the q value at the plateau, q_p, is close to a rational number, for example 4, one can expect an m/n infernal mode (or fat interchange mode) to develop there. This is shown in figure 9.12a.

To explain the EHOs further, let us examine the model MHD equation

$$\frac{d}{dr}\left[\left(\frac{1}{q}-\frac{n}{m}\right)^2-(\omega+n\Omega_{rot})^2\right]\frac{d\xi_m}{dr}+g(r)\xi_m=\sum_{m'}\mathscr{L}_{mm'}\xi_m\xi_{m'}.$$

One can see that at the q plateau, $(1/q-n/m)$ is small. To minimize the inertia effects for localized modes, one has to have $\omega+n\Omega_{rot}=0$. This shows that the $n=1$ mode frequency is close to the rotation frequency Ω_p at the pedestal top, and the mode frequencies ω_n with the toroidal mode number n are $-n\Omega_p$. This is confirmed numerically in figure 9.13. This explains the mode spectrum as shown in figure 9.11.

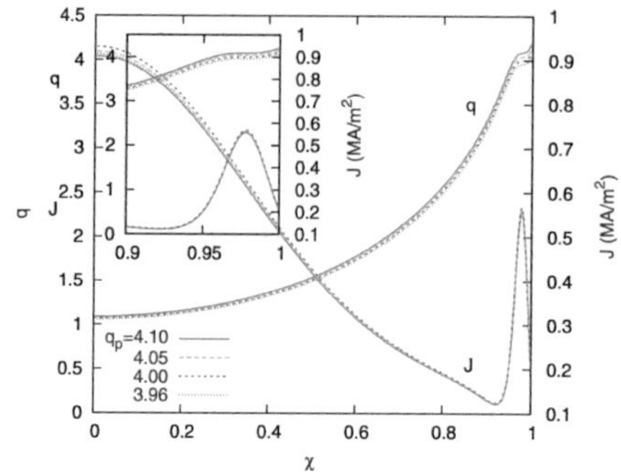

Figure 9.11 Safety-factor profiles versus normalized poloidal magnetic flux. The safety-factor values at the plateau range from 4.1, 4.05, 4, to 3.96 top-down. Reproduced from [23] with the permission of AIP Publishing.

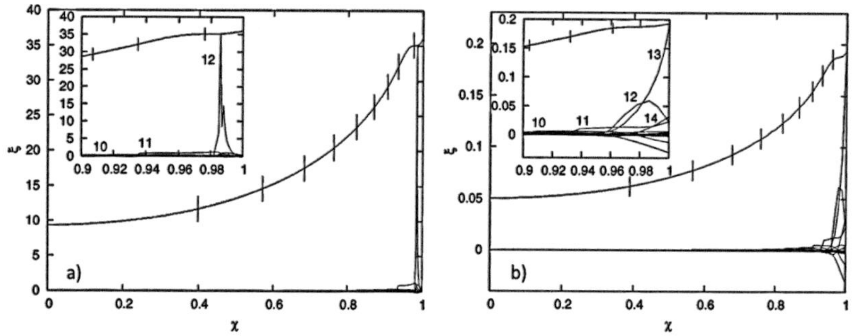

Figure 9.12 The $n = 3$ eigenfunction versus normalized poloidal magnetic flux for the cases with (a) $q_p = 4$ and (b) $q_p = 4.05$. Reproduced from [23] with the permission of AIP Publishing.

Note that the resonance surface $q = m/n$ being above or below q_p makes a difference. This is because the sidband contribution is different between the two cases. When the resonance surface $q = m/n$ is above q_p, the sideband contribution from the core plasma plays an important role so that the frequency multiplying rule cannot be reproduced. This also coincides with the experimental observation. When q_p is slightly larger than 4, as in the case $q_p = 4.05$, the sidebands are still packed near q_p since q is steep at the plasma edge as shown in figure 9.12b. Therefore, the frequency multiplying rule is still preserved in this case. These are confirmed by the numerical results as shown in figure 9.13.

The frequency multiplying rule will be improved if the nonlinear coupling is taken into account. For example, the nonlinear coupling of $n = 1$ modes can give rise to the source with $n = 2$ and $\omega = 2\omega_1$. Since this frequency is close to the $n = 2$ eigenfrequency, it can be quasi-resonantly amplified. Similarly, the nonlinear coupling of

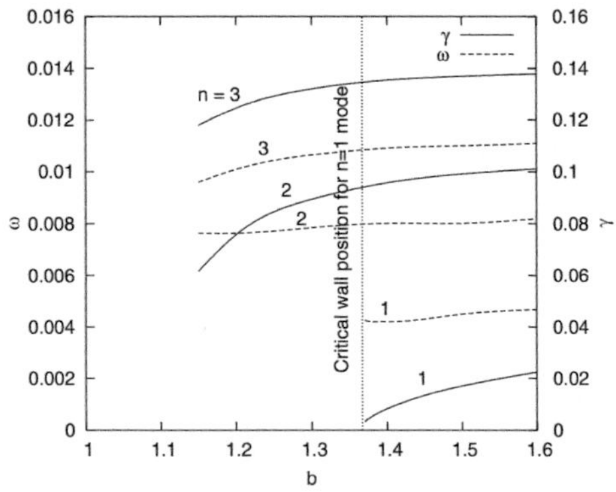

Figure 9.13 The eigenfrequencies ω and growth rates γ for $n = 1-3$ modes versus the wall position b. Reproduced from [23] with the permission of AIP Publishing.

$n = 1, 2$ modes can generate the source with $n = 3$ and $\omega = \omega_1 + \omega_2$. Again, this can be amplified due to the closeness of this nonlinear source frequency to the $n = 3$ linear eigenfrequency.

9.10 FORCED RECONNECTION, THE SWEET-PARKER MODEL, AND BEYOND

In this section, we describe the so-called forced reconnection. Unlike the cases with the reconnection being induced by the instabilities spontaneously, the forced reconnection, as suggested by its name, results from pushing together two plasmas emerging in the oppositely directed magnetic field as shown in figure 9.14. This forces the field lines to reconnect. This phenomenon occurs widely in space plasmas, such as solar winds and Earth magnetictail. We will mainly describe the so-called Sweet-Parker model for the interpretation [8, 9]. This is the most fundamental theory, as proved by the numerical simulations, for example, in reference [24]. Nevertheless, the Sweet-Parker model gives rise to a growth rate that is smaller in order of magnitude than the observations. There are still active investigations into this topic. We will discuss some of them.

Let us now describe the Sweet-Parker model. Suppose the two magnetized plasmas are pushed toward each other as shown in figure 9.14 with the velocity v_{in}. Outside the singular layer $x = 0$, one can use the ideal MHD frozen-in law to express the velocity

$$v_{in} = \frac{E_z}{B_*}. \tag{9.59}$$

Here, E_z is the electric field in the z direction. Using the induction equation, one obtains

$$E_z = \frac{\eta B_*}{\mu_0 \delta}, \tag{9.60}$$

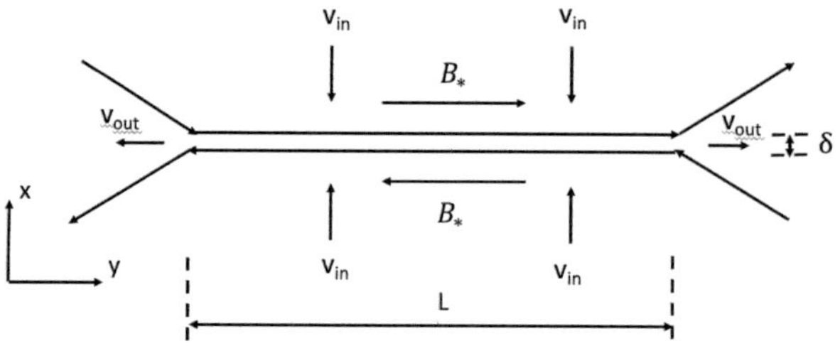

Figure 9.14 The Sweet-Parker reconnection picture.

where δ is the scale length of the singular layer. Suppose the scale length of the singular-layer width is L, and the speed flowing out of the layer is v_{out}. Using the continuity equation, one obtains

$$Lv_{in} = \delta v_{out}. \tag{9.61}$$

Furthermore, noting that the work done by the $\mathbf{j} \times \mathbf{B}$ force is converted to the outflow kinetic energy, one has

$$\frac{B_*^2}{\mu_0} = \rho v_{out}^2. \tag{9.62}$$

With the results in equations (9.59) - (9.62), one can estimate the reconnection rate.
First, we introduce the Lundquist number as given in Section 7.2

$$S = \frac{\mu_0 L v_A}{\eta}, \tag{9.63}$$

where the Alfvén speed is defined as $v_A = B_*/\sqrt{\mu_0 \rho}$. From equation (9.62), one can estimate the outflow speed

$$v_{out} = v_A. \tag{9.64}$$

Inserting equation (9.64) into equation (9.61), the layer thickness can then be obtained

$$\delta = L\frac{v_{in}}{v_A}. \tag{9.65}$$

Note that using equations (9.65), (9.60), and the definition of S in equation (9.63) one obtains

$$\frac{v_{in}}{v_A} \sim \frac{\delta}{L} \sim \frac{1}{L}\frac{\eta B_*}{\mu_0 E_z} \sim \frac{1}{L}\frac{\eta}{\mu_0 v_{in}},$$

one finally obtains

$$\frac{v_{in}}{v_A} \sim S^{-1/2}. \tag{9.66}$$

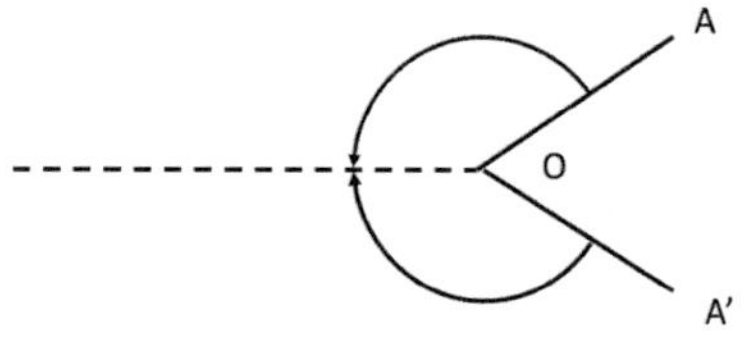

Figure 9.15 Formation of the current sheet in the configuration in figure 9.14.

Suppose all inflow plasmas end up with reconnection. Equation (9.66) shows that the magnetic reconnection rate is scaled as $\tau_A^{1/2}\tau_R^{1/2}$, where the Alfvén and resistive times, τ_A and τ_R, are defined in Section 7.2. Using the Sweet-Parker model to estimate the reconnection time scale of the solar flares, the time scale is roughly a few ten days. However, in reality, the observed time scale is just a few minutes to an hour. A similar phenomenon occurs in the Earth magnetictail during the solar wind. This leads to the exploration of various models in this field. In passing, Kadomtsev's disruption theory as reviewed in Section 9.4 uses a similar framework to estimate the reconnection rate.

Let us now further discuss this issue. In reference [25], S. I. Syrovatskii points out that forcing two plasmas immersed in the opposite magnetic field together can cause the magnetic field lines to rearrange so that an electric current sheet can be formed in the neutral layer. A large amount of magnetic energy can be stored in the layer and can be explosively released during magnetic reconnection. It is thought to power phenomena like solar flares. In Syrovatskii's theory, a complex analysis is used with the conformal transform performed. This can actually be understood from the real space analyses. In subsection 4.6.1, we have seen that at the X-point tip, one can reduce the Grad-Shafranov equation to the Poisson equation, and the resulting equation can be solved in polar coordinates. As shown in figure 9.15, one can use the initial conditions for ψ and $\partial\psi/\partial\theta$ on the surfaces OA and OA' to integrate the equation poloidally counterclockwise and clockwise, respectively. One can expect that the respective solutions meet at the neutral surface with the same ψ. However, one cannot assume that $\partial\psi/\partial\theta$ is the same in general. It can only happen for a specific angle $\angle A0A'$. The discontinuity of $\partial\psi/\partial\theta$ on the neutral surface indicates there must be a current sheet at the neutral surface.

There is another interesting phenomenon at the neutral surface regarding the particle gyromotion. In figure 9.16, an electron orbit is shown. Because of the direction change, the electron does not gyrate in a circle locally. Instead, it drifts in the $-\mathbf{z}$ direction. Similar analyses show that the ions drift in the opposite direction, *i.e.*, in the positive $\mathbf{z}$ direction. This shows that charged particles escape in the direction perpendicular to the magnetic field, and ions and electrons exit in the opposite direction when arriving at the neutral surface. Because of losing the plasma pressure at the neutral layer, the plasmas are sucked into the neutral layer. Furthermore, because ions and electrons drift in opposite directions. A current sheet is formed. The current

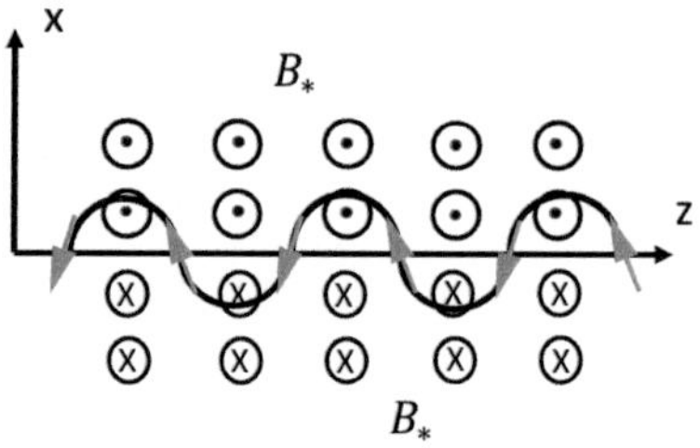

Figure 9.16 Charged particle drifts at the neutral surface $x = 0$.

can induce the reconnection of the magnetic field as discussed in Section 9.1. This gives an alternative picture to the Sweet-Parker model.

9.11 CONCLUSIONS AND DISCUSSION

In this chapter, we have reviewed various nonlinear theories, covering the tokamak disruptions, edge localized modes, off-axis heating, blob transport, edge harmonic oscillations to the solar wind, and Earth magnetotail physics.

The nonlinear development involves multiple physics processes. Although a brute force treatment may work, the best approach is to extract the key physical mechanisms and the fast physical processes. This leads the current reviews to focus on explaining the physics. The materials reviewed in this chapter are especially useful for the interpretation of experimental observations and the control of the damaging instabilities in the magnetically confined plasmas.

In this chapter, we have focused on the explosive instabilities. There are nonlinear instabilities near the threshold [26, 27] as well as the so-called fishbone instabilities [28]. They are reviewed in reference [11]. The readers may also refer to reference [29] for non-ideal MHD theory.

References

1. Wesson J 1997 Tokamaks (Oxford: Clarendon Press).
2. Rutherford P H 1973 Nonlinear growth of the tearing mode, *Phys. Fluids* 16 1903–8. https://doi.org/10.1063/1.1694232
3. Carrera R, Hazeltine R D, Kotschenreuther M 1986 Island bootstrap current modification of the nonlinear dynamics of the tearing mode *Phys. Fluids 29*, 899–902. https://doi.org/10.1063/1.865682
4. Chang Z, Callen J D 1990 Global Energy Confinement Degradation Due to Macroscopic Phenomena in Tokamaks *Nuclear Fusion 30*, 219. DOI 10.1088/0029-5515/30/2/003
5. Hazeltine R D 1976 Review of neoclassical transport theory, *Advances in Plasma Physics vol 6* ed A Simon and W B Thompson (New York: Wiley) p 273

6. Kadomtsev B B 1975 Disruptive instability in tokamaks *Sov. Tech. Phys. Lett.* (Engl. Transl.); (United States), Vol. 1:5; ISSN STPLD
7. Bandyopadhyay I, Igochine V, Sauter O, *et al.* 2025 MHD, disruptions and control physics: Chapter 4 of the special issue: on the path to tokamak burning plasma operation *Nucl. Fusion 65*, 103001. DOI 10.1088/1741-4326/ade7a0
8. Sweet P A 1958 Electromagnetic Phenomena in Cosmical Physics (Cambridge UK: Cambridge University Press).
9. Parker E N 1957 Sweet's mechanism for merging magnetic fields in conducting fluids, *Geophys. Res. 62*, 509–20. https://doi.org/10.1029/JZ062i004p00509
10. Zheng L J, Furukawa M 2010 Current-interchange tearing modes: Conversion of interchange-type modes to tearing modes *Phys. Plasmas 17*, 052508. DOI 10.1088/0957-4484/21/4/045305
11. Zheng L J, Takahashi H, Fredrickson E D 2008 Edge-localized modes explained as the amplification of scrape-off-layer current coupling *Phys. Rev. Lett. 100*, 115001. https://doi.org/10.1103/PhysRevLett.100.115001
12. Takahashi H, Fredrickson E D, Schaffer M J 2008 Scrape-Off-Layer Current Model for Filament Structure Observed during Edge-Localized Modes in the DIII-D Tokamak *Phys. Rev. Lett. 100*, 205001. DOI: https://doi.org/10.1103/PhysRevLett.100.205001
13. Rack M, Wingen A, Liang Y, Spatschek K H, Harting D M, Devaux S, JET-EFDA contributors 2012 A Thermoelectric currents and their role during ELM formation in JET *Nucl. Fusion 52*, 074012. DOI 10.1088/0029-5515/52/7/074012
14. Staebler G M, Hinton F L 1989 Currents in the scrape-off layer of diverted tokamaks *Nucl. Fusion 29*, 1820. DOI 10.1088/0029-5515/29/10/017
15. Zheng L 2015 Advanced tokamak stability theory. (London: IOP Publishing). https://iopscience.iop.org/book/mono/978-1-6270-5423-2
16. Coppi B 1980 Nonclassical transport and the 'principle of profile consistency' *Comments Plasma Phys. Controlled Fusion 5*, 261.
17. Hogeweij G M D, Oomens A A M, Barth C G *et al.* 1996 Steady-State Hollow Electron Temperature Profiles in the Rijnhuizen Tokamak Project *Phys. Rev. Lett. 76*, 632. DOI: https://doi.org/10.1103/PhysRevLett.76.632
18. Bortolon A, Duval B P, Pochelon A, Scarabosio A 2006 Observation of Spontaneous Toroidal Rotation Inversion in Ohmically Heated Tokamak Plasmas *Phys. Rev. Lett. 97*, 235003. DOI: https://doi.org/10.1103/PhysRevLett.97.235003
19. Myra J R, D'Ippolito D A Stotler D P, Zweben S J, LeBlanc B P, Menard J E, Maqueda R J, and Boedo J 2006 Blob birth and transport in the tokamak edge plasma: Analysis of imaging data *Phys. Plasmas 13*, 092509. https://doi.org/10.1063/1.2355668
20. Krasheninnikov S I 2001 On scrape off layer plasma transport *Physics Letters A 283*, 368–370. https://doi.org/10.1016/S0375-9601(01)00252-3
21. Burrell K H *et al.* 2005 Advances in understanding quiescent H-mode plasmas in DIII-D *Phys. Plasmas 12*, 056121. https://doi.org/10.1063/1.1894745
22. Solano E R *et al.* 2010 Observation of confined current ribbon in JET plasmas *Phys. Rev. Lett. 104*, 185003. https://doi.org/10.1103/PhysRevLett.104.185003
23. Zheng L, Kotschenreuther M, Valanju P 2013 Low-n magnetohydrodynamic edge instabilities in quiescent H-mode plasmas with a safety-factor plateau *Nucl. Fusion 53*, 063009. DOI: 10.1088/0029-5515/53/6/063009
24. Biskamp D 1986 Magnetic reconnection via current sheets *Phys. Fluids 29*, 1520–53. https://doi.org/10.1063/1.865670

25. Syrovatskii S I 1971 Formation of Current Sheets in a Plasma with a Frozen-in Strong Magnetic Field *Soviet Journal of Experimental and Theoretical Physics, 33*, 933.
26. Berk H L, Breizman B N, Pekker M 1996 Nonlinear dynamics of a driven mode near marginal stability *Phys. Rev. Lett. 78*, 1256–9. https://doi.org/10.1103/PhysRevLett.76.1256
27. Breizman B N, Berk H L, Pekker M S, F Porcelli F, Stupakov G V, Wong K L 1997 Critical nonlinear phenomena for kinetic instabilities near threshold *Phys. Plasmas 4*, 1559–68. https://doi.org/10.1063/1.872286
28. Chen L, White R B, Rosenbluth M N 1984 Excitation of internal kink modes by trapped energetic beam ions *Phys. Rev. Lett. 52*, 1122–5. https://doi.org/10.1103/PhysRevLett.52.1122
29. Biskamp D 1997 Nonlinear Magnetohydrodynamics (Cambridge, UK: Cambridge University Press).

10 Magnetic fusion concepts and conclusive remarks

This chapter will discuss the major concepts of magnetic confinement for fusion plasmas. We will focus on three major concepts: tokamak, stellarator, and mirror. We will briefly review the development history of these concepts and discuss their advantages, disadvantages, prospects, and key related physics issues.

10.1 TOKAMAK

Tokamak concept was first conceived by Soviet physicists Andrei Sakharov and Igor Tamm. Experiments were constructed from 1951 at Kurchatov Institute in Moscow. Their 1958 T-1 device is sometimes considered the first tokamak [1]. Numerous tokamaks have been built since then, such as TFTR, JET, DIII-D, EAST, etc. Currently, the world's largest operating tokamak by plasma volume and current is JT-60SA in Japan. The primary global effort for fusion research is ITER [2], as shown in figure 10.1. It is intended to demonstrate the scientific and technological feasibility of fusion tokamak concept, by producing 10 times more energy than is required to heat it, *i.e.*, $Q > 10$. In recent years, commercial fusion companies have also joined the tokamak research, for example, SPARC by Commonwealth Fusion Systems.

The tokamak was invented about the same time as the stellarator. Both of them are aimed to confine plasmas in a closed field system by bending the plasma column into a torus. To address the charge separation issue due to the different curvature drifts of ions and electrons, which can cause the $\mathbf{E} \times \mathbf{B}$ particle losses, the field line rotation transform is introduced. In tokamaks, the transformer principle is employed. The Ohmic current is driven toroidally. The field line rotational transform is generated without sacrificing the toroidal symmetry. Besides the limitation of computational ability for coil design in the early days, the tokamak achieved early success simply because of its toroidal symmetry feature. Nowadays, the best confinement parameters achieved are still from tokamak experiments.

The toroidal symmetry in tokamaks matters. From the Hamiltonian mechanics, we know that symmetry gives rise to the particle motion invariance. It constrains the plasma transport and the development of instabilities. This helps the tokamak concept. Besides, the induction of toroidal current in a tokamak can also play the role of Z pinch. This compresses and heats tokamak plasmas. Plasma confinement in tokamaks can be modulated by arranging the Ohmic current profile and heating power. The off-axis heating during the current ramping-up process can help generate a pressure profile that exhibits the so-called internal transport barrier feature.

However, the Ohmic current is transient. This limits the tokamak discharge time. Besides, the Ohmic current raises the plasma energy. Consequently, the

DOI: 10.1201/9781003522133-10

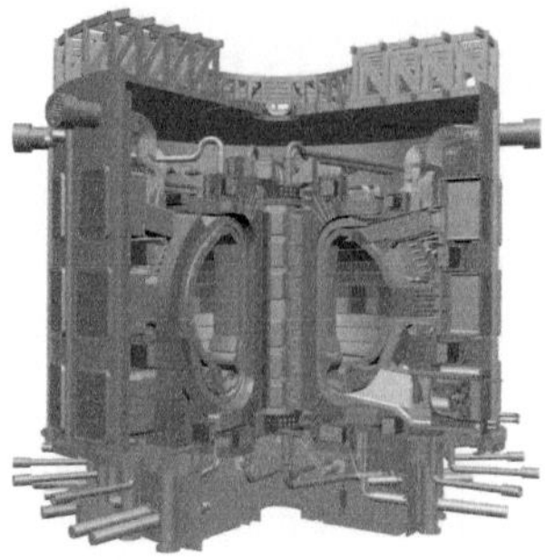

Figure 10.1 ITER: the international tokamak device under construction in France (U.S. Department of Energy from United States, Public domain, via Wikimedia Commons).

current-driven instabilities result. The harmful plasma disruption is uniquely tied to the tokamak confinement. Because of the current gradient, the pressure-driven interchange modes in tokamaks interchange not only the plasma pressure, but also the current. This results in the coupling between the interchange and tearing modes in tokamaks. Consequently, the electron transport in tokamaks is usually larger than that for ions due to the excitation of the current-interchange tearing modes. This enhances the radial transport in tokamaks.

In discussing tokamak physics, one cannot skip the so-called high-confinement mode (H-mode) [3]. It is a confinement mode in which the particle and energy confinements are significantly enhanced. Tokamak plasmas usually exhibit a confinement degradation feature. The plasma temperature and pressure do not increase with the heating power (such as neutral beam injection) applied. Such a case is usually referred to as the low mode (L mode) discharge. However, it was first observed in the ASDEX tokamak that above a certain heating power threshold, the plasma discharge suddenly switches to the H-mode with a dramatic improvement in particle and energy confinement of the plasma. The L and H mode pressure profiles are shown in figure 10.2.

Traditionally, the H mode is interpreted as the formation of the transport barrier at the pedental, where the plasma pressure gradient is steep near the plasma edge. It is thought that because the $\mathbf{E} \times \mathbf{B}$ rotation shear rate is larger than the instability growth rate at the pedestal, the turbulent eddies are broken up and a transport barrier is created that prevents energy leakage. Consequently, plasma pressure and temperature are allowed to build up. However, the experimental observations show some exceptions. H-mode confinement is found to be rather quiet. It is not the turbulence being torn away by the rotation shear, but there seems to be no instabilities at all. Also, both numerical simulation and experimental observation show that the $\mathbf{E} \times \mathbf{B}$ rotation shear rate in the negative triangularity tokamak is much higher. Nonetheless, the H mode is hardly achieved in the negative triangularity configuration. Theoretically, it is also shown that in the system with magnetic shear, the turbulence eddies are not distorted by the rotation shear. The turbulence eddies are instead decorrelated by the

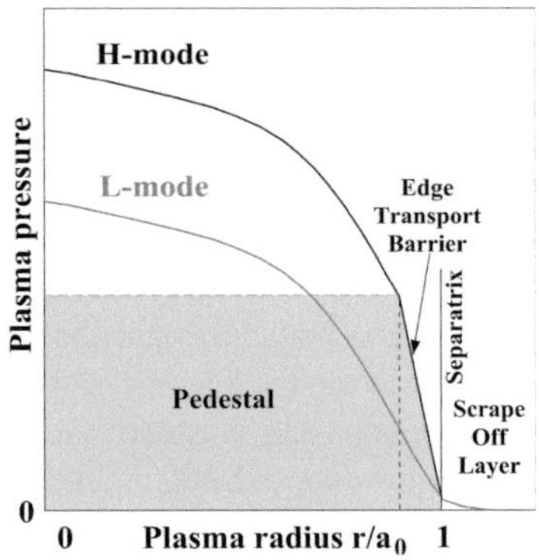

Figure 10.2 Pressure profiles for L and H modes. Reproduced from reference [4] with the permission of IAEA.

rotation curvature [5]. Perhaps, more or other physics mechanisms are involved. The L-H transition is consistent with the two stability regimes in the ballooning stability diagram. In this picture, the L mode is in the first stability regime, while the H-mode pedestal is in the second stability regime, with the internal region remaining in the first stability regime. Because there is an unstable parameter region between the first and second regimes, sufficient heating power is required to run over from the first to the second regime. This explains that there is a heating power threshold for the L-H transition. The I mode is just in the status near the second stability boundary.

There is another rotation-induced stability mechanism. The flute modes in a magnetically confined plasma are tied to the charge separation due to the different curvature drifts between the ions and electrons. In the presence of an edge electric well, the plasma rotation is usually not of isorotation, *i.e.*, the rotation velocity along the magnetic field line is not constant. This can lead to a short-circuit effect on the charge separation that drives the flute modes. This effect is stranger in the positive triangularity tokamaks than in the negative ones.

The bootstrap current also plays a crucial role. From the ballooning mode stability diagram, one can see that without counting its contribution to reducing the magnetic shear, the magnetic shear is too high so that the first and second stability regimes are widely separated, and transition to H mode is hard. The QH mode is even quieter. The picture based on the $\mathbf{E} \times \mathbf{B}$ rotation shear rate is even harder to work. However, the modification of the q profile successfully interprets the edge harmonic oscillation observed in the QH mode discharge. The experiment does observe a current ribbon at the pedestal top as discussed in Section 9.9.

H-mode confinement is good. However, it is often tied to the damaging edge localized modes (ELMs), which discharge the particle and heat from the pedestal to the scrape-off layer (SOL) and eventually to the diverters. Divertor materials are hardly able to sustain such a discharge. The ELMs can be viewed as a certain kind of microdisruption at the plasma edge. The coupling of SOL current plays a significant role [6]. Various approaches have been proposed to mitigate them. The word "mitigate" is used here instead of "suppress". This is to emphasize that the suppression of mild instabilities in this case can only cause a bursting of stronger instabilities. This thought motivates the way to constrain the damaging ELMs.

Let us discuss one of the possible ways for mitigating ELMs by applying the so-called resonant magnetic perturbations (RMPs). Because the applied perturbed magnetic field resonates with the field at the pedestal top, a stochastic field structure is formed so that the transport is enhanced and the pedestal pressure gradient is reduced. Consequently, the MHD modes become stable there, and ELM excitation is avoided. Reference [7] proposes an alternative RMP approach. Because in the presence of X points, the edge safety exhibits a dual q feature, one can simply apply axisymmetric coils to resonate the magnetic field at the X point to generate a stochastic field line area to mitigate the ELMs. This is simpler in engineering.

Another topic for tokamak physics is the negative triangularity tokamak. The negative triangularity tokamak is originally proposed as a possible solution for the divertor heat load issue especially during the ELM burst. When the divertors are moved from the high field side in the positive triangularity case to the low field side in the negative triangularity case, a larger separatrix wetted area results. This is simply because the larger major radius of the divertor location is larger in the negative triangularity case. Also, a larger room is yielded for divertor engineering design in the negative triangularity case.

Traditionally, one is concerned about the low stability beta limit in the negative triangularity case. The H-mode confinement can hardly be achieved in the negative triangularity configuration. Only the recent TCV and DIII-D experiments found that the L-mode discharges in the negative triangularity experiments can reach about the same level of normalized beta as the H-mode confinement in the positive triangularity tokamaks [8, 9]. This revives the interest in the negative triangularity tokamaks.

There is an important development for the negative triangularity tokamak research. Reference [10] showed numerically that the safety factor in the negative triangular configuration is generally lower than that in the positive triangular configuration. Also, the bootstrap current in the negative triangularity configuration is bigger. It is found that even when the bootstrap current fraction reaches as high as 95%, the safety factor value in the edge region remains reasonably low. This has motivated them to explore the advanced steady-state scenario with the negative triangularity. It is further found that the negative triangularity configuration with high bootstrap current fraction, high poloidal beta, and peaked pressure profiles can achieve higher normalized beta for low-n modes than the positive triangularity case. In a certain parameter domain, the normalized beta can reach about twice the extended Troyon limit, while the same computation indicates that the positive triangularity configuration is indeed constrained by the Troyon limit. This good stability property is obtained since the equilibrium is characterized by the steep pressure gradient appearing in the core region, where the magnetic shear is negative, while the safety factor near the edge region remains low. This discovery is important to tokamak research. The

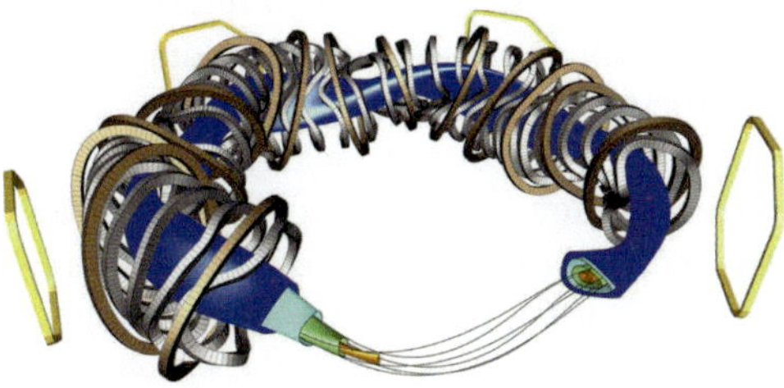

Figure 10.3 Wendelstein 7-X stellarator in Germany (Sunn Pedersen, et al. and The Wendelstein 7-X Team, CC BY 4.0, via Wikimedia Commons).

toroidal symmetry is preserved, while no Ohmic current is needed, and even a high beta confinement is achieved.

These are brief prospect reviews of tokamaks. Readers are referred to reference [11]. Reviews of nonlinear theories in Chapter 9 are also mainly related to the tokamak physics interpretations.

10.2 STELLARATOR

The stellarator was invented by L. Spitzer in 1951 [12]. Much of its early development was carried out by Spitzer's team, which later became the Princeton Plasma Physics Laboratory.

To date, several stellarators have been built, such as LHD in Japan and Wendelstein 7-X in Germany, which is shown in figure 10.3. A stellarator creates the field line rotational transform by twisting the coils and/or the coil axis. This is a thoughtful idea and has a great advantage over tokamaks in that there is no need to induce the toroidal plasma current to generate the field line rotational transform. Historically, the stellarator was invented even earlier than the tokamak, or about the same time. But, the tokamak was the first to achieve success. Today, we know that the original stellarator design has not addressed the particle orbit loss issue. Twisting the coils and/or the coil axis indeed generates the field line rotational transform. However, it cannot assure the existence of perfect magnetic surfaces and the confinement of trapped particles, even assuming that there are nested magnetic surfaces. The Grad-Shafranov equation is derived for a symmetric system, and in the toroidally symmetric tokamaks, the canonical momentum is conserved, which assures the confinement of charged particles.

One of the most important contributions of the stellarator concept is the invention of the so-called quasihelically symmetric stellarators as reviewed in reference [13] and also in Section 4.4. It is impossible to have an axisymmetric stellarator. However, one can seek the particle confinement with the invariance in the guiding center picture. It is proven that, using the Boozer coordinates, the dependence of the Lagrangian on the magnetic field in the guiding center description is only through its magnitude. Therefore, if there is a symmetry for $B = |\mathbf{B}|$, one can have an invariance of guiding center motion according to the Hamiltonian equations. This assures the confinement of trapped particles in the guiding center motion time scale. The symmetry of magnetic field magnitude in the Boozer coordinates requires just $B = B(\psi, M\theta - N\zeta)$, i.e., B is a two-dimensional function.

Another important contribution is the concept of minimization of geodesic curvature, the mnigeneous stellarator [13]. The geodesic curvature defines the guiding center excursion of charged particles across the magnetic surfaces. If the geodesic curvature is minimized to be vanishingly small, the banana width of trapped particles vanishes, and the trapped particles are certainly confined. This solves not only the trapped particle confinement issue, but also reduces the neoclassical transport of the stellarator system.

There are tremendous efforts to optimize the stellarator concepts, especially along these two lines. However, the existence of perfect nest magnetic surfaces remains a challenge. So far, the existence of magnetic islands has only been avoided by requiring the safety factor to be a ratio of big rational numbers to avoid the resonances. This leads the design to usually have a flat safety-factor profile. In this sytem the magnetic shear nearly vanishes. This is a significant drawback, since the magnetic shear plays an imporant role in the stabilization of MHD modes.

The other issue with the current stellarotr design is about the determination of the coil structure. It was known in the early days that both the magnetic flux and the gradient of the flux should be continuous on the plasma-vacuum interface. This is explained in subsection 4.3.1. But sometimes in stellarator design, only the magnetic flux continuity is imposed. Consequently, the magnetic islands appear at the plasma-vacuum interface. In tokamaks, there is toroidal current. Application of the vertical magnetic field can control the equilibrium Shafranov shift. In stellarators, controlling the Shafranov shift needs to be particularly noted. Note that the Shafranov shift increases with plasma beta. The outward shooting method for determining the coil structure described in Section 4.3 is usual for stellarator coil design and feedback control.

10.3 MIRROR

The magnetic-mirror concept was proposed by Gersh Budker at the Kurchatov Institute, Russia, and Richard F. Post at the Lawrence Livermore National Laboratory in the US in the early 1950s. The concept and its history were reviewed in reference [14]. The advantage of the mirror concept lies in its simplicity and conceptual beauty—the symmetry. The challenges of the concept are two-fold: it has open ends in the axial direction and a bad curvature in the radial direction. Due to funding resource limitations, the mirror research has been nearly frozen in favor of the tokamak concept for years. Only in recent years have the mirror researchers begun to revive.

Although the mirror geometry is simple, controlling hot plasma for nuclear fusion is never trivial. The key to the success of the mirror concept lies in a deep understanding of the underlying physics. To improve the mirror equilibrium and stability without impeding its symmetry is the key issue. Even if the geometry is simple, one may not overlook the proper coil design. The limiter confinement is not an option.

One has to calculate the free boundary equilibrium with a vacuum gap between the hot plasma and the wall. The similar issues dealt with in tokamaks may need to be addressed in mirrors, especially the feedback control.

Perhaps, the most urgent issue for mirror research currently is to demonstrate that the mirror concept is able to achieve the same level of temperature and confinement time as tokamaks. This is not a trivial task, noting that the achieved temperature and confinement time in the current mirror experiments are much lower than those of tokamaks. In tokamaks, the neutral beam heating is commonly used. In mirrors, one has to consider the symmetry-breaking issue. One can hardly make the perpendicular neutral beam heating symmetric. Note that a tokamak generally benefits from Ohmic heating, which is also a kind of Z pinch to the plasma column. Analyses for what would happen if the Ohmic heating power is used to pinch the mirror plasma need to be implemented. In tokamaks, plasma is heated through Ohmic heating. In mirrors, the magnetic moment invariant leads to the ions and electrons being heated up directly for perpendicular temperature. With the same power, one can expect that the mirror theta pinch process is much more effective. Indeed, like a tokamak, the pinch effect tends to be transient. However, after the theta pinch process, the charged particles tend to precess azimuthally. An additional adiabatic invariant from particle precession helps the mirror plasma confinement and stabilization. This research can provide a quick and direct assessment of whether the same level of temperature and confinement time as modern tokamaks can be achieved in a mirror configuration. The early theta-pinched mirror experiment Scylla IV-P showed encouraging signs. The instability spectrum is much cleaner than in tokamaks [15]. The so-called 'wobble' instabilities are non-explosive. Very likely, they are just caused by the symmetry defectiveness in the azimuthal direction. We believe that it would be quite different if the pinch time could last as the Ohmic current ramping-up time in tokamaks, and the symmetry issue is properly addressed. It should also be pointed out that the tokamak current is carried by plasma, but the mirror current is carried by coils. With superconducting coils, the final magnetic field in mirrors can be sustained. It is not a transient confinement. The theta pinch effects are only used for compression and heating.

10.4 CONCLUSIVE REMARKS

In physics, one searches for the first cause—the first principle, and then deduces from the fundamental level of established laws of physics to explain more complicated phenomena. With advanced parallel computers, one may think of the possibility of simulating multiple billion particles from the Newtonian law of motion and Maxwell equations. However, the motion of the particles depends not only on the differential equations but also on the initial conditions. In the restricted region, it depends on the boundary conditions as well. This makes the first principle in describing many particles system different from what one may think in brute force simulation efforts. One simply is unable to specify the initial and boundary conditions for many charged particle systems. Various conservation laws for multiple particle systems derived from the fundamental physics laws have to be considered as part of the first principle. The

MHD equations are actually the set of conservation laws: the density, momentum, and energy conservation.

The importance of MHD theory lies also in the fact that charged particles tend to move together with the magnetic field lines in the perpendicular direction. This makes the MHD fluid description possible, even considering that charged particles have a long correction length. Certainly, beyond the MHD description, one still needs to consider the kinetic effects in the direction parallel to the magnetic field and finite Larmor radius effects in the perpendicular direction. MHD description also excludes the coupling of charge-separation-induced electrostatic modes. Nevertheless, the MHD theory remains the most important theory in magnetic fusion and magnetically confined plasma physics. Nowadays, for magnetic fusion experiments, it is a consensus that the system cannot survive unless it is in MHD equilibrium and is stable to MHD modes. With modern computing power and advanced numerical schemes, many theoretical models can be implemented numerically. This can further advance the MHD mode research.

For decades, controlled nuclear fusion has been considered to be the holy grail of clean and lasting energy. With the advancements in experiments and theoretical insights in recent years, we see hope, but we also have to be frank about the reality. 'In the sweat of thy face shalt thou eat bread'... Only 'the truth shall make you free'. The simplicity and beauty are the Creator's wisdom: 'it is so', 'it is good'. The challenge of controlling nuclear fusion lies in losing simplicity. The importance of MHD lies exactly in its simplicity and beauty. MHD theory is also among the earliest developments of plasma physics, such as the MHD energy principle, the singular layer theory, etc. Even in this case, the MHD theory keeps evolving. With a generation's effort, many pioneering contributions were achieved in this field. It is hoped that this book helps in passing these on...

'They that sow in tears shall reap in joy.'

References

1. Shafranov V D, Bondarenko B D, Goncharov G A 2001 On the history of the research into controlled thermonuclear fusion, *Physics - Uspekhi 44*, 835–65. https://fire.pppl.gov/rf˙shafranov.pdf
2. Ikeda K 2007 Progress in the ITER Physics Basis, *Nucl. Fusion 47*, E01. DOI 10.1088/0029-5515/47/6/E01
3. Wagner F *et al* 1982 Regime of Improved confinement and high beta in neutral-beam-heated divertor discharges of the ASDEX tokamak *Phys. Rev. Lett. 49*, 1408–12. https://doi.org/10.1103/PhysRevLett.49.1408.
4. Lang P T *et al.* 2013 ELM control strategies and tools: status and potential for ITER, *Nucl. Fusion 53*, 043004. doi:10.1088/0029-5515/53/4/043004
5. Zheng L 2015 Advanced Tokamak Stability Theory (London: IOP Publishing & San Rafael, CA: Morgan & Claypool). https://dx.doi.org/10.1088/978-1-6270-5423-2
6. Zheng L J, Takahashi H, Fredrickson E D 2008 Edge-localized modes explained as the amplification of scrape-off-layer current coupling *Phys. Rev. Lett. 100*, 115001. https://doi.org/10.1103/PhysRevLett.100.115001

7. Linjin Zheng, Kotschenreuther M T, Waelbroeck F L, and Austin M E 2025 X-point effects on the ideal MHD modes in tokamaks in the description of dual-poloidal-region safety factor, *Phys. Plasmas 32*, 012501. https://doi.org/10.1063/5.0227541
8. Camenen Y. et al. 2007 Impact of plasma triangularity and collisionality on electron heat transport in TCV L-mode plasmas *Nucl. Fusion 47*, 510–16. DOI 10.1088/0029-5515/47/7/002.
9. Austin M E et al. 2019 Achievement of Reactor-Relevant Performance in Negative Triangularity Shape in the DIII-D Tokamak *Phys. Rev. Lett. 122*, 115001. DOI: https://doi.org/10.1103/PhysRevLett.122.115001.
10. Linjin Zheng, Kotschenreuther M T, Waelbroeck F L, Austin M E, Rowan W L, Valanju P, Liu X 2024 Prospects of negative triangularity tokamak for advanced steady-state confinement of fusion plasmas in MHD stability consideration *Fundamental Plasma Physics 10*, 100051. https://doi.org/10.1016/j.fpp.2024.100051
11. Wesson J 1997 Tokamaks (Oxford: Clarendom Press).
12. Spitzer L 1951 A proposed stellarator, Technical report, Project Matterhorn.
13. Lise-Marie Imbert-Gérard, Elizabeth J P, Wright A M 2025 An Introduction to Stellarators: From magnetic fields to symmetries and optimization, (SIAM, Philadelphia). https://epubs.siam.org/doi/book/10.1137/1.9781611978223
14. Post R, 1987 The magnetic mirror approach to fusion, *Nucl. Fusion 27*, 1579. DOI 10.1088/0029-5515/27/10/001
15. Freidberg J P, Pearlstein L D 1978 Rotational instabilities in a theta pinch *Phys. Fluids 21*, 1207-17. https://doi.org/10.1063/1.862359

Index